Microbiological Sensors for the Drinking Water Industry

Microbiological Sensors for the Drinking Water Industry

Edited by
Torben Lund Skovhus and Bo Højris

IWA PUBLISHING

Published by **IWA Publishing**
Alliance House
12 Caxton Street
London SW1H 0QS, UK
Telephone: +44 (0)20 7654 5500
Fax: +44 (0)20 7654 5555
Email: publications@iwap.co.uk
Web: www.iwapublishing.com

First published 2018

British Library Cataloguing in Publication Data
A CIP catalogue record for this book is available from the British Library

ISBN: 9781780408682 (Paperback)
ISBN: 9781780408699 (eBook)
ISBN: 9781780409559 (ePub)

Contents

Editorial preface

As population in and around major cities grows exponentially, the responsibilities of the associated drinking water suppliers increase accordingly. In each city, thousands, hundreds of thousands or even millions of citizens rely on a continuous supply of clean, healthy drinking water every day of the year. From the supplier's point of view, delivering safe drinking water is an endless battle against pollution and degrading water quality: a battle fought with source protection, continuous risk assessment and monitoring. This is a battle that must be won, because the alternative is unthinkable.

Nevertheless, as occasionally or quite frequently seen from a global perspective, people become ill from drinking water that was supposed to be safe. In many of these cases, people end up being affected before the supplier realizes there is a problem. One might expect that some monitoring scheme, set up in accordance with local legislation, could prevent such outbreaks. However, the *consumer* is often the fastest-responding microbiological sensor compared to time-consuming laboratory analyses of planned grab samples, etc. Some water suppliers actually use shopping data from local pharmacies as "early" warnings of potential outbreaks of waterborne contamination.

In 1883, the German doctor and bacteriologist Robert Koch published an article entitled "About Detection Methods for Microorganisms in Water". Ever since, growth-based methods have been applied as the gold standard for assessing microbiological conditions in drinking water. Culture-based methods are still widely used within the drinking water industry for detection of general and specific bacterial presence. However, there are two major problems with these methods. First, not all bacteria can be cultivated in the laboratory, as they may be quite viable in their natural habitat but not viable under laboratory conditions, and will thus not show up in the analyses, thereby misrepresenting or under representing the community. Secondly, it takes between one and several days to culture bacteria, depending on the specific type, which means that the response times of the analyses are much slower than the rate at which the drinking water is distributed and consumed. Hence, the consumer remains the fastest microbiological sensor.

Developments over the past few decades have created an arsenal of rapid molecular microbiological methods (MMM) that are capable of detecting microorganisms, general or specific, within hours. These are, however, still laboratory based and require sampling, which consequently limits the number of analyses and still leads to relatively long response times. What the drinking water industry really needs, to live up to their responsibilities, are fast, accurate online (real-time) methods for monitoring changes in water quality; and that is what this book is about.

Though it might take a while before the general legislative authorities become aware of it, there are already online water quality methods available. Unfortunately, the objectives of the various stakeholders in the drinking water quality community are not well aligned with one another. Legislation is falling behind technology development; maintenance personnel lack education on water quality issues; installed sensors produce loads of data that are not used in a very intelligent manner, etc. These difficulties are generally caused by the inability of the different stakeholders to communicate on the same level, i.e., using the same language.

The intent of this book is to gather the different stakeholders to a common starting point, bringing all of the issues to the table and creating a united foundation for future development in the field of sensor technology and drinking water safety. Therefore, this book is not solely intended for drinking water suppliers. It is also meant to be a technical resource and inspiration for the people responsible for setting the legislative standards of tomorrow, and as an aid in the education of future drinking water professionals, including engineers, chemists, microbiologists, field technologists and managers.

The book is made up of 16 individual chapters in 5 sections. While the reader may choose to start at a specific topic based upon past experience and specific interests, the book is structured in a logical way so it can be studied from first to last page with a progression of topics in the different sections. **Section 1**, on background and perspectives, contains Chapters 1 to 3. Here some of the most experienced educators and scientists in the field of drinking water microbiology provide their perspective on the past, present and future of monitoring microbiological water quality. This part of the book is meant as an introduction to the field, as well as inspiration for future development. **Section 2**, on industry needs, is made up of Chapters 4 and 5, where two significant frontrunner companies within the drinking water supply world provide their views on, and experience with, online monitoring of microbiological water quality. Then follows **Section 3** on sensor technologies (Chapters 6 to 11) with presentations of existing technologies that are already being applied for online monitoring in the field today. **Section 4** deals with the testing and evaluation of online monitoring methods (Chapter 12) and covers an important aspect that is often neglected: the handling all of the collected data and appropriate data processing (Chapter 13). **Section 5** (Chapters 14 to 16) sets all of the technology into perspective by dealing with (i) risk assessment and critical control points in drinking water systems, (ii) educational aspects related to current and future water safety professionals, and (iii) the challenges that lie within a legislative system that is only slowly responding to these new technologies.

The editors wish to thank all of the authors and reviewers who contributed their time, knowledge, and expertise to this book, making it an invaluable resource for many years to come.

Please enjoy reading, studying and discussing the content of the book, which we truly hope will benefit and stimulate utilization of online monitoring as well as further research and developments in the field of sensor technology and drinking water safety.

Torben Lund Skovhus
VIA University College
Horsens, Denmark

Bo Højris
Grundfos Holding A/S
Bjerringbro, Denmark

Foreword by Eberhard Morgenroth

HOW TO MAKE GOOD USE OF MORE INFORMATION FOR SAFE DRINKING WATER

How would we change our approaches to provide safe drinking water if we had reliable and abundant information in real time about water quality parameters in the drinking water system? Drinking water quality depends on the interplay of source water protection, treatment, and the water distribution system. Technological and management approaches to water systems have evolved over centuries. Over the years, our empirical understanding has been enhanced by scientific understanding of the underlying processes. However, design, operation, and management today are still largely based on empirical guidelines and without *feeling the pulse* of how the water system is performing in real time. The current book explores, from a range of perspectives, how microbiological sensors can improve the way we manage our drinking water systems.

Our current drinking water systems were developed so that they work well with limited or no online sensors. Why change? There are new challenges and new opportunities that require us to rethink our approaches. In many regions of the world, water is scarce and water supply has to make use of more and more challenging water sources. More challenging water sources require more advanced treatment and closer process control. Treatment processes are being developed that require less energy and less chemicals while achieving better product water qualities – again, operation of these processes requires close monitoring of source waters and the processes themselves. Aging water distribution systems in industrial countries, in some cases constructed more than a century ago, will require significant investments. Retrofitting and managing these aging systems should be informed by monitoring the actual performance of the system. Finally, in many parts of the world, centralized water systems are augmented with decentralized systems producing potable drinking water or non-potable water for irrigation, toilet flushing, or other non-potable water uses in households and industry. Ensuring microbial water safety for this diversity of water systems is a future challenge we have to address.

If we want to address these new challenges, how can we make use of additional information from online sensors? Simply introducing online sensors into existing systems and generating additional data will not improve much. We need to learn how reliable information from sensors can lead to better water quality or reduced costs and effort. Are online sensors primarily used as an early warning to trigger manual sampling and inspection? Or can we sufficiently trust online sensors to guide operation directly? Will reliable monitoring lead to improved design and management approaches? There is significant promise – but we still have very limited understanding of the microbial processes that occur in the distribution system and that are affected by water quality and system operation. The current book addresses these questions from the perspective of researchers, technology providers, operators, and regulators. We need researchers to develop and test new sensors. We need practitioners and regulators to gain experience and confidence working with information provided based on sensor signals. We need to develop tools and routines to work responsibly with big data sets. To what extent are engineers today educated to take advantage of increasing availability of sensor data? How do we utilize these new possibilities responsibly?

The state of a system can often be inferred only indirectly from sensor measurements. For example, if a sensor provides particle counts as proxy for microbial contamination, it is essential for the user to understand under what conditions this proxy is valid or not. Effective management of large data sets requires skills that need to be introduced in professional training and university education of environmental engineers. Deriving valuable information from large datasets requires a systematic approach for outlier detection and for creating data streams and databases that can be trusted. This systematic approach must be based on a fundamental understanding of processes within the system and the sensors themselves. In addition, it needs to take advantage of machine learning approaches to help identify outliers or flag unexpected behavior.

I am convinced that microbiological sensors in drinking water systems will help improve our current water services and allow us to extend water services to places that are currently not served. But sensors alone will not be sufficient to improve our water systems. It requires innovative scientists, creative engineers, responsible operators, and engaged regulators to jointly explore and advance water systems. I wish that the current book will help to advance this joint learning process and guide future developments.

Eberhard Morgenroth
ETH Zürich and Eawag
Switzerland

Foreword by Koen Huysman

HOW TO LINK SENSORS WITH YOUR GOALS FOR OPERATIONAL EXCELLENCE

There is a long tradition of monitoring the microbiological quality of drinking water during production and distribution stages. Classical microbiological detection methods like plate counts are widely criticized for not being representative compared with the total treated water volume or the loss of time between sampling and availability of the results. On top of that, it is not easy to cover a continuous process like drinking water production because of practical issues like working hours and routines.

It is also difficult to relate biological data to other parameters such as iron, ammonium, manganese, etc. Changes in biological quality are mostly more difficult to interpret owing to variations over time with several possible explanatory variables. Microbiological monitoring devices often seem complementary with other tools already in use, but almost never seem to replace the existing tools.

Microbiological sensors for the drinking water industry are in full development, with great possibilities in the relatively near future. It will not be easy to make a choice between the different upcoming tools. Having to use many parallel sensors each with their own needs and follow-up requirements is neither optimal nor convenient.

After reaching consensus within a company to gather extra information about the production and distribution processes, the next hurdle to take is defining what kind of information you want for what purpose. Only then can a choice be made about the desirable measurement devices, taking into account the available budget. For example, do you want to gather information that is more general or something very specific? Is it for public trust, threat of terrorism, detecting (fecal) contamination, part of a water safety plan, process optimization, etc…?

At Pidpa, a Flemish drinking water company treating deep groundwater, microbiological sensors are considered complementary to the classical approaches, since it is currently not feasible to fulfil all legislative requirements (e.g. sampling frequency and number of parameters) with only one type of sensor. Nonetheless, a more general approach where the company takes measures based on its own risk analysis is highly appreciated by legislative instances.

A clear example of the usefulness of microbiological sensors is the following. Traditionally, as in most water companies, Pidpa used chlorine combined with UV disinfection in order to have a disinfection barrier to keep the drinking water safe for use. However, the dosage was so low that it would not solve a major contamination problem. It merely resulted in a false sense of safety and masked possible failures within the production and distribution facilities. After removal of the chlorine-dosing units, Pidpa decided to install online sensors for total particles and bacterial counts at the entrance of the distribution network and at a few water towers in order to detect possible treatment process failures. So, in this case, sensors were introduced for operational control and early warning in order to obtain operational excellence in the future.

The market for microbiological sensors is currently in full development, but is not yet mature and many challenges still have to be managed in the near future. Each sensor has its own advantages and disadvantages, with a total cost of ownership depending on sampling frequency, amount of produced data and use of chemicals. Placing several sensors in a distribution network in a way that one can follow changes throughout the system can only happen when relatively cheap and almost maintenance-free sensors become available. There are already sensors available for specific bacteria. However, measuring different specific bacteria (Coliforms, Enterococci, Giardia, Cryptosporidium, Legionella, etc.) online and simultaneously at very low levels is desirable for drinking water purposes, but is not yet available.

Within the product range of microbiological sensors, it is difficult to compare the output of one sensor to another even when they target the same parameter, owing to different measurement techniques. This can become frustrating for customers who want to compare measurements from different sensors to make the right choice for their application. Furthermore, the difficulty of comparing the results of the sensors with traditional laboratory methods makes it hard to for sensors to be accepted as a replacement. When legislators make the drinking water companies more responsible for looking at risks, a reduction of the number of classical grab samples should be possible if a water company can show that there is a promising global approach in the market.

Either way, it is not an excuse to stay on the sidelines. For the time being, and for our case at Pidpa, a more general approach where the total bacterial activity is monitored seemed more appropriate. In the beginning, it is important that someone with a lot of practical knowledge uses sufficient time to analyze the data in order to recognize events and effects of events on the sensor values, without wanting or needing to explain each change in value. Each company will have to decide which events and values are considered normal or not, taking into account different parameters. First, sensors will be implemented in a production environment where it is somewhat easier to couple sensor values with other data from a SCADA system (backwash of a filter, capacity change, replacement of pumps, etc.). It will certainly be challenging to use sensors in a distribution network with thousands of kilometres of piping and many more parameters, known or unknown, like residence time, change of valve position or water direction, not being continuously monitored.

In the coming period, the combination of sensors and big data analyses will certainly help make progress in this area. Algorithms will have to be developed to recognize patterns and events coupled with the loads of other data available from your SCADA system.

Koen Huysman
Pidpa
Belgium

About the Editors

Torben Lund Skovhus is Researcher and Project Manager at VIA University College in the Centre of Applied Research & Development in Building, Energy & Environment (Horsens, Denmark). He graduated from Aarhus University, Denmark, in 2002 with a Master's degree (cand.scient.) in Biology. In 2005 he finished his Ph.D. from Department of Microbiology, Aarhus University. In 2005, Torben was employed at Danish Technological Institute (DTI) in the Centre for Chemistry and Water Technology, where he was responsible for the consultancy activities for the oil and gas industry around the North Sea. Torben was heading DTI Microbiology Laboratory while he was developing several consultancy and business activities with the oil and gas industry. He founded DTI Oil & Gas in both Denmark and Norway where he was Team and Business Development Leader for five years. Thereafter Torben worked as Project Manager at DNV GL (Det Norske Veritas) in the field of Corrosion Management in both Bergen and Esbjerg.

Torben is currently chair of NACE TEG286X and ISMOS TSC an organization he co-founded in 2006. He is an international scientific reviewer and the author of 60+ technical and scientific papers and book chapters related to industrial microbiology, applied biotechnology, corrosion management, oilfield microbiology, water treatment and safety, reservoir souring and biocorrosion.

Dr. Bo Højris is a Lead Researcher at Grundfos Holding A/S in the Global Technology & Innovation department (Bjerringbro, Denmark). He received his Ph.D. in environmental engineering from Aalborg University for studies done at the Center for Biofilm Engineering at Montana State University. He has spent the past 20+ years as researcher, consultant, and lecturer within the interdisciplinary field spanned by materials, environments, and microorganisms. Bo has authored or co-authored 40+ scientific and technical papers. For the past 8 years, he has been with Grundfos, the world's largest pump manufacturer, pioneering in the development of sensors for water quality monitoring and processes for water treatment. In 2014, Bo and two of his colleagues won an Innovation award for the development of a 10-minute resolution online bacteria monitor.

List of Contributors

Finn Mollerup
VCS, Denmark

Ditte A. Søborg
Centre of Applied Research & Development in Building, Energy & Environment, VIA University College, Denmark

Frederik Hammes
Department of Environmental Microbiology, Eawag: Swiss Federal Institute of Aquatic Science and Technology, Switzerland

Bofu Li
Centre for Water Resources Studies, Dalhousie University, Canada

Hans-Jørgen Albrechtsen
Department of Environmental Engineering – DTU Environment, Technical University of Denmark, Denmark

Henrik Braathen
Colifast, Norway

Joep Appels
microLAN B.V., The Netherlands

Lars S. Olesen
Grundos Holding A/S, Denmark

Loren Ramsay
Centre of Applied Research & Development in Building, Energy & Environment, VIA University College, Denmark

Mathis Dahlqvist
Grundfos Holding A/S, Denmark

Patrick A. Whalen
LuminUltra Technologies Ltd, Canada

Sarah C. B. Christensen
HOFOR A/S, Water Quality Section, Denmark

Sophie Courtois
SUEZ, CIRSEE – Centre International de Recherche sur l'Eau et l'Environnement, France

Syed A. Imran
Aquastructure Solutions Inc., Canada

Torben Lund Skovhus
Centre of Applied Research & Development in Building, Energy & Environment, VIA University College, Denmark

Section 1

Background and perspectives

Chapter 1

History and perspective – the challenge of the gap between legislation and scientific achievements: The new era of enlightenment

Hans-Jørgen Albrechtsen

1.1 INTRODUCTION

Controlling microbial water quality is crucial to ensure safe drinking water – and this subsequently calls for proper monitoring and documentation of the occurrence of microorganisms in drinking water.

Since the end of the 1880s, monitoring has been achieved by various culture-based methods, in which the microorganisms must proliferate to be detected. Today these principles still form the basis for almost all legally binding guidelines and standards relating to microbial drinking water quality (e.g. EU, 1998; US EPA, 2017; US SDWA, 1974; WHO, 2017; EN ISO 6222:2000; EN ISO 9308-1:2014; EN ISO 9308-2:2012).

Recently, a range of different sensors has emerged with abilities to measure microbial quality aspects of drinking water. Furthermore, we are entering the age of "Industry 4.0" (Lu, 2017; Sedlak, 2015), and automation and digitalization is playing an increasingly important role. Also for this reason, the use of sensors becomes essential. However, one major challenge for the implementation of sensors for microbial water quality is that they often are based on measuring principles that are quite different from the classical culture-based methods.

Therefore, one of the major challenges in the implementation of sensors in modern management of microbial drinking water quality is to overcome the clash between signals from the sensors and the standards based on classical measuring methods.

Editors: Torben Lund Skovhus and Bo Højris
doi: 10.2166/9781780408699_0003

1.2 THE HISTORICAL DEVELOPMENT OF CULTURE-BASED METHODS

Already in 1881, Robert Koch had demonstrated the use of culturing bacteria on solid media, i.e. the application of agar or gelatin to solidify the bacterial culture medium containing growth substrates such as ammonium salts, yeast residues and a variety of sugars. Within a few years Percy Frankland had – in 1885 – applied this method to measure the number of bacteria in water, e.g. for evaluation of the efficiency of filtration. However, this raised a discussion on which medium is best? How many bacteria are acceptable? How good is this measure for understanding the occurrence of pathogens? (Payment *et al.*, 2003).

A few years later, in the 1890s, Theobald Smith proposed *Escherichia coli* as a drinking water indicator since it is part of the normal bacterial flora of mammals and warm-blooded animals (Prescott & Winslow, 1915). In the absence of a specific test for *E. coli*, surrogate tests for *E. coli* were developed, e.g. the 'fecal coliform test' proposed by Eijkman in 1904, based on the principle that *E. coli* is more thermotolerant than other members in its family. This method was further modified to detect the 'total coliform', which in addition to *E. coli* included a range of bacteria for which growth is not restricted to the mammalian colon. In 1914 the US Treasury Department proposed this test to be a standard for drinking water safety (Edberg *et al.*, 2000) and these principles have been the foundation for monitoring microbial drinking water quality since.

Several methods have been implemented to detect and quantify the number of *E. coli*. A simple version of the Most Probable Number or MPN procedure was implemented in 1914 (Ashbolt *et al.*, 2001). This is a statistical-based assay where a sample is diluted until extinction, and the serial dilutions are inoculated into broth media. The number of bacteria in the sample is estimated based on the distribution of the positive and negative responses (expressed as growth or biochemical reactions such as gas production during fermentation) among the dilutions.

E. coli can also be detected as colony formation in solidified selective media either on the top of agar (spread plate) or by mixing the water sample with melted agar (pour plate method). During the 1950s membrane filtration became an alternative to the MPN approach (Ashbolt *et al.*, 2001). This method is based on filtering the water sample through a membrane which retains the bacteria; after filtering the filter is incubated on top on a solid agar plate where the bacteria on the filter form colonies. The advantage with this approach was the increased sensitivity of the method, since a larger volume (100 mL or more) can be concentrated on the filter. This is still an accepted method for enumeration of *E. coli* (EN ISO 9308-1:2014).

In the early 1990s the MPN methods were further developed into a miniaturized setup, where detection is based on an enzymatic reaction in the so-called Colilert® technique (IDEXX), which is the state-of-the-art method (EN ISO 9308-2:2012).

However, all these methods require growth of the investigated cells, either to provide enough cells to make a visible colony or to produce enough enzyme to catalyze transformation of the substrate to produce a detectable signal. Furthermore, these methods require incubation time and subsequent relatively long response times, to ensure that the numbers are sufficiently low.

1.3 ADVANTAGES AND LIMITATIONS OF THE CLASSICAL CULTURE-BASED METHODS

All culture-based methods have this strong limitation – the incubation time – since it takes time for the microorganisms to grow to such a quantity that they can be detected, whether that is as a colony or a biochemical reaction. Although the methods have improved substantially, as exemplified by the state-of-the-art and standard method for *E. coli* and coliforms (Colilert® from Idexx), they still need 18 hours of incubation to provide certainty whether a given water sample is safe – although a contaminated sample may give a positive signal earlier. When monitoring drinking water distribution systems, the water might have been used or consumed before the result finally is reported.

On the other hand culture-based methods all have one clear advantage – the cells which are detected are obviously viable, otherwise no colony would have been formed. The use of *E. coli* provides an extremely sensitive analytical approach. Since the concentration of *E. coli* in wastewater is on the order of 10^6–10^7/100 mL (Lucas *et al.*, 2013; Andersen, 2015), and the level of detection for the standard method for use in drinking water is <1/100 mL, this method is able to detect traces of wastewater even after 1–10 million times of dilution.

However, the reliability of the viability on the other hand is also a challenge for this measuring principle, since e.g. *E. coli* is more sensitive to UV radiation and disinfection chemicals such as chlorine than many other pathogens (Chang *et al.*, 1985). In this way the absence of the indicator microorganism may be a false negative result, because the given water sample may still be infectious.

Despite these aspects, *E. coli* is still seen by many as the best biological drinking water indicator for public health protection (Edberg *et al.*, 2000).

1.4 MEASURING GENERAL MICROBIAL POPULATIONS OR MICROBIAL INDICATORS

The methods described above can be used in two quite different principles. The first principle – to count the number of unspecific bacteria in the water – is referred to as heterotrophic plate counts (HPC). This can be done in many different ways, but a common method for drinking water is described by EN ISO 6222:2000.

The second principle is to use indicator microorganisms (*E. coli, Clostridium perfringens*, intestinal enterococci, etc.), which are indicators of the risk of infection, i.e. the presence of pathogens. Since the majority of waterborne diseases are related to fecal contamination of the water, the philosophy is based on the selection of an indicator microorganism which always is present in the digestive system of mammals. So, if this microorganism is detected in a given water sample, the sample is likely contaminated with feces and as such likely to contain pathogens. This approach is quite effective, since the presence of an indicator organism not only reflects the risk of presence of bacteria, but also viruses and protozoans, which may occur in polluted water. On the other hand, pathogens that are not colon-related, such as *Legionella,* may also occur in water. Colon-related indicator microorganisms would never be able to reflect their occurrence.

A third approach could then be to analyze for each of the specific pathogens. Unfortunately, it is a problem that this would require a very broad array of analyses to include all pathogens, including viruses and protozoans. Furthermore, this approach would also require that methods are available for all relevant pathogens – and although methods are emerging (Bosch *et al.*, 2008), analysis for viruses in drinking water is not routine yet. On top of that, the approach of direct analysis for pathogens would require that we know all the pathogens to analyze for. Most of these specific analyses require molecular-based methods – an area in very rapid development (e.g. Clark *et al.*, 2011).

1.5 MONITORING FOR COMPLIANCE WITH GUIDELINES VERSUS PROCESS CONTROL

Most often, drinking water is monitored for compliance with legal guideline values. Here, *E. coli* is the 'golden standard', and most often this is monitored at end of pipe, where the water is available for the consumer. Since the resulting microbial drinking water quality is affected by many processes such as the source water quality, treatment efficiency including disinfection, storage and transport, monitoring for process control is highly relevant. In this respect, the indicator microorganisms are not that relevant, because they may not be present in the raw water (e.g. groundwater), are more sensitive to disinfection than other target microorganisms, or unable to represent aftergrowth, where various (non-pathogenic) bacteria are growing in and deteriorate the water during storage and transport. In these cases, a generic measurement of the microbial communities may be relevant, and here, heterotrophic plate count (HPC) has been used historically (Payment *et al.*, 2003). This was originally seen as a measure for the total microbial population, but due to the selectivity of all culture media, it is obvious that this is not the case, and standard methods only estimate a fraction as little as on the order of 0.01% (WHO, 2003). This phenomenon has been called 'The Great Plate Count Anomaly' (Staley & Konopka, 1985).

Some of the bacteria may be dead, or in a state where they are alive but currently unable to grow on agar media – the so called 'Viable But Not Culturable' state (VBNC). However, the vast majority (over 99%) of bacteria have never been cultured (Achtman & Wagner, 2008), so it is unlikely that culturing will ever be able to detect or quantify all the bacteria present.

Other methods than culture-based ones can be used to quantify this group of bacteria, including direct counting in the microscope, quantitative polymerase chain reaction (qPCR) quantification using primers specific for bacteria, or flow cytometry and they are challenging the conventional heterotrophic plate counts for routine microbial drinking water monitoring (Van Nevel *et al.*, 2017) – with potential for being developed into sensors.

1.6 SENSORS

Currently, a wide range of sensors for microbial drinking water quality is emerging on the market, and ideally the sensors should fulfill these criteria:

- in-line or at-line sampling
- continuous measurements
- results in real time
- on-line rapid transmission of data

Some sensors (e.g. BACTcontrol from MicroLan) can monitor coliforms and *E. coli*, basically using the same enzymatic reaction as the standard method Colilert® (Idexx): i.e. coliforms use β-galactosidase to metabolize ortho-nitrophenyl–beta-galactoside (ONPG) to produce a yellow color, and *E. coli* uses β-glucuronidase to metabolize 4-methylumbelliferyl-beta-D-glucuronide (MUG) and create fluorescence.

However, most of the sensors are based on measuring principles different from the culture-based methods used in the guidelines such as the WHO Guidelines for Drinking-Water Quality or EU Drinking Water Directive (Tatari *et al.*, 2016; WHO, 2017; EU, 1998). This includes biochemical measurements based on enzymatic reactions such as alkaline phosphatase (Bactcontrol from MicroLan) (Albrechtsen *et al.*, 2017), biochemical constituents such as ATP (Delahaye *et al.*, 2003; Hammes *et al.*, 2010; Vang *et al.*, 2014) which has been fully automated in the EZ-ATP analyzer (AppliTek, Promega), flow cytometry (Hammes *et al.*, 2012) or optical detection of bacterial cells (Grundfos BACMON) (Højris *et al.*, 2016).

For these sensors there are no guidelines yet, or records of values for safe drinking water – and there are seldom simple correlations to known microbial or process parameters.

1.7 OPPORTUNITIES OFFERED BY SENSORS

Sensors may offer a much higher sampling frequency – if not continuous monitoring – than ordinary monitoring based on grab sampling. This gives the opportunity to detect pulse contaminations which may quickly pass through a system and which may not be detected by grab sampling (Figure 1.1). Although a pulse contamination may only last for a short period, it may still cause disease in people drinking the contaminated water. It would be ideal if sensors could provide an alarm to prevent the spread of the contaminated water, even though they are detecting something other than specified as the microbial water quality parameters (EU, 1998; US EPA, 2017).

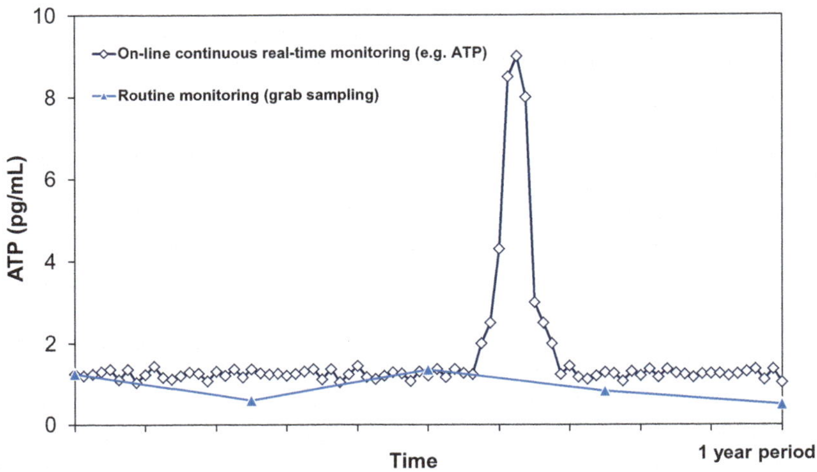

Figure 1.1 Conceptual illustration of a pulse contamination passing a measuring point in a distribution network, which will be detected by continuous monitoring, but which could pass unnoticed between two grab samples. The contamination would be detected by continuous or high-frequency sampling – here illustrated by ATP-measurements – whereas the peak may be overlooked by grab sampling when collected with time intervals of months (Vang, 2013).

Sensors may also provide a faster response (hours, or maybe even minutes) which may allow the water utility to be pro-active and prevent the spread of the contamination. This is in strong contrast to traditional methods, which require at least 18 hours for a response.

Sensors can be installed strategically to monitor locations which are particularly critical or sensitive. This could be to monitor the freshly produced drinking water when it leaves the waterworks to monitor for a potential breach in the hygienic barrier in the treatment. Water supply systems without disinfection residual are particularly sensitive to contaminants entering via leaks in a water tower or a reservoir (bird feces, surface water) or via renovation of the distribution network (dirt, surface water, sewage). Such a threat could be monitored by sensors installed at the outlet from water towers and reservoirs, or at major pipe connections/ branches. Sensors could also be located at the inlet to sensitive consumers, such as hospitals, pharmaceutical and food industry plants or at high-risk industrial plants or wastewater treatment plants (Figure 1.2). The same locations may have been selected for grab sampling in traditional monitoring, but usually not at the same sampling frequency.

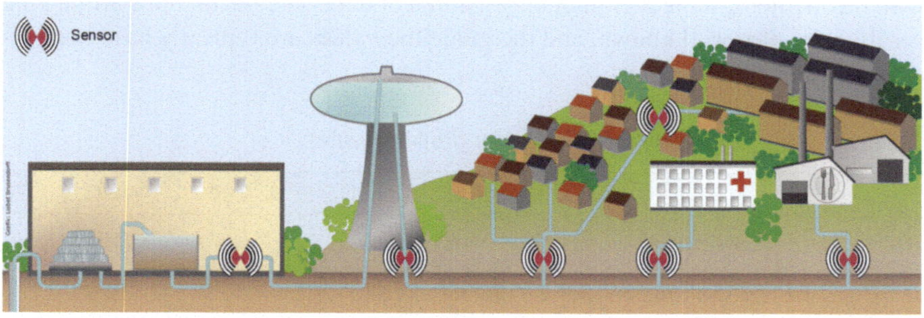

Figure 1.2 Potential locations of on-line sensors providing real-time analysis for continuous monitoring of microbial drinking-water quality in a distribution system (Vang, 2013).

More frequent sampling is in accordance with World Health Organization (WHO) recommendations which long have stated that "Water must be examined regularly and frequently because pollution is often intermittent and may not be detected if examination is limited to one or only a small number of samples. For this reason, it is better to examine drinking water frequently by means of a simple test rather than less often by several tests or a more complicated test" (WHO, 1996).

1.8 NEW INSIGHTS THROUGH MONITORING WITH SENSORS

First of all, the use of new methods always provides new insights since microbial methods usually supplement each other – they offer different views through different windows into the complex microbial world.

Continuous monitoring may also provide a better understanding of the monitored system. As an example, monitoring of the microbial water quality in a slaughterhouse by the optical sensor Grundfos BACMON (Højris *et al.*, 2016) showed a daily peak with an increase by a factor of ten of total bacteria. A thorough investigation revealed that the peak was caused by a combination of stagnant water in an unsuitable pipe layout and the onset of the morning cleaning. So new data could lead to optimization of the piping system.

1.9 THE POSITION OF REGULATORS AND UTILITIES

The fact that many new sensors measure other parameters than those regulated by legal guidelines (Table 1.1) such as EU Drinking Water Directive (EU, 1998) or US Safe Drinking Water Act (US SDWA, 1974) is a severe challenge. Many regulators and utilities are often very reluctant to shift from a well established and well known traditional measuring principle which seems to have been a useful tool to document the safety of the provided water. Besides the long experience with the methods, there are often long records of data. The fluctuations around the baseline are also well known, and the guideline values are typically based on this experience.

Table 1.1 Microbial parametric value for drinking water.

	EU, 1998*	US EPA, 2009	WHO, 2017
E. coli, Fecal coliform	0/100 mL	0/L	0/100 mL
Coliforms	0/100 mL	0/L	0/100 mL
Enterococci	0/100 mL	–	–
Clostridium perfringens (including spores)	0/100 mL†	–	–
Heterotropic plate count 22°C	No abnormal change	n/a, but recom. <500/mL	–
Cryptosporidium	–	0/L	–
Giardia lamblia	–	0/L	–
Legionella	–	0/L	–
Viruses (enteric)	–	0/L	–

*Under revision by 2018.
†Measured for water originating from or is influenced by surface water.

However, many of the existing, regulated microbial parameters are imperfect to some degree, so implementation of alternative parameters may contribute to improvement of the water quality.

The implementation of the principle of Water Safety Plans (WHO, 2005) may also call for use of sensors. Here, the principle is to go upstream in the water production line to prevent any deterioration of the water quality as early as possible, and generally to be proactive and prevent deterioration of the water quality instead of waiting to remediate when it has happened. To achieve this, any useful parameters can be used. This is in contrast to most legally regulated monitoring programs today, which have a strong focus on compliance with defined guideline values for classical parameters, monitored at the end-of-pipe at the consumer with methods where the results seldom are available before the investigated water has been used.

With this background, there is a strong need to establish baselines and experience with the new parameters offered by the sensors.

Potentially, the analytic approaches can be combined, so when a microbial event is registered by a sensor, a grab sample could be collected automatically for later analysis by conventional methods to be compared with legal standard values.

To take advantage of all these new opportunities to further improve the surveillance of water quality and to increase the water safety, we have to reduce the gap between legislation and all the recent scientific achievements that the array of new sensors are offering. Although the technology development is fast, the regulators have to be open minded and consider other parameters than *E. coli*. Although *E. coli* over more than a century has proven its values, it also has several limitations, and as long the regulators are awaiting sensors to be developed to detect *E. coli*, a wealth of opportunities to use other parameters may be wasted. The Internet of Things (IoT) and Big Data have shown that unusual combinations of very different data can provide highly valuable information. Operators should also learn the meaning of the parameters provided by the new sensors, and they should use the sensors as a support in the daily management of water treatment and supply, and not be restricted to parameters regulated by legislation. This process could be facilitated by further scientific investigations and analysis of the obtained data from the sensors, as well as systematic documentations of the advantages and disadvantages of the different sensors.

Well documented, concrete and successful cases can be very persuasive.

1.10 REFERENCES

Achtman M. and Wagner M. (2008). Microbial diversity and the genetic nature of microbial species. *Nature Reviews Microbiology*, **6**(6), 431–440.

Albrechtsen H.-J., Arnedo M. J., Appels J., Baquero D., Galofré B., Lindhardt B., Puigdomènech C. and Wagner F. B. (2017). On the way to on-line monitoring of microbial drinking water quality. *Abstract from Water Quality Technology Conference (WQTC) 2017*, AWWA, Portland, United States.

Andersen S. J. (2015). Quantitative Health and Microbial Risk Assessment for Public Warning During Flood. PhD Thesis, Technical University of Denmark, Kgl. Lyngby. Denmark.

Ashbolt N. J., Grabow W. O. K. and Snozzi M. (2001). Indicators of microbial water quality. In: Water Quality: Guidelines, Standards and Health, L. Fewtrell and J. Bartram (eds), World Health Organization (WHO), IWA Publishing, London, UK. ISBN: 1 900222 28 0.

Bosch A., Guix S., Sano D. and Pintó R. M. (2008). New tools for the study and direct surveillance of viral pathogens in water. *Current Opinion in Biotechnology*, **19**(3), 295–301.

Chang J. C., Ossoff S. F., Lobe D. C., Dorfman M. H., Dumais C. M., Qualls R. G. and Johnson J. D. (1985). UV inactivation of pathogenic and indicator microorganisms. *Applied and Environmental Microbiology*, **49**(6), 1361–1365.

Clark S. T., Gilbride K. A., Mehrvar M., Laursen A. E., Bostan V., Pushchak R. and McCarthy L. H. (2011). Evaluation of low-copy genetic targets for waterborne bacterial pathogen detection via qPCR. *Water Research*, **45**, 3378–3388.

Delahaye E., Welté B., Levi Y., Leblon G. and Montiel A. (2003). An ATP-based method for monitoring the microbiological drinking water quality in a distribution network. *Water Research*, **37**, 3689–3696.

Edberg S. C., Rice E. W., Karlin R. J. and Allen M. J. (2000). *Escherichia coli*: the best biological drinking water indicator for public health protection. *Journal of Applied Microbiology*, **88**(Supplement: S), 106S–116S.

EN ISO 6222:2000 (2000). Water quality – Enumeration of culturable micro-organisms – colony count by inoculation in a nutrient agar culture medium.

EN ISO 9308-2:2012 (2012). Water quality – Enumeration of *Escherichia coli* and coliform bacteria – part 2: most probable number method.

EN ISO 9308-1:2014 (2014). Water quality – Enumeration of *Escherichia coli* and coliform bacteria – part 1: membrane filtration method for waters with low bacterial background flora.

EU (1998). Drinking Water Directive (Council Directive 98/83/EC of 3 November 1998 on the quality of water intended for human consumption). 1998L0083 - EN - 27.10.2015 - 003.001–1.

Hammes F., Goldschmidt F., Vital M., Wang Y. and Egli T. (2010). Measurement and interpretation of microbial adenosine tri-phosphate (ATP) in aquatic environments. *Water Research*, **44**(13), 3915–3923.

Hammes F., Broger T., Weilenmann H.-U., Vital M., Helbing J., Bosshart U., Huber P., Odermatt R. P. and Sonnleitner B. (2012). Development and laboratory-scale testing of a fully automated online flow cytometer for drinking water analysis. *Cytometry Part A*, **81A**, 508–516.

Højris B., Christensen S. C. B., Albrechtsen H.-J., Smith C. and Dahlqvist M. (2016). A novel, optical, on-line bacteria sensor for monitoring drinking water quality. *Science Report*, **6**, 23935.

Lu Y. (2017). Industry 4.0: A survey on technologies, applications and open research issues. *Journal Industrial Information Integration*, **6**, 1–10.

Lucas F. S., Therial C., Gonçalves A., Servais P., Rocher V. and Mouchel J. J. (2013). Variation of raw wastewater microbiological quality in dry and wet weather conditions. *Environmental Science and Pollution Research*, **21**(8), 5318–5328.

Payment P., Sartory D. P. and Reasoner D. J. (2003). The history and use of HPC in drinking-water quality management. In: Heterotrophic Plate Counts and Drinking-water Safety, J. Bartram, J. Cotruvo, M. Exner, C. Fricker and A. Glasmacher (eds), World Health Organization (WHO), IWA Publishing, London, UK, pp. 20–48. ISBN: 1 84339 025 6.

Prescott S. C. and Winslow C.-E. (1915). Elements of Water Bacteriology, 3rd edn. John Wiley, New York, USA. [cited in Edberg *et al.* 2000].

Sedlak D. (2015). Water 4.0. The Past, Present, and Future of the World's Most Vital Resource. Yale University Press, US. ISBN: 9780300212679.

Staley J. T. and Konopka A. (1985). Measurement of in situ activities of nonphotosynthetic microorganisms in aquatic and terrestrial habitats. *Annual Review of Microbiology*, **39**, 321–346.

Tatari K., Corfitzen C. B., Albrechtsen H.-J. and Christensen S. C. B. (2016). Sensors for microbial drinking water quality. Technical University of Denmark, DTU Environment, Kgs. Lyngby. p. 43.

US EPA (2009). National Primary Drinking Water Regulations (NPDWR). EPA 816-F-09-004.

US EPA (2017). Analytical Methods Approved for Compliance Monitoring under the Revised Total Coliform Rule. US Environmental Protection Agency, Office of Water (MS – 4606 M), EPA 821-F-17-004.

US SDWA (1974). Safe Drinking Water Act. Title XIV of The Public Health Service Act: Safety of Public Water Systems (Safe Drinking Water Act), and subsequent amendments. www.epa.gov/sdwa/title-xiv-public-health-service-act-safety-public-water-systems-safe-drinking-water-act (accessed 4 May 2018).

Van Nevel S., Koetzsch S., Proctor C. R., Besmer M. D., Prest E. I., Vrouwenvelder J. S., Knezev A., Boon N. and Hammes F. (2017). Flow cytometric bacterial cell counts challenge conventional heterotrophic plate counts for routine microbiological drinking water monitoring. *Water Research*, **113**, 191–206.

Vang O. K. (2013). Development of on-line sensor for microbial quality of drinking water based on ATP-measurement. PhD Thesis, Technical University of Denmark, Kgl. Lyngby. Denmark.

Vang O. K., Corfitzen C. B., Smith C. and Albrechtsen H.-J. (2014). Evaluation of ATP measurements to detect microbial ingress by wastewater and surface water in drinking water. *Water Research*, **64**, 309–320.

WHO (1996). Guidelines for Drinking-Water Quality. Second edition. Volume 2 – Health Criteria and other Supporting Information. International Programme on Chemical Safety, World Health Organization Geneva, Switzerland. ISBN 92 4 154480 5 (v. 2).

WHO (2003). Expert consensus. In: Heterotrophic Plate Counts and Drinking-Water Safety - the Significance of HPCs for Water Quality and Human Health, J. Bartram, J. A. Cotruvo, M. Exner, C. R. Fricker and A. Glasmacher (eds), IWA Publishing on behalf of the World Health Organisation, London, UK.

WHO (2005). Water Safety Plans. Managing Drinking-Water Quality from Catchment to Consumer. WHO Geneva, Switzerland.

WHO (2017). Guidelines for Drinking-Water Quality. Fourth edition incorporating the first addendum. World Health Organization. ISBN 978-92-4-154995-0.

Chapter 2

The need for speed – rapid evolution of microbiological testing in drinking water

Patrick A. Whalen, Dan Kroll and Peter Gallant

2.1 INTRODUCTION

The detection of microorganisms in water has occurred for millennia. Initially, it happened incidentally – ancient cultures observed that when water was boiled prior to consumption, the incidence of sickness was dramatically reduced. They did not fully understand why this occurred, but the practice has continued to this very day as a reliable method for the disinfection of water. In the 19th century, the roots of modern water microbiology were laid through the work of John Snow in the City of London. Snow's work enabled others to conclusively link the discharge of sewage from overloaded cesspools into the Thames river to outbreaks of cholera and typhoid fever, dispelling the popular belief that these outbreaks were rooted in clouds of sickness descending upon the city, otherwise known as Miasmatic Theory (Halliday, 2001). Simultaneously, cities around the world were installing water filters to remove sediment from water sources prior to its delivery for consumption, mainly to improve aesthetic qualities (primarily suspended sediment, taste and odor) but also indirectly removing at least some microbiological contaminants. At the same time, scientists including Robert Koch and Louis Pasteur were pioneering the field of microbiology via microscopes and culture tests – and thus the science of microbiological sensors

doi: 10.2166/9781780408699_0015

was born. Into the 20th century, the widespread adoption of microbiological testing, coupled with water filtration and later chlorination, resulted in the most significant decline in mortality rates in modern history (Cutler & Miller, 2005).

Despite early rapid advances, microbiological testing methods such as plate-growth methods remained largely unchanged until the mid-20th century, when biochemical and molecular-based test methods were first developed. The reasons for these developments were many: a desire for more rapid results, greater specificity, greater objectivity and ease-of-use by non-skilled workers, and the continual search for the "holy grail" – sensors embedded on-line in the water distribution system that can detect and characterize bacteriological targets in real-time, all the time. By the early 21st century, science had entered the golden age of microbiological test method development with a multitude of different technology options for achieving one or more of the above-mentioned goals. However, scientists have still not found the "perfect" sensor technology, and it is not obvious that a single technology is emerging as the leading contender for widespread deployment in public water supply networks. Furthermore, there is a wide gap between the available technology, end-user capability to deploy and manage sensor networks, and the nature of the regulatory compliance environment such that even if the perfect microbiological sensor existed, it would be a struggle to deploy rapidly across the industry.

This chapter will provide both a historical review of microbiological detection technologies as well as an overview of selected biosensor technologies for water quality monitoring, with a focus on the challenges that must be overcome to ensure the successful deployment of advanced microbiological sensor technologies in water supply networks.

2.2 ANCIENT AND MEDIEVAL TIMES – EARLY MICROBIOLOGICAL SENSING

Without water, life cannot exist. It is therefore no surprise that human civilization developed close to sources of fresh water. Though water quantity was typically the deciding factor in where communities were founded, records indicate that our forerunners realized the benefits of a high-quality water supply thousands of years ago. For example, Sanskrit writings document the use of charcoal (now activated carbon) (Enzler, 2018) and Egyptian hieroglyphs mention the use of alum (USEPA, 2000), both of which are still in use today (Figure 2.1). Greek writings and even the Bible mention the use of filters to remove impurities (APEC Water, 2013). In general, these treatments were used to improve the aesthetic quality of water and led to the theory that if it is clean, it is safe to drink.

Unfortunately, that theory was not necessarily always true due to the hidden microbiological threats that lurked beyond view of the human eye. Indeed, other cultures such as the Chinese inadvertently discovered how to make water microbiologically safe through other means – that being, tea. By boiling water to make tea, they also disinfected the water. In Medieval Europe, there was at least an anecdotal awareness of microbiological contamination risks in even clean-appearing water that led many to quench their thirst with wine or beer (Harris & Grigsby, 2009). In some ways, these actions were in response to results gleaned from the first microbiological sensors – gross observation of the impact contaminated water had on other people.

Figure 2.1 Drawings on the walls of Egyptian rulers Amenophis II and Rameses II (APEC Water, 2013).

Gross observation also led civilizations to seek sources of water elsewhere when waterborne illness occurred. Similarly, when water quantities became limited and began to restrict the growth of cities, methods were developed to bring in additional sources of safe water. In Rome a series of aqueducts (Figure 2.2) were developed over a 500-year period and allowed it to become the largest city of its time, far larger than the water resources within its periphery were able to sustain both from a perspective of quantity and quality (Enzler, 2018).

During the industrial revolution, western civilization grew at an exponential rate – especially in urban centres. This put pressure on these major population centres to supply sufficient quantities of clean water to sustain that population growth. Some cities resorted to building extensive water supply networks to bring water from far away locations, similar to the Roman aqueducts. Others began experimenting with methods that could convert contaminated water into clean water, such as slow sand filters which were first deployed in Scotland in the early 1800s (Blake, 1956). Neither of these solutions, however, brought forth a direct and impactful public awareness to the risks of microorganisms in drinking water supplies and their linkage to the water cycle. Ironically, it was the birth of the water closet that did exactly that.

Figure 2.2 Roman Aqueducts (Cartwright, 2012).

2.3 19TH CENTURY – LINKING THE WATER CYCLE TO HUMAN HEALTH

In 19th century England, the cesspit was a ubiquitous piece of infrastructure designed to capture and store human waste underground (Morris, 2009). The next phase of waste-handling technology, the water closet – invented in the late 18th century (Hardy, 1984) – was becoming more popular. To handle the elevated water flows from this new apparatus, city planners and engineers began to install modern sewer systems to allow these cesspits to drain more rapidly and not overflow. The drainage point of these systems were natural bodies of water, such as the Thames river.

Starting in the 1830s and continuing through to the 1860s, large swathes of London were overcome by outbreaks of cholera (Morris, 2009). The leading theory of the day was that a cloud of sickness had descended upon the city. This Miasmatic Theory (Halliday, 2001) had been used to explain such epidemics in the past, which seemed to come and go with the changing of the seasons. This time, however, the outbreaks did not stop despite seasonal change.

The physician John Snow began to explore the situation with great interest. His research began in 1849 and came to a head with the 1854 Soho epidemic (Morris, 2009). He found that most instances of illness and resulting fatalities occurred near certain cisterns. He further found that fatalities not in that specific area often correlated with families that obtained water from the same cisterns, mainly on the basis that they preferred the taste of that water. He thus concluded that this was not a case of miasma outbreaks; rather, it was a waterborne outbreak.

After several months of investigation, two sources of contamination were identified. First, the integrity of some cesspits had failed, with their contents leaching into cisterns within their proximity. Second, the water source of some cisterns (namely, those being fed by draw points from the Thames river used by the Southwark and Vauxhall Waterworks Company) were located downstream of sewer system discharge points. It was thus concluded that the source of these outbreaks was human faecal material. Snow's study (Figure 2.3) was a major turning point in the history of public health and is regarded as the founding event of the science of epidemiology (Morris, 2009).

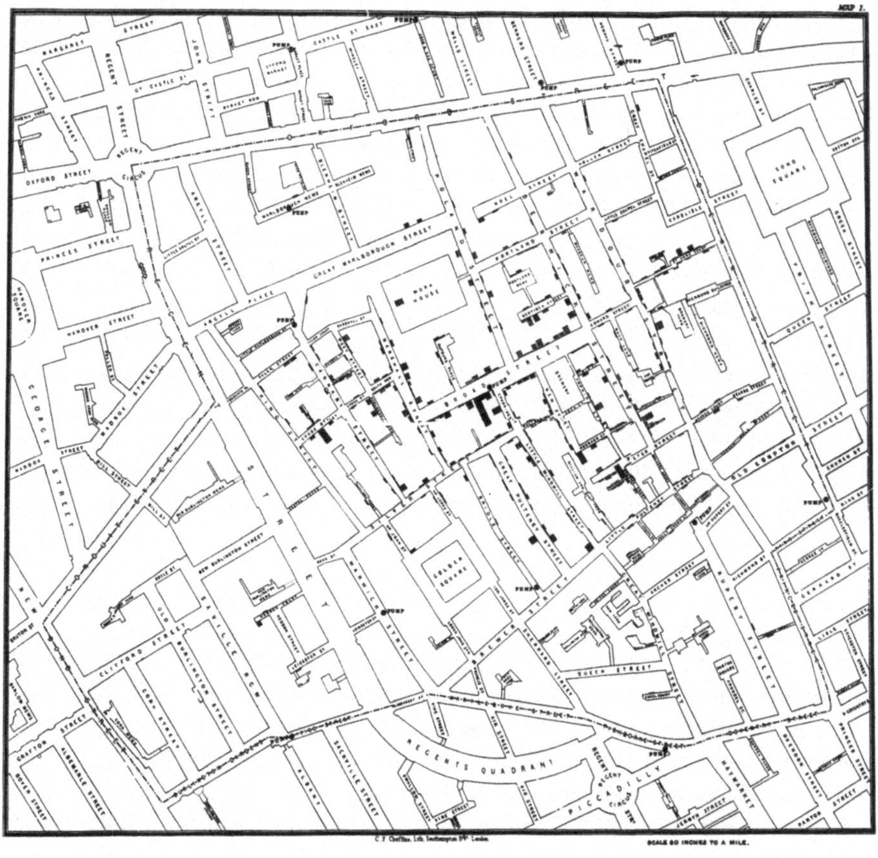

Figure 2.3 Original map of cholera cases in the Soho epidemic of 1854 (Barton, 2018).

Snow's revolutionary work instigated significant political controversy, and his theories were not widely accepted by the time of his death in 1858. However, further debate and study resulted in full acceptance even by his most prominent opponents such as William Farr in 1866. In the following years, work done by other prominent researchers including Louis Pasteur (France) and Robert Koch (Germany) led to the birth of today's most common microbiological sensors: microscopes and growth-based culture tests. It came to be in the 1880s that the Germ Theory of disease overtook Miasmatic Theory as the leading explanation for human infection (USEPA, 2000). A "golden era" of bacteriology ensued, in which the theory quickly led to the identification of the actual microorganisms that cause many diseases, waterborne or otherwise.

On the back of these discoveries, several advances in water and wastewater treatment gained increasing adoption. Filtration was identified as a means to improve water quality prior to human consumption. Slow sand filtration was developed in the United Kingdom in the early 1800s and improved upon in the United States as rapid sand filtration in the late 1800s (Melosi, 2000). At the same time, mechanisms to contain human waste and prevent contamination of drinking water sources were developed, such as the activated sludge process in the early 1900s (Beychok, 1967). These factors combined to provide a multi-barrier approach against the threat of microbiological contamination in drinking water supplies, but they alone were not enough to mitigate the risk to acceptable levels.

Chlorination was first proposed in 1902 and was met with heavy resistance from the public, which feared the concept of 'doping' their drinking water (Enzler, 2018). Resistance to this was all but overcome in 1916 when the city of Milwaukee lost chlorination for a period of 7 hours due to operator error, resulting in 60 fatalities and over 100,000 illnesses (Becker, 1974). The resulting widespread adoption of chlorination by villages, towns and cities in the United States reduced the incidence of cholera and typhoid from 100 to 0.1 cases per 100,000 inhabitants over the first half of the 20th century, which is often referred to as the single greatest achievement in public health improvement in the 20th century (CDC, 2012; Cutler & Miller, 2005).

2.4 20TH CENTURY – ESTABLISHMENT OF REGULATORY FRAMEWORKS

By the early 1900s, public awareness about the risks associated with microorganisms in drinking water supplies had been established, as had the understanding of the water cycle and its impact on water quality. The major causes of waterborne disease had been identified and robust solutions for their prevention had been developed along with the means to confirm their effectiveness. All the tools were therefore available to solve the problem; it was now a matter of ensuring wide adoption through laws and regulations. The first example of such regulations was laid in the mid-19th century when the City of London passed the Metropolis Water Act of 1852 to ensure that all water supplied to the city would be filtered (City of London, 1852).

In 1914, the Congress of the United States passed the Clean Water Act, which mandated certain minimum provisions around water treatment and management (USEPA, 2000). In the 1960s, these regulations were strengthened primarily in response to deteriorating source water quality caused by waste discharged from rapidly growing industrial production. This led to a major overhaul of water regulations in the form of the Safe Drinking Water Act in 1974 (USEPA, 2000). At this point microbiological quality issues had given way to mostly chemical issues as the primary concern, such as industrial wastewater discharge causing contamination in the form of eutrophication. Later it was recognized that an over-reliance on chlorination to address microbiological concerns had created another problem in the form of disinfection by-products, which are created by the interaction of disinfectants such as chlorine with organic matter present in source water.

In general, modern regulations prescribe the following preventative measures and testing methods to guard against microbiological contamination in most countries around the world.

2.4.1 Pressure

Water authorities are required to verify the maintenance of a positive pressure within their water supply at all times to prevent the infiltration of external contamination. However, it is recognized that maintenance of a positive pressure at all points within a water distribution system at all times is an impossible objective; hence, the quality of water within the distribution system must also be monitored on a constant basis (Snoeyink, 2006).

2.4.2 Turbidity

Turbidity is used as a general indication of contamination from soil or other debris that may imply the introduction of microorganisms. Though this technique can be deployed online to deliver results continuously and in real-time, it is not economically possible to cover all parts of the distribution system, nor is it possible to distinguish microbiological constituents from non-biological constituents (Schilz, 2018).

2.4.3 Disinfectant residual

Many operators carry out their duties under the assumption that if one maintains a measurable disinfectant residual, then microorganisms cannot exist. While there is logic to this thought, it is not a panacea – owing to the nature of microorganisms, it is possible to have both a measurable disinfectant residual and living microorganisms in a given sample. Furthermore, depending on the disinfectant used, water chemistry may have a substantial impact on its efficacy, such as the mechanisms by which chlorine is impacted by pH (Hydro Instruments, 2010). Finally, not all disinfectants provide a residual (e.g. ultraviolet radiation), nor can all residuals be easily measured (e.g. non-oxidizing biocides), and this philosophy cannot cover situations where disinfectants are not used in the water system (e.g. private wells, small towns, and certain countries).

2.4.4 Faecal indicators

Once it became known that waste was the primary source of waterborne illness, it became easier to design microbiological tests that would detect the presence of microorganisms associated with faecal matter. The work of Koch, Pasteur, and others served as the basis for early test methods, including the Heterotrophic Plate Count (Bartram *et al.*, 2003). As use became more widespread, however, it became apparent that the rate of false positives in indicating faecal contamination was unacceptably high. This was due to the fact that this category of microorganism is one of the more prevalent in the natural environment and hence in water in general. Therefore, researchers developed more specific test methods such as tests for only coliform bacteria or even more specific still, tests for only *E. coli* (which is most closely linked to faecal matter). These tests are the primary two relied upon today by water consumers around the world to indicate safe water and are embodied in easy-to-use methods from a variety of manufacturers that are readily commercially available. One example is the IDEXX Colilert-18 test that can detect and quantify *E. coli* and coliform bacteria after incubation for 18 hours (IDEXX, 2018).

Despite the robust evolution of water cycle management, treatment processes, monitoring programs and regulatory frameworks, access to clean drinking water is not yet universal – and where it is not readily available, the impact on public health can be devastating. The World Health Organization estimates that 2.1 billion people lack access to safe drinking water, twice that many lack adequate sanitation, and that contaminated drinking water causes over 500,000 diarrhoeal deaths each year (WHO, 2017). Clearly, civilization still has significant progress to make in terms of achieving 100% coverage in clean water and sanitation services – and not just in bringing it to those parts of the world where it does not currently exist.

2.5 21ST CENTURY – A NEW PARADIGM

Much has changed since the days of John Snow. Most of the world's population has access to clean and safe drinking water, with a significant proportion paying very affordable rates to have that clean water delivered directly to their home. Treatment processes are more affordable and accessible, and the basic monitoring tools are simple enough to be used by almost anyone. However, even where sophisticated drinking water infrastructure and management programs exist, there are some fundamental problems rapidly approaching that are forcing the sector to re-evaluate priorities.

2.5.1 Aging infrastructure and shifting demand

Water infrastructure in the western world is aging and subsequently deteriorating. This is leading to greater and greater rates of water loss in modern cities. Not only is this a waste of water and therefore money, it also represents a significant hazard. If water can leak out of the distribution network, contaminant-laden water can also potentially enter. Increasing ingress will invariably lead to a deterioration in water quality, which will lead to an increased rate of biofilm formation, the inevitable consequences of which will include microbiologically influenced corrosion, aesthetic (i.e. taste and odor) issues, or worse, human health impacts (Qureshi & Shah, 2014).

Furthermore, population growth is slowing down in some areas of the world while speeding up in others. Not only does this mean that water distribution infrastructure is being put under different and sometimes conflicting pressures, but utilities are also being forced to seek alternative water sources of questionable microbiological quality, such as direct potable reuse (Gale, 2018) or seawater desalination (Gleick, 2018). Put simply, systems that were designed for one purpose in the past may not be fit for their required purpose at present, not to mention that new infrastructure will be required in new major population centres. These factors all combine to create major operating and capital budgetary challenges (Qureshi *et al.*, 2014).

2.5.2 Changing workforce

The demographics of the water sector are changing rapidly (Brueck *et al.*, 2010). Many experienced water operators and engineers are retiring and taking with them substantial institutional relating to operation and data interpretation best practices. This leaves water utilities in a difficult situation given ever-constrained operating budgets as to how they secure equivalent knowledge moving forward.

2.5.3 Consumer awareness

Consumers are becoming more aware of and informed about the products they use and consume, including the water they use and consume. There is however no established educational resource to teach the public about what they should expect from their water supply outside of the obvious (i.e. water should be available when needed and it should be safe to use). Consumers can also connect with one another instantly via social media. This means that if a customer perceives a failure in water quality – whether they are correct or not – this information or misinformation can be disseminated widely nearly instantaneously.

Thus, improved engagement with and education of the public is of paramount importance. This problem will only become more pressing as sensor technology becomes less and less expensive and becomes integrated with point-of-use filtration systems and appliances, as many consumers will not be sufficiently informed in how to understand the data. Given these facts, it is inevitable that elevated consumer awareness will influence the evolution of regulatory frameworks in terms of the quantity, frequency and subsequent reporting of measurements made, which will add to water utility operating costs.

2.5.4 Evolving science

Over the past several decades, scientists, engineers and regulators have recognized that there is a myriad of waterborne threats to human health in addition to coliforms and *E. coli* (Krewski *et al.*, 2004). These include organisms such as cryptosporidium, giardia, enterococci, algae, amoeba, legionella and more. Additionally, there are several nuisance microorganisms that can affect water quality and aesthetics, such as nitrifying bacteria, corrosion-causing bacteria, and those contributing to odoriferous compounds such as geosmin and 2-methylisoborneol. Owing to the rapid advances in genomic technologies, this list is growing exponentially, and it is inevitable that new threats will be discovered that need to be monitored in the future, again adding to the costs of water utility operation.

While the incidence of waterborne disease has been dramatically reduced in many parts of the world through basic water cycle management, treatment, and monitoring, there are fundamental shifts in infrastructure availability, demographics, and awareness that create new challenges. Evolving science and consumer awareness will require the sector to advance its monitoring capabilities to cover a broader spectrum of contaminants more frequently, but this must be done in the context of limited budgets and workforce turnover all while continuing to drive adoption of the fundamentals in underrepresented parts of the world. While difficult, this is not an impossible task given the great many technology advancements made in recent times to enable civilization to increase productivity – in other words, to accomplish more given less time.

2.6 THE ECONOMICS OF TIME IN DRINKING WATER

There are many economic challenges facing the water sector in the early 21st century, including both capital and operating cost constraints. Given these pressures and recognizing that regulatory framework modifications will take time, the advancement and adoption of microbiological sensors requires a firm economic argument. In that vein, there is perhaps no more obvious area that the sector can benefit from the adoption of new microbiological sensors than in improving productivity and thus available time.

Current practices for microbiological problem recognition and resolution are reactive in nature. This is in large part due to the industry standard microbiological test being the culture test (WHO, 2006), which requires microorganisms to grow from presumptive single cells to entire colonies so that they can be counted, either through colorimetric indication or counted by the human eye. At best, the timeframe to obtain results from such tests is 1 day, but in many cases can be as much as a week. Furthermore, these tests tend to be designed for specific threats (e.g. faecal contamination) and will subsequently ignore most microorganisms.

Waiting at least a day to obtain results creates a few several side effects. Firstly, microbiological contamination is unlikely to remain static while waiting for results – contamination may grow, relocate or even dissipate while waiting for results. This means that the results once obtained may not reflect present state in the process. Secondly, the delay in obtaining results may lead to an ineffective use of resources such as with operators being prevented from doing further work or with water supplies remaining shut during the delay, such as in the case of line breaks and repairs. The following sections outline several benefits that water utilities would attain by having more rapid microbiological testing results.

2.6.1 Water and chemical conservation

Knowing sooner as to the location and extent of contamination could help to ensure that operators apply just the right amount of remedial treatment, for instance additional chemical disinfectant or flushing or both. This approach can save in the consumption of chemicals and the wasting of water, both of which have a real cost for the utility. For example, optimization of annual flushing activities to reduce their duration to when the objective (i.e. scouring of biofilm) has been achieved – rather than running for a prescribed period as done in most cases – has been shown to equate to $54,000 in annual savings for a utility serving 20,000 customers (Whalen *et al.*, 2014).

2.6.2 Time and travel conservation

Being able to have results on-site and in near-real-time could save on operator time and travel expense costs for water operators to carry out investigations and/or apply remedial treatment, such as in instances of customer complaints about aesthetic issues. A shorter time-to-result could also enable shorter time-to-action, thus avoiding having to go back-and-forth to the location of contamination. Furthermore, performing testing on-site would eliminate concerns around changes in microbiological characteristics of samples during transportation and storage. This could provide real reductions in operating costs through primarily less operator time and transportation fuel costs being used in problem-solving exercises. One reference has identified the potential for annualized savings of $50,000 for a utility serving an average-size municipality serving 10,000 customers (Whalen, 2016).

2.6.3 Boil Water advisory mitigation

Adopting a more proactive approach to drinking water management could improve the ability to mitigate waterborne outbreaks such as those in Walkerton, Ontario in 2000 (Salvadori *et al.*, 2009) and Milwaukee, Wisconsin in 1993 (Gradus, 2014) through reducing their duration and thus their impact, or by their outright prevention. Boil water advisories can result in a massive expense to water utilities directly through the cost of remedial actions and substitute supplies as well as indirectly through reputation impacts or worse actual outbreaks of disease. Presuming a significant boil water advisory to be a 1-in-20-year event, at a cost of $1.3 per person per day in substitute supply costs for 3 days, and $418 per illness at a 10% illness rate (Wagner *et al.*, 2005; all figures adjusted to 2018), preventing boil water advisory for an average (10,000 customer) municipality represents a $23,000 annual savings to the utility.

2.6.4 Infrastructure preservation

Staying ahead of microbiological contamination could minimize impact on infrastructure from issues like corrosion. This would lead to longer infrastructure life thus delaying capital cost expenditures. In the USA alone, forecast drinking water infrastructure investment requirements are $370B over 20 years across the country (New England USEPA, 2008). Based on 35,000 operating drinking water utilities, this equates to approximately $530,000 in annual capital expenditure requirements, on which conservative interest payments would be at least $20,000 per year, assuming a 4% interest rate).

Bringing these opportunity costs together for an average municipal operator serving 10,000 customers results in substantial available savings, as shown in Table 2.1.

Table 2.1 Savings opportunities from advanced microbiological sensors for a utility serving 10,000 customers.

Aspect	Annual Savings ($USD)
Water & chemical conservation	$27,000
Time & travel conservation	$50,000
Boil water advisory mitigation	$23,000
Infrastructure financing	$20,000
TOTAL OPEX	**$120,000**
Infrastructure deferral	$530,000
TOTAL CAPEX	**$530,000**

Clearly, having faster results leading to faster and more targeted action without delay results in significant economic efficiency and benefits to society. Over the past 150 years, the water sector has gone from taking years to days to solve faecal-based microbiological contamination in drinking water processes. Available technology now affords the sector the opportunity to further reduce this from days to minutes, in addition to offering the opportunity to expand beyond faecal contamination to additional health risks and operational concerns.

2.7 UPGRADING THE TOOLBOX

Looking to other sectors for inspiration, there are several technologies that could be leveraged for enhanced detection spectrum and faster response time. Sectors like food, personal care product, and pharmaceutical manufacturing leverage technologies such cell counting, optical sensing, rapid methods, and genomics to improve product quality and reduce inventory holding times to streamline the supply chain and reduce costs. The water sector has the same opportunity to leverage these newer technologies to improve product quality, reduce disruptions, and save time. The following sections discuss three categories of sensors that could improve detection spectrum and reduce response times.

2.7.1 Online cell counting and sensing

As mentioned in Section 2.4.2, the water sector routinely makes use of turbidity to detect intrusion of particles in drinking water processes. This technique is broad spectrum, but perhaps too broad in that it does not differentiate between biological and non-biological material. In the pharmaceutical sector for example, instrumentation such as laser-refraction and in-line optical technology has been used for many years to provide a 24/7/365 warning against specific microbiological incursions in raw, process, and final product water streams.

Recently, such tools have been introduced into the drinking water sector. One example is BACMON (Grundfos, 2017), an in-line camera-based microscope technology that photographs particles in cells and compares against a reference database to confirm the presence of microorganisms. Another is flow cytometry (Van Nevel *et al.*, 2017), a cell counting and sorting technique that has long been a mainstay in the medical sector, though tends to interrogate a small sample size and may require staining to increase specificity and to distinguish live cells from dead cells.

Such techniques can be very beneficial in that they may be always on and are sensitive to low levels of microbiological contamination, though depending on the technology they may not be particularly precise and can be quite expensive. Deployment of such technology would be ideal at strategic locations within a water distribution system such as at reservoirs and booster stations and are an ideal first line of defence when placed alongside other online sensors such as disinfectant residual and turbidity. These sensors would save utilities considerable time in providing alerts to contamination as quickly as possible so that operators could be dispatched to investigate.

2.7.2 Portable and rapid microbiological methods

Rapid, portable methods have been a mainstay of the food processing sector for over 30 years, where having a fast and on-the-spot indicator of biological contamination to verify surface cleanliness at any point in the manufacturing process to ensure no accumulation of biological matter that could lead to the manifestation of pathogenic microorganisms. These methods have completely changed the economics of the food sector in terms of risk avoidance and inventory management (IMMR-4, 2014).

One example of such a technique is the 1st Generation ATP test method, which has also been applied to limited success in certain water applications such as in commercial and industrial cooling systems. These methods however were never designed for water systems, where greater robustness and a quantitative cell count result was required. Over the past decade, a 2nd Generation ATP test method has become commercially available and is designed specifically for water systems. The ATP test is useful in that it detects all living cells and the results are easy to interpret. However, this method is also non-specific and must be used in the context of understanding that water systems are not sterile and there is an underlying background level of microbiological contamination.

Portable tools such as these provide operators with the ability to take the test into the field and identify where and how much microbiological contamination exists at any point in the treatment and distribution system as fast as they can sample, which can lead to a near-real-time problem recognition and correction cycle. They also enable on-site re-testing when suspect results are obtained, which is not possible with laboratory testing due to the process environment having changed by the time results are returned. These methods are useful for guiding and confirming remedial actions given that these actions are typically broad-spectrum and serve as an ideal complement to the online sensors mentioned in 2.7.1 and in that regard are unlikely to ever be replaced. However, these methods do not fulfil the requirement for specificity in the types of microorganisms detected.

2.7.3 Microorganism identification

Since its original discovery, deoxyribonucleic acid (DNA) has been the most widely studied biological cell component. This is not surprising, given that it represents the code for all life. Genomics technology provides the most specific indication possible of the type of biological cell present in the sample – even more specific than a culture test. Indeed, whereas there are only several thousand developed culture media formulations for specific categories of microorganisms, the development of genomics technologies has led the sector to estimate that in total there are over one trillion species of microorganisms on Earth, only 0.001% of which have been identified to date (Bakalar, 2016).

At present, genomics-based methods are mainly focused on DNA measurement, and generally come in two types: quantification (via polymerase chain reaction, or PCR), and identification (for example via next-generation sequencing). In the former, a 'primer' is used to provide a template to locate DNA of the same sequence in the sample, hence one must know what is being looked for before starting the test. In the latter, all individual DNA strands are read and mapped against a reference database to translate into actual microbiological species (a process called bioinformatics). Due to its sensitivity to the target microorganisms, great care must be taken in testing execution to not provide erroneous results.

In present form and compared to current industry standard test methods, these tools are extremely powerful, but can be expensive and time-consuming. In many cases, knowing the specific microorganisms present in the sample can assist with troubleshooting the source of process failures, but may not necessarily change the immediate remedial action. The field of DNA measurement is evolving rapidly, and while these methods are currently best executed in the laboratory, portable and online versions are under development and likely to come to market soon (Juhler *et al.*, 2012). Furthermore, one of the longest-standing deficiencies of genomics-based tests has been an inability to distinguish living from dead cells – and methods to overcome this limitation are being rapidly developed. This increases the likelihood that genomic technology will eventually replace culturing as the go-to mechanism of identifying specific microorganisms. For now, performing genomics-based tests in the laboratory are a powerful tool for utilities to use for troubleshooting complex problems or optimizing unit operations.

These technologies can provide real, tangible benefits for the water sector, all centred on saving time – that is, in minimizing the time to realize, interrogate, and correct contamination events. Each is fit for a different purpose, and none provide the same information as regulatory-approved culture-based methods – hence, they would be complementary. They also all require different skills to execute and to interpret data. All these points require a different way of thinking to achieve adoption in the water sector.

2.8 THE FUTURE AND THE HOLY GRAIL

For years, the water sector has been waiting for the perfect biological sensor to come along and advance its abilities to determine when and where microbiological contamination occurs (Tatari *et al.*, 2016). In contemplating what constitutes the perfect sensor, the following factors may be considered:

1. **Instantaneous time-to-result** – the sensor should provide near-immediate results.
2. **Continuous monitoring** – the sensor should be able to monitor 24/7/365.
3. **Sensitive to incursions** – the sensor should be sensitive enough to detect an incursion at its earliest possible stages.
4. **Specific to known microbiological threats** – the sensor should be specific enough to look for all the threats known at the time.
5. **Goes beyond health-based targets** – providing the capability to measure more than just a few specific microorganisms.
6. **Accuracy** – the sensor should provide valid results across the wide range with no false positives or negatives, including the ability to distinguish live cells from those that have been killed even if the cells are just recently dead.

7. **Reliable and low maintenance** – the sensor should be as reliable as possible and require no more maintenance than commonly used sensors.
8. **Wide coverage and/or portability** – the sensor should be able to be deployed anywhere and everywhere throughout the water management cycle.
9. **Easy to interpret** – the sensor should output results and insights that are easily understood by water process stakeholders.
10. **Affordability** – the sensor should be operationally cost effective and offer a reasonable return on investment of capital.

The above checklist is a very challenging set of criteria for any one technology to meet. This will require the sector to think differently about how to achieve its goals – such as looking at whether a combination of sensors could give the same benefits. This approach would require operators to be educated to think that all microbiological tests measure different characteristics, and that they need to seek multiple perspectives. Table 2.2 shows how the three technologies mentioned in the previous section perform against these goals:

Table 2.2 A comparison of three advanced molecular microbiology techniques against all three under a combined approach to incumbent culture methods.

Technology	Cell Sensing	Portable & Rapid Tests	Genomics	Combined	Culture
Instantaneous	X	X		**X**	
Continuous	X			**X**	
Sensitive	X	X	X	**X**	**X**
Specific			X	**X**	**X**
Inclusive	X	X	X	**X**	
Accurate		X	X	**X**	**X**
Low maintenance	X	X		**X**	
Wide coverage		X	X	**X**	**X**
Interpretation	X	X		**X**	**X**
Affordability	X	X		**X**	**X**

As seen in Table 2.2, an approach where these three types of sensors are deployed together in the drinking water system can provide all the attributes sought. However, all this new data poses a challenge to acceptance by the water community. Fortunately, one of the major advancements of the 21st century is the growth of cloud-based computing and intelligent computer-based decision support systems. All the data from these platforms could be integrated into a single smart system (Figure 2.4) that through basic provision of metadata by the operator could provide them with insights and recommended actions, rather than just random data points.

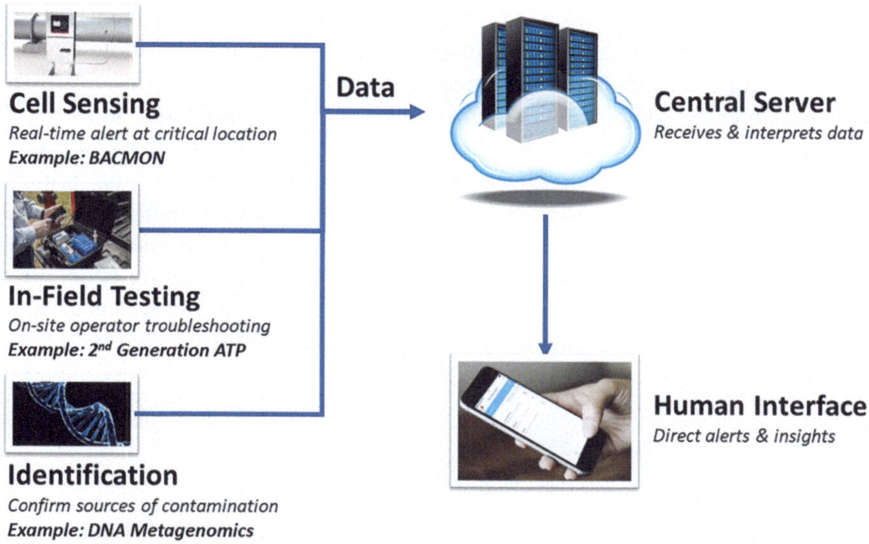

Figure 2.4 Combining three technologies to provide value to the water sector.

Taking the approach outlined here would flip microbiological management from a reactive to a proactive stance, thus enabling the savings of time, resources, and ultimately, money. As such, it is not necessary to wait for the 'perfect' biological sensor – the sector only need apply the tools at its disposal and integrate them into smart decision support systems to reap the benefits.

Achieving acceptance in this regard would require market education. Current training provided to water operators in microbiology is minimal – most are aware only of the required health-based testing and that if those targets are achieved, no further action is required. Taking the next step will require a change in how operators are educated on the risks and opportunities posed by a wider range of microorganisms, as well as adequate adoption drivers – whether regulatory or economic. As the former takes a long time to change, it is likely that adoption would occur faster by focusing on the latter – namely, the return on investment available through the adoption of more advanced monitoring tools such as those outlined in 2.6.

2.9 CONCLUSION

Over the course of 150 years, the water sector went from having an anecdotal awareness of microbiology in water to having robust, regulatory-enforced water management programs that were established help to ensure that drinking water is safe to drink. However, these regulatory frameworks have not kept up with societal changes nor the needs of utilities to stay ahead of emerging risks and address the operational challenges posed by microorganisms. The technology exists today to take a quantum leap in water microbiology management, and its adoption simply requires a change in approach by water utilities to go beyond what is required by regulations. Given the economic pressures these utilities are encountering and given the emergence of more and more unforeseen threats, it is imperative that the sector works to change the landscape in a proactive rather than a reactive way. The need for speed is borne from finding faster ways of detecting and addressing microbiological contamination, and this can be done through technology available today. It is not just about time to result – it is about time to action, and it is about incorporating economics into the traditional health-based conversation to provide a robust, sustainable framework that can overcome microbiological risks both today and tomorrow in sustainable ways.

2.10 REFERENCES

APEC Water (2013). The History of Clean Drinking Water. https://www.freedrinkingwater.com/resource-history-of-clean-drinking-water.htm (accessed 11 March 2018).

Bakalar N. (2016). Earth May Be Home to a Trillion Species of Microbes. *The New York Times*, 23 May, 2016.

Barton M. (2018). John Snow and the 1854 Cholera Outbreak. *Past Medical History*. https://www.pastmedicalhistory.co.uk/john-snow-and-the-1854-cholera-outbreak/ (accessed 15 March 2018).

Bartram J., Cotruvo J., Exner J., Fricker C. and Glasmacher A. (2003). Heterotrophic Plate Counts and Drinking-water Safety: The Significance of HPCs for Water Quality and Human Health. IWA Publishing, London, UK.

Becker E. (1974). A Century of Milwaukee Water, A Historical Account of the Origin and Development of the Milwaukee Water Works. Milwaukee Water Works, Milwaukee, USA.

Beychok M. R. (1967). Aqueous Wastes from Petroleum and Petrochemical Plants. John Wiley & Sons Ltd., New York, USA.

Blake N. M. (1956). Water for the Cities: A History of the Urban Water Supply Problem in the United States. Syracuse University Press, Syracuse, USA.

Brueck T., Isbell M., O'Berry D. and Brink P. (2010). Water Sector Workforce Sustainability Initiative. Water Research Foundation, Denver, USA.

Cartwright M. (2012). Aqueduct. *Ancient History Encyclopedia*. https://www.ancient.eu/aqueduct/ (accessed 15 March 2018).

CDC (2012). A Century of US Water Chlorination and Treatment: One of the Ten Greatest Public Health Achievements of the 20th Century. https://www.cdc.gov/healthywater/drinking/history.html (accessed 11 March 2018).

City of London (1852). An Act to make better Provision respecting the Supply of Water to the Metropolis. 15 & 16 Vict. C.84.

Cutler D. and Miller G. (2005). The role of public health improvements in health advances: the twentieth-century United States. *Demography*, **42**(1), 1–22.

Enzler S. M. (2018). History of water treatment: Significant historical events forming the basis for today's water treatment systems. https://www.lenntech.com/history-water-treatment.htm (accessed 11 March 2018).

Gale S. F. (2018). Battling Water Scarcity: direct potable reuse poised as future of water recycling. *WaterWorld*. http://www.waterworld.com/articles/print/volume-29/issue-9/editorial-features/battling-water-scarcity.html (accessed 15 March 2018).

Gleick P. (2018) Why don't we get our drinking water from the ocean by taking the salt out of seawater? *Scientific American*. https://www.scientificamerican.com/article/why-dont-we-get-our-drinking-water-from-the-ocean/. (accessed 15 March 2018).

Gradus S. (2014). Milwaukee, 1993: The largest documented waterborne disease outbreak in US history. Water Quality & Health Council. https://waterandhealth.org/safe-drinking-water/drinking-water/milwaukee-1993-largest-documented-waterborne-disease-outbreak-history/ (accessed 13 March 2018).

Grundfos (2017). Grundfos Bacmon: Real-time bacteria monitoring. Grundfos Holding product flyer. https://net.grundfos.com/Appl/ccmsservices/public/literature/filedata/Grundfosliterature-5605901.pdf (accessed 15 March 2018).

Halliday S. (2001). Death and miasma in Victorian London: an obstinate belief. *British Medical Journal*, **323**, 1469–1471.

Hardy A. (1984). Water and the search for public health in London in the eighteenth and nineteenth centuries. *Medical History*, **28**(3), 250–282.

Harris S. and Grigsby B. L. (2009). Misconceptions About the Middle Ages. Routledge, London, UK.

Hydro Instruments (2010). *Basic Chemistry of Chlorination*. Telford, USA.

IDEXX (2018). Colilert-18 Product Page. https://www.idexx.com/en/water/water-products-services/colilert-18/ (accessed 11 March 2018).

IMMR-4 (2014) Industrial Microbiology Market Review, Fourth Edition: Global Review of Microbiology Testing in the Industrial Market. Strategic Consulting Inc., Woodstock, USA.

Juhler S., Skovhus T. L. and Vance I. (2012). A practical evaluation of 21st century microbiological techniques for the oil and gas industry. Energy Institute, London, UK.

Krewski D., Balbus J., Butler-Jones D., Haas C., Isaac-Renton J., Roberts K. and Sinclair M. (2004). Managing the microbiological risks of drinking water. *Journal of Toxicology and Environmental Health*, **67**, 1591–1617.

Melosi M. V. (2000). The Sanitary City: Urban Infrastructure in America from Colonial Times to the Present. Johns Hopkins University Press, Baltimore, USA.

Morris R. D. (2009). The Blue Death: Disease, Disaster, and the Water We Drink. Harper Collins, New York, USA.

New England USEPA (2008). Water Infrastructure Funding Options. United States Environmental Protection Agency, New England Branch.

Qureshi N. and Shah J. (2014). Aging infrastructure and decreasing demand: A dilemma for water utilities. *AWWA Journal*, **106**, 51–61.

Salvadori M. I., Sontrop J. M., Garg A. X., Moist L. M., Suri R. S. and Clark W. F. (2009). Factors that led to the Walkerton tragedy. *Kidney International. Supplement*, **112**, S33–4. doi: 10.1038/ki.2008.616.

Schilz T. (2018). Turbidity measurement key to drinking water treatment. *WaterWorld*. http://www.waterworld.com/articles/print/volume-20/issue-5/awwa-exhibitors/turbidity-measurement-key-to-drinking-water-treatment.html (accessed 15 March 2018).

Snoeyink V. L. (2006). Drinking Water Distribution Systems: Assessing and Reducing Risks. National Research Council. The National Academies Press, Washington, USA.

Tatari K., Corfitzen C. B., Albrechtsen H.-J. and Christensen S. C. B. (2016) Sensors for microbial drinking water quality. DTU Environment, Department of Environmental Engineering, Technical University of Denmark.

USEPA (2000). The History of Drinking Water Treatment. USEPA Office of Water.

Van Nevel S., Koetzsch S., Proctor C. R., Besmer M. D., Prest E. I., Vrouwenvelder J. S., Knezev A., Boon N. and Hammes F. (2017). Flow cytometric bacterial cell counts challenge conventional heterotrophic plate counts for routine microbiological drinking water monitoring. *Water Research*, **119**, 191–206.

Wagner M. M., Wallstrom G. L. and Onisko A. (2005). Issue a Boil-Water Advisory or Wait for Definitive Information? A Decision Analysis. AMIA Symposium.

Whalen P. A. (2016). 2nd Generation ATP Technology: Real-Time Microbiological Decision Support. BlueTech Forum, Dublin, Ireland.

Whalen P. A. and Tracey D. R. (2014). A Novel Solution to Flushing Optimization Through Rapid Microbiological Field Testing. Singapore International Water Week, Singapore.

WHO (2006). Guidelines for Drinking-Water Quality, First Addendum to Third Edition. World Health Organization, Geneva, Switzerland.

WHO (2017). Drinking Water Fact Sheet. http://www.who.int/mediacentre/factsheets/fs391/en/ (accessed 11 March 2018).

Chapter 3

Microbiological sensors for drinking water: Terminology and central concepts

Loren Ramsay

3.1 INTRODUCTION

Monitoring microbiological parameters in drinking water is an interdisciplinary field. Many different employees – including field and laboratory personnel, technicians, engineers, microbiologists and chemists – work together to create, design, implement and operate microbiological sensors as well as to interpret the resulting data. Microbiological monitoring methods are undergoing rapid change, as new innovations in analytical methods, electronics, engineering, computer capacity and other fields are constantly extending boundaries.

In such fields, it is important to develop a common terminology. Common terminology provides a framework for the different employees to engage in succinct communication, but also promotes a more vivid clarity for creative thinking about complex themes. Such terminology allows us to communicate central concepts, to distinguish between similar and unique techniques, to categorize and generalize phenomena and to provide a united foundation for triggering new ideas for future development. Common terminology helps avoid ambiguity, misconceptions and fuzzy thinking in general.

In this chapter, central concepts related to microbiological sensors are identified and a common terminology is developed.

3.2 MICROORGANISMS

Microbiological sensors focus on the measurement of organisms which are too small to be seen by the human eye without the aid of a microscope. These organisms

Editors: Torben Lund Skovhus and Bo Højris
doi: 10.2166/9781780408699_0035

include prokaryotes (unicellular organisms that lack a membrane-bound nucleus) such as bacteria and archaea as well as some eukaryotes (uni- or multicellular organisms with a membrane-bound nucleus) including protists, algae and fungi. Viruses will be excluded in this chapter as they are not normally considered to be microorganisms. It is estimated that there are on the order of one trillion species of microorganisms (Loceya & Lennona, 2017).

The diversity of microorganisms does not stop with taxonomy but is also exhibited in properties such as life stage, mobility and effect on humans and infrastructure. These properties are discussed in this chapter. Owing to this diversity, the job cut out for microbiological sensors is enormous.

It should perhaps be noted here that the living conditions for microorganisms in drinking water are quite unique. Often, conditions in drinking water are oligotrophic (low nutrient content) and aerobic (free oxygen is present). Temperatures may vary greatly according to the season (surface water sources) or be more stable throughout the year (groundwater sources, depending on the drinking water's residence time in the distribution network). Drinking water may contain disinfectant residuals although this is not the case in countries such as Denmark and The Netherlands (Hijnen *et al.*, 2018). In general, these living conditions contrast with living conditions in the human body. This means that patterns of growth and survival of microorganisms may be quite different in drinking water than under other conditions.

3.2.1 Dead or alive?

Due to the risk of disease, the main interest in microbiological contamination of drinking water is related to living microorganisms. Therefore, it is advantageous for a microbiological sensor to be capable of distinguishing between living and dead microorganisms. However, a simple division into living and dead microorganisms is inadequate for interpreting biosensor results. Several categories are described in Figure 3.1.

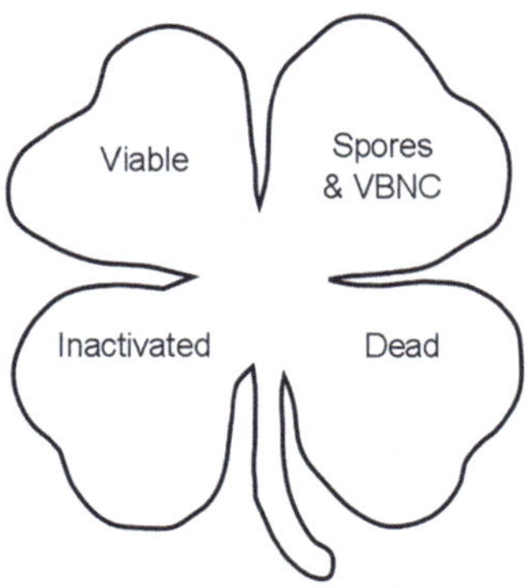

Figure 3.1 Different life states of microorganisms.

Viable: Viable microorganisms have metabolic activity and are capable of multiplying. Many of these may be detected on standard laboratory media. An interesting category of viable microorganisms which may affect their resistance to disinfection includes those growing inside other organisms, such as bacteria living in protozoa (Berry *et al.*, 2006).

Spores & VBNC: Under adverse conditions, some bacteria are capable of producing highly resistant structures known as spores. These dormant structures can return to normal growth when conditions improve. Microorganisms unable to form spores may enter a state called Viable But NonCulturable, or VBNC. These microorganisms do not grow on standard laboratory media although they are characterized by a low metabolic activity. As with spores, they are capable of revival under the right conditions (Oliver, 2010).

Inactivated: Microorganisms that have passed through a disinfection unit at the waterworks that utilizes ultraviolet light may have become inactivated. This means that they are still alive, but incapable of replication. Inactivated microorganisms are generally not of concern, as they typically lack potential to cause disease. However, some studies have indicated that there may be a risk of reactivation and regrowth following the practical application of disinfection (Basu *et al.*, 2007).

Dead: Dead microorganisms are incapable of multiplying and eventually degrade as their cell membranes lyse, and their content is eventually spilled. Following death, the stability of ATP (Nescerecka *et al.*, 2016) and enzymes is limited, as is the shape and electrical properties of the cell. These degradation processes progress at various rates (Rice & Bayles, 2008).

Different microbiological sensors have different abilities to distinguish between these categories.

3.2.2 Planktonic or sessile?

Microorganisms may also be categorized according to their mobility, see Figure 3.2. The cells of planktonic microorganisms are individually suspended in the water phase and are transported either passively with the water flow or actively through locomotion (e.g. flagella, the small appendage protruding from a cell with the purpose of cell movement). Planktonic microorganisms often make up only a small minority of the microorganisms found in drinking water distribution systems (Liu *et al.*, 2013).

Microorganisms may also be found in biofilms in the drinking water distribution system. A biofilm is a combination of cells of microorganisms and a slimy secretion of the cells called extracellular polymeric substance (EPS) which coats the surface (Flemming & Wingender, 2010). A biofilm may be considered a coordinated and functional community of microorganisms. When cells are attached in a biofilm, they are called sessile (immobile, attached to a surface), and may become physiologically distinct from planktonic cells of the same species. Biofilm may be found as a coating on the walls of pipes in the distribution system or as a coating on individual inorganic loose deposits (iron oxides, calcium carbonate, sand, silt, etc.) which have settled in the distribution system. When there is more than 1 g/m^2 of these loose deposits, the number of bacteria in the loose deposits (particle-associated bacteria or PAB) may exceed the number of bacteria in the pipe wall biofilm (Batté *et al.*, 2003).

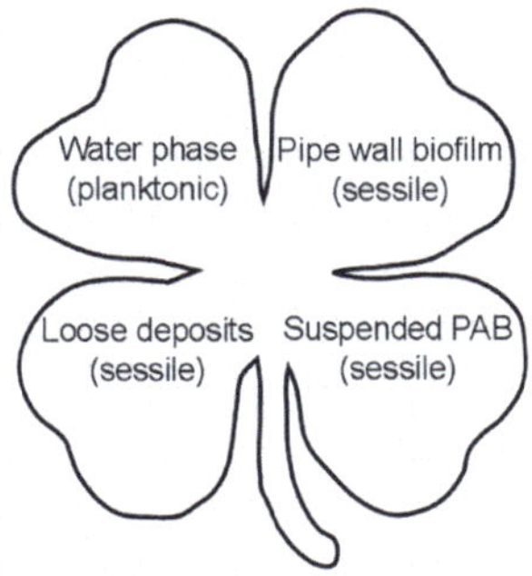

Figure 3.2 Categories of locations of microorganisms in drinking water distribution systems.

A special category of bacteria is suspended PAB. In distribution systems, particles in loose deposits may be re-suspended due to hydraulic peaks, or pieces of wall biofilm may detach and become suspended in the water phase. PAB provide special challenges as they may resist disinfection, impart a greater risk of bacteria reaching the customers' taps, and provide a source for bacterial regrowth downstream (Liu *et al.*, 2016). Especially relevant for this chapter is the fact that certain monitoring methods are unable to quantify PAB (for example plating methods would count a particle with many bacteria as a single colony).

3.2.3 Pathogenic or harmless?

A final categorization of microorganisms in this chapter is related to the effect of the microorganism, see Figure 3.3.

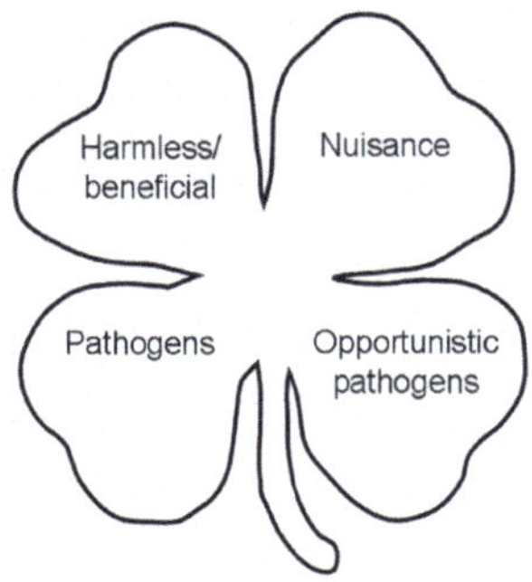

Figure 3.3 Categories of microorganisms in relation to their drinking water effect.

Many microorganisms are harmless to humans and to infrastructure and may even have a beneficial function. In the human body, microorganisms fulfil many important functions including microorganisms on our skin and in our gut (Enders, 2015). In drinking water treatment systems with biofilters, bacteria are the workhorses of the purification processes. Recent efforts have begun to identify bacteria in biofilters (Gülay *et al.*, 2016). The function of many bacteria in drinking water is unknown and until further insight is gained, they may simply be considered harmless, collateral microorganisms. In general, harmless and beneficial microorganisms do not need to be removed before water consumption. This approach is practiced in Denmark, where disinfection is not used at any point in the treatment plant or the distribution network.

Other microorganisms create a nuisance for drinking water providers and consumers. They may cause water quality issues and/or infrastructure damage and compromise consumer confidence. Examples of nuisance problems caused by microorganisms include production of foul-smelling/tasting compounds such as geosmin and methylisoborneol, infrastructure damage from microbial-induced corrosion, drinking water discoloration, and accelerated chloramine decay.

A pathogen is a microorganism that can cause disease. Prior to the implementation of drinking water chlorination in developed countries in the early 1900s, waterborne diseases such as cholera and typhoid fever were widespread (Sedlak, 2014). The total number of potential waterborne pathogens is very large. A short list of some of the best-known waterborne pathogens is shown in Table 3.1 (Leclerc *et al.*, 2002).

Table 3.1 Partial list of pathogens that may cause waterborne disease.

Category	Species	Disease
Bacteria	*Campylobacter jejuni*	Campylobacteriosis
Bacteria	*Vibrio cholerae*	Cholera
Bacteria	*Salmonella typhi*	Typhoid fever
Bacteria	*Shigella dysenteriae*	Dysentery
Protozoa	*Cryptosporidium parvum*	Cryptosporidiosis
Protozoa	*Giardia lamblia*	Giardiasis

Opportunistic pathogens are microorganisms that do not cause disease under normal circumstances but can cause disease when presented with the opportunity. These opportunities include a patient's weakened immune system, a penetrating injury and lack of competition (such as a disrupted gut flora). Important members of this group of microorganisms include *Legionella pneumophila, Staphylococcus aureus* and *Pseudomonas aeruginosa.* Some opportunistic pathogens have been observed to colonize plumbing fixtures in buildings (Falkinham, 2015).

Since the number of different pathogens is large, it is not feasible to measure each pathogen individually with a microbiological sensor. One approach is therefore to utilize key indicator microorganisms as surrogates for the presence of pathogens.

3.3 MICROBIOLOGICAL CONTAMINATION

For this chapter, microbiological contamination of drinking water is defined as:

> *The presence of any pathogenic or non-pathogenic microorganisms in drinking water above baseline values or the presence of microorganisms that cause the water to be less suitable due to technical, organoleptic or health concerns.*

Several aspects of this definition warrant closer inspection. Firstly, this definition includes the situation in which the number of microorganisms has increased above the baseline values, even if the number is below a critical level for the country in question. This is not the case in all definitions of microbiological contamination – some definitions only consider distinct events above a critical level to be contamination. Secondly, the definition considers non-pathogenic as well as pathogenic organisms to be contamination – as long as they are above baseline values or have an actual effect on technical, organoleptic or health aspects of the drinking water. This means that an unwanted increase of bacteria with no health concerns may also be considered contamination. Finally, although definitions of drinking water quality generally contain references to both safety and consumer trust, the definition of contamination above excludes a lack of consumer trust as true contamination. Consumer trust then becomes an issue that can be treated separately.

In this section, microbiological contamination will be discussed with respect to its causes, its location within the drinking water production chain from source to tap and various typical temporal patterns of microbiological contamination.

3.3.1 Sources of microbiological contamination in drinking water

In some water supplies, microbiological contamination is a frequent or even chronic problem, whereas other supplies have never experienced problems, but are still concerned about the risk of an accidental or deliberate contamination event. But where does this contamination stem from? This section divides potential causes into four categories and provides examples of various causes of contamination. A list and illustration of these causes is shown in Figure 3.4.

(1) *Source water.* Microbiological contamination may be intrinsic in source water that is utilized for drinking water production. Surface water sources are generally more vulnerable to microbiological contamination than

groundwater sources. Perhaps the most famous example of contamination from surface water sources is the Milwaukee *Cryptosporidium* outbreak in 1993. This massive outbreak of watery diarrhea caused 400,000 illnesses, including 69 fatalities and was due to *Cryptosporidium* oocysts in the source water that passed through the filtration system of one of the city's water-treatment plants (MacKenzie *et al.*, 1994).

(2) *Ingress.* Ingress is the unwanted entrance of microorganisms from a source outside of the supply system. Examples of ingress are widely varied. One example is the ingress of sewage through a leak in a water pipeline. Since pipelines are generally pressurized, however, the greatest likelihood of ingress is when there is a pressure loss, such as when a pump fails or when pipe repairs are carried out. Distribution systems in hilly areas may even experience a powerful vacuum on hill tops in this situation, as water in the pipelines drains towards lower-lying areas. Another example of ingress may occur during flooding following heavy rainfall, where contaminated water finds its way into the water supply system through unprotected groundwater wells. Human error resulting in improper pipe connections may also allow contaminated water to enter the distribution system through backpressure or a siphoning effect. Finally, ingress may occur intentionally as a terrorist act.

(3) *Mobilization.* Microorganisms are omnipresent in distribution systems, whether the system utilizes disinfection or not. These microorganisms are found in the biofilm of pipe walls as well as on the loose sediments in the pipes. Pressure and flow transients in the pipeline can cause these sessile microorganisms to slough off (biofilm) or be re-suspended (loose sediments), thereby being transported to the consumer. A special case of mobilization may occur in the water treatment process immediately following the backflush of a sand filter. The first hour following backflush may result in the release of particles with their associated microorganisms into the distribution network.

(4) *Growth.* Microbiological contamination may be created within the distribution system through the growth of microorganisms. This is often related to the microbiologically influenced corrosion (MIC) of pipe materials or leaching of chemicals from the pipeline material. Under the right circumstances, bacteria promoting MIC may proliferate in cast iron (Hu *et al.*, 2018) and copper (Pizarro & Vargas, 2016) pipelines. New plastic pipelines may leach organic compounds such as alkylphenols that are used as additives (Brocca *et al.*, 2002), promoting the growth of a biofilm on the walls of the pipe. A long transportation time in the distribution network (on the order of days) combined with elevated water temperatures can result in the aftergrowth of planktonic bacteria, depending on the amount of assimilable organic carbon in drinking water that is neither disinfected nor biologically stable (Prest *et al.*, 2016).

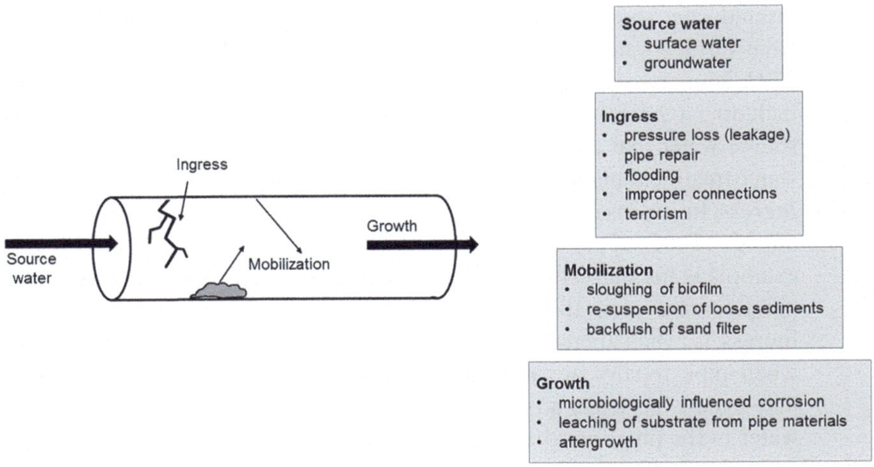

Figure 3.4 Sources of drinking water contamination with microorganisms (left, illustration and right, list).

3.3.2 Location of contamination sources

It can be conceptually advantageous to categorize microbiological contamination of drinking water by its location, since the cause, severity and remediation may vary, depending on where in the water supply system, the contamination has occurred. Figure 3.5 identifies a number of locations.

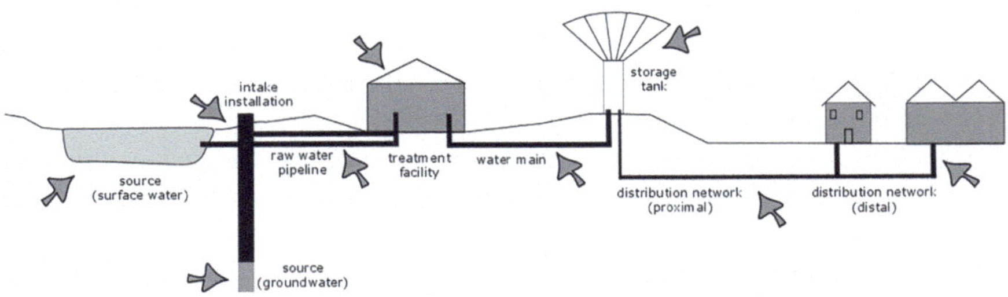

Figure 3.5 Location of potential sources of microbiological contamination.

As can be seen from the figure, there are many potential locations for contamination. However, the source of serious microbiological drinking water contamination frequently stems from the surface water or the distal distribution network. One of the most urgent goals of a contamination investigation is to identify the original location of the contamination.

3.3.3 Contamination patterns

Different microbiological contamination events exhibit different temporal patterns, ranging from steady, continual, low levels of contamination to sudden spikes with extremely high, but short-lived contamination. These temporal patterns are very important, since they dictate the required monitoring frequency as well as giving us valuable clues for identifying the source of contamination. In this chapter, five conceptual temporal patterns are defined, see Figure 3.6.

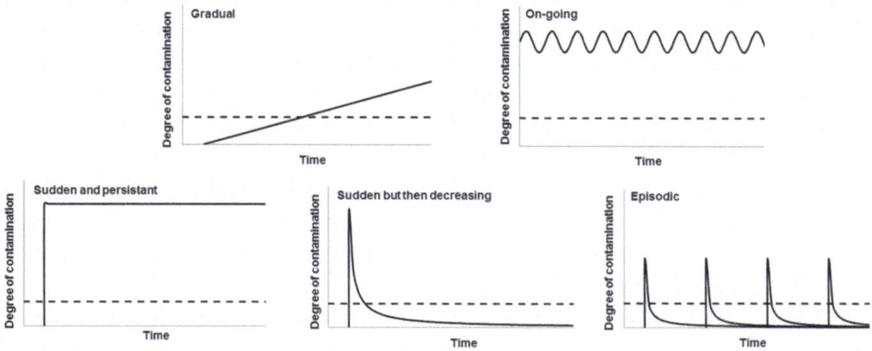

Figure 3.6 Conceptual temporal patterns of microbiological contamination in drinking water (dashed line: level of concern).

Gradual: Gradually increasing biological contamination may start well below levels of concern and increase over a period of days or weeks. One possible cause for this temporal pattern is growth (see Section 3.3.1). This pattern may be seen in the distal area of the distribution network in an area with summer tourists at the beginning of the off-season. At the end of the tourist season, water demand drops, the transportation time in the network increases and high late-summer temperatures may still be prevalent. This combination of the water age and temperature can lead to aftergrowth of microorganisms. Other possible causes include the infestation of higher organisms such as snails, beetles or worms that contain microorganisms in their gut. This temporal pattern requires a monitoring system with a high sensitivity so that the trend can be identified before it reaches the level of concern. The frequency of measurement may be relatively low, on the order of days to weeks.

On-going: On-going contamination may show similar results through extended time periods, such as the entire summer season or even longer. The cause for this pattern may for example be low-quality source water, inadequate water treatment, dirty pipelines or dead-end pipeline sections with very low flow giving rise to aftergrowth. This is perhaps the only pattern for which the current monitoring practice of infrequent grab sampling is suitable.

Sudden-and-persistent: Contamination incidents that appear suddenly and then persist are relatively rare. Such a contamination pattern could, for example, occur in connection with a deterioration of the source water quality or the ingress of contaminated water through a newly arisen pipe rupture, or long-term flooding. The monitoring requirements for this temporal pattern are focused on a high sampling frequency and less on the sensitivity of the method.

Sudden-then-decreasing: Sudden contamination, which then decreases with time, is more common than the previous pattern. This pattern could occur as a result of human error, which is immediately discovered and rectified, or as a result of a terrorist act. It could also be due to ingress of contaminated water from a short-lived flooding episode. Yet another potential cause is the running-in of new pipelines. Leaching of organics from the plastic material may cause the rapid formation of a young, poorly anchored biofilm of microorganisms living on these nutrients. In these early stages of biofilm formation, bacteria are more easily torn loose by transient flow/pressure, resulting in high bacteria counts in the water for a number of days until a more solid biofilm is formed. There is a risk that a sudden-then-decreasing pattern may go completely undetected unless the monitoring system has a high frequency of sample collection.

Episodic: This pattern involves recurring short-lived spikes of elevated numbers of microorganisms. These incidences may occur at random intervals rather than the even pattern shown in Figure 3.6. This conceptual pattern may be the most common contamination pattern found in drinking water supplies. The cause of episodic contamination may be related to heavy precipitation events or mobilization of contaminated pipe sediment due to changes in flow velocity. The pattern may also be related to contamination in connection with pipe repairs. Episodic contamination is typically a great challenge to the monitoring system. A low frequency of sample collection can render identification of the temporal pattern impossible. Only through a high sample collection frequency and high sensitivity can the true pattern be recognized and bring information value to the monitoring budget.

3.4 MICROBIOLOGICAL SENSORS

Recently, there has been much focus in the literature regarding technology advances in developing microbiological sensors for drinking water monitoring (Dejus *et al.*, 2017; Banna *et al.*, 2014; Lee *et al.*, 2012; Storey *et al.*, 2011; Hall *et al.*, 2009). In this section, various aspects of these sensors are discussed.

Before proceeding, however, two definitions are required. The first definition involves water quality monitoring in a water supply system:

> *The on-going execution of any set of observations or measurements of a physical, chemical or biological characteristic of water with the purpose of early warning, regulatory compliance, operational control, investigation of a contamination event or other water quality management objective.*

This definition requires that observation/measurement is a recurring activity, being executed over a time interval. These on-going measurements may be automated (for example the logging of conductivity values every 10 seconds in a pipeline) or manual (for example collection of groundwater samples and laboratory analysis for pesticides once every 5 years).

The second definition involves microbiological sensors, which are used for a specific type of water quality monitoring. For the purposes of this chapter, a microbiological sensor is:

> *Any on-line, automatic device which measures a parameter related to the microbiological water quality in real-time or near real-time, converts the resulting impulse to a signal and sends the signal to a display, data processor or data acquisition unit.*

This definition requires that the device which measures, converts and sends a signal is automated. This means that a method requiring manual steps may be very useful for water quality monitoring but does not fulfil the definition of a microbiological sensor. The definition describes sensors as on-line, meaning that samples are not removed from the premises prior to measurement. More specifically, a differentiation may be made between in-line measurement (where a probe placed directly in the path of the water) or at-line measurement (where a subsample is removed from the main stream but measured automatically on the premises). Note that the definition above would include a turbidity meter as a microbiological sensor, since it is on-line and automatic and since there often is a relationship – albeit tenuous – between turbidity and the microbiological water quality (Mohammed *et al.*, 2018).

3.4.1 Goals of monitoring with microbiological sensors

Microbiological sensors may not be suitable for every situation. For example, the need for sampling during the investigation and mapping of a contamination that has spread widely in a drinking water distribution system may suggest that it is more efficient to collect samples manually in a great many locations and subsequently analyse the samples in a central location, rather than installing microbiological sensors in each sampling location (Storey *et al.*, 2011).

In situations where microbiological sensors are deemed relevant, it is important to start by articulating the goal of the monitoring. Only after determining the goal can the most suitable microbiological sensor be selected and installed. And since no single microbiological sensor is suitable for all goals, more than one sensor may be required.

If possible, multiple benefits should be obtained if monitoring is to be sustained in the long run. If the monitoring goal is simply obtaining a warning if a rare terrorist act occurs, monitoring may be too expensive if no other benefits are involved. However, if the same sensor also can give information regarding operational control, it may be another matter.

Five major goals of monitoring with microbiological sensors are shown in Figure 3.7. A description of each of these goals is given in the following paragraphs.

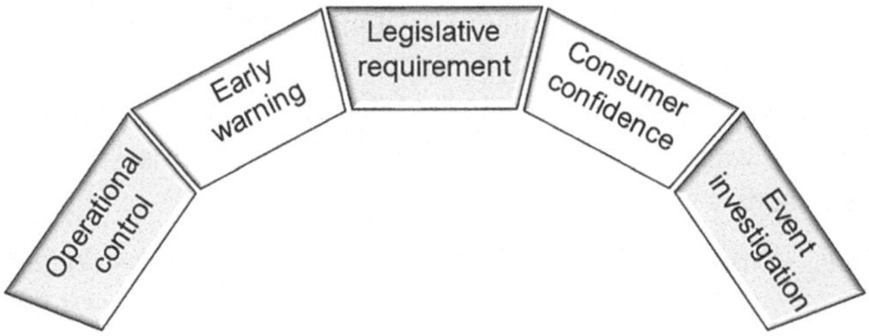

Figure 3.7 Five major goals of monitoring drinking water with microbiological sensors.

Operational control: The purpose of operational monitoring is to determine the effect of some action such as changes in the source water, changes in the treatment process (i.e. disinfectant dosage or commissioning new sand filters) or changes in the distribution network (i.e. commissioning new pipeline sections). An additional purpose can be related to investigations undertaken for a broader research or development activity, especially those that require identification of trends over time. Operational control monitoring also provides the development of a water quality baseline which may be useful for later comparisons during a contamination event.

Early warning: The purpose of early warning monitoring is to discover a contamination event in time to reduce risk to the consumer. A contamination event may be intentional or accidental and result in a rapid change in water quality, especially if it takes place in the distribution network. To be suitable for this goal, therefore, a microbiological sensor must be automated and rapid to allow for a timely response. Water security concerns have greatly increased since the terrorist attack in the USA on 11 September 2001. However, the frequency of accidental contamination is much greater than that of intentional contamination.

Legislative requirement: The purpose of legislative requirement monitoring is to document compliance. Regulations often define many monitoring aspects including sampling location, sampling frequency and even the analytical method to be used. Obviously, the selected microbiological sensor must be in alignment with the requirements to be useful for this goal. In many cases, use of microbiological sensors may not be applicable, as is the case where plate count methods are specified. It should be noted, however, that legislative requirements may change over time. For example, plate count methods currently in the legislation will likely eventually give way to faster methods (Van Nevel *et al.*, 2017).

Consumer confidence: Microbiological water quality is a major issue for many consumers. Although consumers often lack insight into actual contamination risks faced by a local drinking water supplier, they are still concerned with water quality. Knowing that the microbiological quality is monitored day and night can therefore increase consumer confidence. Consumer confidence is internationally recognized as a legitimate goal of drinking water production and is often stated as such. The goal described by the Bonn Charter, for example, is "Good safe drinking water that has the trust of consumers" (International Water Association, 2004).

Event investigation: Immediately following the onset of a contamination event, event investigations are often called for. These investigations may include a wide variety of activities such as source identification, contamination mapping and characterization of the contamination. To be useful in these activities, microbiological sensors must be mobile so that they can be installed quickly at a variety of locations in the distribution network.

3.4.2 Sensor attributes

Water quality deterioration in connection with drinking water production and distribution may appear quite suddenly. Microbiological contamination in distribution pipelines is an example of this, with deterioration sometimes occurring within minutes. Other changes in water quality may happen much more slowly. Figure 3.8 illustrates the time involved in water quality deterioration for different scenarios.

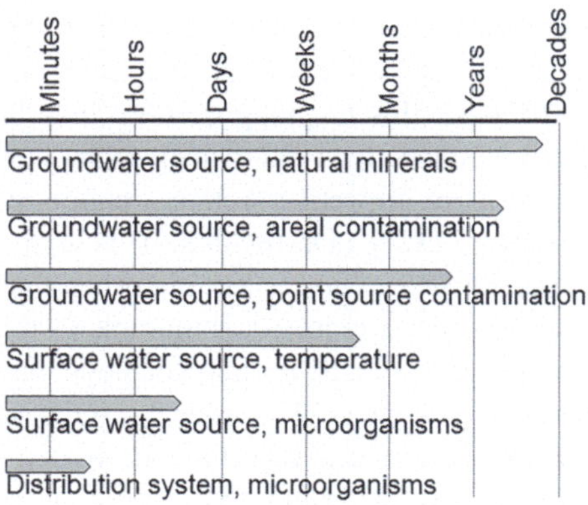

Figure 3.8 Typical time interval of water quality deterioration for selected drinking water parameters in various scenarios.

Fast measurement: Perhaps one of the most valuable attributes of a microbiological sensor is the capability of making fast measurements. Especially in connection with early warning of microbiological contamination, for example, fast measurements are of the essence. Two other attributes become important as a consequence of fast measurement. The first of these is frequent sampling. A fast measurement loses much of its value if samples are not collected frequently.

Therefore, a combination of frequent sampling and fast measurement is a desired attribute for microbiological sensors. The total time should preferably be on the order of minutes or seconds. The second attribute that is a consequence of fast measurements is automation. If sampling and measurement are to take place on the order of minutes, manual work would require an operator around the clock (including night-time, weekends and holidays) dedicated to monitoring. Therefore, automation in microbiological sensors is highly desirable. Fast measurements, frequent sampling and automation provide an opportunity for instigating a rapid response such as closing of valves to prevent the unchecked spread of the contamination in the distribution system, flushing a pipeline, and warning the consumers to boil their water before drinking.

Sensitivity: In many countries, one of the main regulatory requirements for microorganisms is the determination of *E. coli*. The criterion is often set at <1 CFU/100 mL (CFU stands for Colony Forming Unit). This criterion often coincides with the detection limit of the analytical measurement. This is not a desirable situation, since there is no warning when the concentrations begin to approach the criteria. It is therefore preferable that the sensitivity of the measurement is a factor 10 or more below the criteria. Sensitivity is therefore an important attribute for microbiological sensors.

Cost and mobility: When a contamination event is discovered, there is often the need for mapping of the extent of the contamination in the distribution network. This activity requires taking measurements at many different locations throughout the network that could not have been predicted prior to the detection of the contamination event. A practical approach for this mapping exercise may be to collect samples manually at the many different locations and to analyse the samples centrally. Here, an attribute of mobility is more important so that a temporary laboratory can be set up for the duration of the event investigation. This approach requires only a single sensor, whereby the attribute of cost is less important than an approach where multiple permanent installations are required. If multiple installations are required, the sensor must be economically suitable for mass deployment.

Specificity: Some methods such as ATP may be quite sensitive to the total number of microorganisms but have a very low sensitivity to specific bacteria. Others are sensitive to a specific bacterium, but insensitive to the total number of microorganisms. For the investigation of a contamination event, characterization of the type of contamination is helpful as it can assist in identifying the source and the health risk. In these cases, a sensor with the capability of specificity, i.e. being able to identify the dominating species of bacteria is needed.

These examples demonstrate clearly, that microbiological sensor attributes are closely connected with the goal of monitoring. Therefore, a single listing of desirable attributes given in a prioritized order without reference to the monitoring goal does not make much sense. Table 3.2 below is just one interpretation of which attributes are most important for which monitoring goal. Desired attributes should be selected for each individual situation.

Table 3.2 Examples of which sensor attributes (rows) are most important for which monitoring goal (columns).

Attributes	Operational Control	Early Warning	Legislative Requirement	Consumer Confidence	Event Investigation
Fast response		×		×	
Frequent sampling		×		×	
Automation	×	×		×	
Specific			×		×
Sensitive			×		×
Low cost		×			
Low maintenance					
Complies with regulations			×		
Robust					
Telemetry		×		×	
Few false negatives			×	×	×
Few false positives					×
No waste stream					
Easy interpretation		×	×		
Battery driven					
Wide measuring range					
Multiple parameters					
Mobility					×

3.4.3 Metrics and measurement principles

The use of concentration as a metric for gauging chemical water quality is commonplace. Whether speaking of calcium which is important for bones and teeth or naturally occurring arsenic which can cause cancer after an extended latency period, the concentration metric with units of mg/L or μg/L is used. Drawing an analogy between chemical and microbiological measurements in drinking water can be very instructive. Chemical measurements encompass a long list of completely different parameters, many of which are familiar and some which are not. A similar situation holds true for microbiological parameters. Here, there are a great number of genera and species of microorganisms to be measured, some of which have familiar names and many which do not. Microorganisms may include the bacterium *Nitrobacter* that are involved in the desirable removal of nitrite and the protozoan pathogen *Cryptosporidium*, which can cause gastrointestinal illness. In neither the chemical nor microbiology case would a complete description of the quality of the water be obtained by measuring a single parameter. Through this analogy it is abundantly clear that the search for a single one-size-fits-all microbiological sensor for drinking water is futile, just as it would be futile to search for a single chemical sensor that will analyse for all chemical compounds. A single sensor can never give a full picture.

With chemical measurements, the focus is typically quantitative analysis. With microbiological measurements, the situation can be more complicated. In many instances, a quantitative analysis of the number of organisms is of interest. In other instances, however, qualitative information is the goal. Next Generation Sequencing is an example of a method that provides qualitative information regarding the dominating species or identifies similarities between different microbiological communities (Eckert & Skovhus, 2018).

Overall, measurements can be divided into culture-dependent and culture-independent techniques. Interestingly, there seems to be a gap between the results of these techniques with very little overlap of taxa (Stefani *et al.*, 2015).

The classical culture-dependent technique is the heterotrophic plate count or HPC (Allen *et al.*, 2004). With this technique, microorganisms in the water sample are incubated and allowed to grow into visible colonies on a plate with nutrient medium. The results, which are given in units such as colony-forming units per volume are highly dependent on the precise conditions of the method. This is especially true with respect to nutrient medium, inhibitors, incubation time and temperature. This dependency on the conditions used means that HPC results are worthless unless conditions are stated, leading to the need for standardized conditions (APHA/AWWA/WEF, 2012). Since the HPC has been used for decades, a lot of experience has been gained in interpreting results. However, the method is difficult to automate, it is time-consuming and particles with multiple bacteria attached may go unrecognized and count as a single colony. The method also results in a measurement of a certain subgroup of microorganisms that are not directly related to drinking water quality or pathogens. Since culture-dependent techniques are time-consuming and require skilled operators, they are less suitable for use in a microbiological sensor.

Culture-independent techniques include a wide range of measurement principles. Widespread measurement principles in chemical analyses include gravimetry, electrochemistry, spectrometry, thermal analysis and many others. Many of these principles are less suited for microbiological analyses, due to lack of sensitivity, specificity, ease of automation or other requirements. The remaining field of techniques is, however, still very broad as seen in the literature (Zourob *et al.*, 2008; Moldenhauer, 2008; Sherchan, 2017). Fluorescence spectrometry is a widespread culture-independent measurement principle for microbiological parameters. It can be utilized to rapidly determine concentrations of ATP and enzymes contained in specific groups of bacteria. A disadvantage is that the fluorescing molecule must often be added to the system, requiring that the method includes addition of chemicals. Image analysis methods and impedance (cytometry) methods are generally fast, but not specific, measuring all intact cells. It appears that microbiological sensors in general require a trade-off between desired attributes, often between a fast response and microorganism specificity.

3.4.4 Monitoring strategies

In a perfect world, a microbiological sensor would excel at all the attributes in Table 3.2 such that the sensor would be suitable for every conceivable monitoring goal. In the extreme, this means that the perfect microbiological sensors would be so low cost and low maintenance that it could be installed at every tap in every home (point of use sensing), where it would measure all relevant and specific microbiological parameters. Individual decisions would not need to be made, since telemetry would ensure that every measurement is collected in real time, sent to an automatic algorithm that determines and initiates the appropriate remedial action immediately.

In the real world, however, there are many constraints. This opens the door for developing more or less intelligent monitoring strategies. A number of reports (Besmer *et al.*, 2017; Storey *et al.*, 2011; Banna *et al.*, 2014; Lee *et al.*, 2012) emphasize that it is currently not feasible to identify and quantify the many different parameters and that sensors should therefore "detect anomalous changes in the baseline water quality without specific regard to precision, accuracy or identification of the contaminant" (Hall *et al.*, 2009).

Such monitoring strategies may be visualized as shown in Figure 3.9 in which there are three separate monitoring tasks: alarm, sampling and characterization. By separating these activities, it is possible to find mature technologies today, even though the perfect sensor has not been developed. These technologies must simply be integrated into a functional system solution.

Figure 3.9 Practical monitoring strategy showing three separate tasks.

The three tasks of this practical monitoring strategy are described briefly below:

Task 1: Alarming. This task requires a sensor that automatically carries out fast and frequent measurements in-line or at-line. The sensor must trigger a clear alarm if a significant change in an indicator parameter, related to microbiological water quality, is noted. It is naturally preferable that an alarm is sufficiently trustworthy to trigger remedial action, i.e. that there are few false positives and no false negatives.

Task 2: Sampling. The alarm from Task 1 sets automatic sampling in motion. This means that appropriate samples are collected, regardless of the time of day. These may include filtered or non-filtered samples as well as grab or time-proportional samples.

Task 3: Characterization. In this task, the automatically collected samples are manually gathered and analysed, typically at a central location. Due to the complex nature of some of the analytical methods, they may be time-consuming, and samples may have to be express mailed to a specialist laboratory. Typically, these analyses are specific in order to clarify the precise cause of the triggered alarm.

It should be noted, that each of the above tasks may require multiple methods. Therefore, a combination of microbiological sensors, a combination of various types of samples and a combination of various analytical methods for characterization will often produce better results than a single sensor.

3.5 CONCLUDING REMARKS

Water quality monitoring for microbiological parameters in drinking water is complex. Microorganisms may be found in a variety of states and they produce very different effects depending on species. The sources, pathways and temporal patterns of contamination vary greatly. To afford the best protection, monitoring points should be strategically located as close to the point of use as feasible.

Microbiological sensors are emerging as valuable tools for drinking water monitoring, filling the serious shortcomings of culture-based analysis of grab samples. As with chemical sensors, no one microbiological sensor is endowed with all of the attributes that are desirable for all monitoring goals. To alleviate this technical burden, monitoring strategies may divide monitoring into three tasks: alarm, sampling and characterization. In addition, multiple methods working in combination for each of these tasks may be required to provide the highest level of consumer safety and trust.

Although numerous microbiological sensors are now developed, widespread implementation in the drinking water industry is yet to come. Widespread implementation of microbiological sensors requires that adequate drivers are in place. Currently, the legislative driver is totally inadequate, since it is still based on infrequent grab samples. Meanwhile, implementation must be driven by the drinking water supplier's desire to use best practice or by public pressure.

3.6 REFERENCES

Allen M. J., Edberg S. C. and Reasoner D. J. (2004). Heterotrophic plate count bacteria – what is their significance in drinking water? *International Journal of Food Microbiology*, **92**, 265–274.

APHA/AWWA/WEF (2012). Standard Methods for the Examination of Water and Wastewater. APHA/AWWA/WEF, Washington DC, USA, p. 541.

Banna M. H., Imran S., Francisque A., Najjaran H., Sadiq R., Rodriguez M. and Hoorfar M. (2014). Online drinking water quality monitoring: review on available and emerging technologies. *Critical Reviews in Environmental Science and Technology*, **44**, 1370–1421.

Basu S., Page J. and Wei I. W. (2007). UV disinfection of treated wastewater effluent: influence of color, reactivation and regrowth of coliform bacteria. *Environmental Engineer: Applied Research & Practice*, **4**, 1–8.

Batté M., Appenzeller B. M. R., Grandjean D., Fass S., Gauthier V., Jorand F., Mathieu L., Boualam M., Saby S. and Block J. C. (2003). Biofilms in Drinking Water Distribution Systems. *Review in Environmental Science and Biotechnology*, **2**, 147–168.

Berry D., Xi C. and Raskin L. (2006). Microbial ecology of drinking water distribution systems. *Current Opinion in Biotechnology*, **17**, 297–302.

Besmer M. D., Hammes F., Sigrist J. A. and Ort C. (2017). Evaluating monitoring strategies to detect precipitation-induced microbial contamination events in karstic springs used for drinking water. *Frontiers in Microbiology*, **8**, 2229.

Brocca D., Arvin E. and Mosbæk H. (2002). Identification of organic compounds migrating from polyethylene pipelines into drinking water. *Water Research*, **36**, 3675–3680.

Dejus S., Neščerecka A. and Juhna T. (2017). On-Line Drinking Water Contamination Event Detection Methods. Environment. Technology. Resources. *Proceedings of the International Scientific and Practical Conference*, [online] 1, p. 77. Available at: <http://journals.ru.lv/index.php/ETR/article/view/2627>.

Eckert R. B. and Skovhus T. L. (2018). Advances in the application of molecular microbiological methods in the oil and gas industry and links to microbiologically influenced corrosion. *International Biodeterioration & Biodegradation*, **126**, 169–176.

Enders G. (2015). Gut – the Inside Story of Our Body's Most Underrated Organ. Scribe, Melbourne, Australia.

Falkinham J. O. (2015). Common features of opportunistic premise plumbing pathogens. *International Journal of Environmental Research and Public Health*, **12**, 4533–4545.

Flemming H. C. and Wingender J. (2010). The biofilm matrix. *Nature Reviews Microbiology*, **8**, 623–633.

Gülay A., Musovic S., Albrechtsen H. J., Al-Soud W. A., Sørensen S. J. and Smets B. F. (2016). Ecological patterns, diversity and core taxa of microbial communities in groundwater-fed rapid gravity filters. *ISME Journal*, **10**, 2209–2222.

Hall J., Szabo J., Panguluri S. and Meiners G. (2009). Distribution System Water Quality Monitoring: Sensor Technology Evaluation Methodology and Results. EPA 600/R-09/076, US EPA, Washington DC, USA, pp. 1–60.

Hijnen W. A. M., Schurer R., Bahlman J. A., Ketelaars H. A. M., Italiaander R., van der Wal A. and van der Wielen P. W. J. J. (2018). Slowly biodegradable organic compounds impact the biostability of non-chlorinated drinking water produced from surface water. *Water Research*, **129**, 240–251.

Hu J., Dong H., Xu Q., Ling W., Qu J. and Qiang Z. (2018). Impacts of water quality on the corrosion of cast iron pipes for water distribution and proposed source water switch strategy. *Water Research*, **129**, 428–435.

International Water Association (2004). The Bonn Charter for Safe Drinking Water, p. 20. Available at: <http://www.iwa-network.org/downloads/1447946928-IWA bonn charter.pdf>.

Leclerc H., Schwartzbrod L. and Dei-Cas E. (2002). Microbial agents associated with waterborne diseases. *Critical Reviews in Microbiology*, **28**(4), 371–409.

Lee A., Francisque A., Najjaran H., Rodriguez M. J., Hoorfar M., Imran S. A. and Sadiq R. (2012). Online monitoring of drinking water quality in a distribution network: a selection procedure for suitable water quality parameters and sensor devices. *International Journal of System Assurance Engineering and Management*, **3**, 323–337.

Liu G., Verberk J. Q. J. C. and Van Dijk J. C. (2013). Bacteriology of drinking water distribution systems: an integral and multidimensional review. *Applied Microbiology and Biotechnology*, **97**, 9265–9276.

Liu G., Ling F. Q., van der Mark E. J., Zhang X. D., Knezev A., Verberk J. Q. J. C., van der Meer W. G. J., Medemal G. J., Liu W. T. and van Dijk J. C. (2016). Comparison of particle-associated bacteria from a drinking water treatment plant and distribution reservoirs with different water sources. *Science Reports*, **6**(1). dx.doi.org/10.1038/srep20367.

Loceya K. J. and Lennona J. T. (2017). Scaling laws predict global microbial diversity. *Proceedings of the National Academy of Sciences of the USA*, **113**, 5970–5975.

MacKenzie W. R., Hoxie N. J., Proctor M. E., Gradus M. S., Blair K. A., Peterson D. E., Kazmierczak J. J., Addiss D. G., Fox K. R., Rose J. B. and Davis J. P. (1994). A massive outbreak in Milwaukee of Cryptosporidium infection transmitted through the public water supply. *The New England Journal of Medicine*, **331**, 161–167.

Mohammed H., Hameed I. A. and Seidu R. (2018). Comparative predictive modelling of the occurrence of faecal indicator bacteria in a drinking water source in Norway. *The Science of the Total Environment*, **628–629**, 1178–1190.

Moldenhauer J. (2008). Overview of rapid microbiological methods. In: Principles of Bacterial Detection: Biosensors, Recognition Receptors and Microsystems, M. Zourob, S. Elwary & A. P. F. Turner (eds), Springer, New York, USA, pp. 49–79.

Nescerecka A., Juhna T. and Hammes F. (2016). Behavior and stability of adenosine triphosphate (ATP) during chlorine disinfection. *Water Research*, **101**, 490–497.

Oliver J. D. (2010). Recent findings on the viable but nonculturable state in pathogenic bacteria. *FEMS Microbiology Reviews*, **34**, 415–425.

Pizarro G. E. and Vargas I. T. (2016). Biocorrosion in drinking water pipes. *Water Science and Technology: Water Supply*, **16**(4), 881–887.

Prest E. I., Hammes F., van Loosdrecht M. C. M. and Vrouwenvelder J. S. (2016). Biological stability of drinking water: controlling factors, methods, and challenges. *Frontiers in Microbiology*, **7**, 1–24.

Rice K. C. and Bayles K. W. (2008). Molecular control of bacterial death and lysis. *Microbiology and Molecular Biology Reviews*, **72**, 85–109.

Sedlak D. (2014). Water 4.0: The Past, Present, and Future of the World's Most Vital Resource. Yale University Press, New Haven, Connecticut, USA.

Sherchan S. P. (2016). Real-time monitoring of microorganisms in potable water using online sensors. In: Handbook of Online and Near-real-time Methods in Microbiology, M. Lackner, W. Grabow and P. Stadler (eds), CRC Press, Boca Raton, Florida, USA.

Stefani F. O. P., Bell T. H., Marchand C., De La Providencia I. E., El Yassimi A., St-Arnaud M. and Hijri M. (2015). Culture-dependent and -independent methods capture different microbial community fractions in hydrocarbon-contaminated soils. *PLoS ONE*, **10**(6), 1–16.

Storey M. V., van der Gaag B. and Burns B. P. (2011). Advances in on-line drinking water quality monitoring and early warning systems. *Water Research*, **45**, 741–747.

Van Nevel S., Koetzsch S., Proctor C. R., Besmer M. D., Prest E. I., Vrouwenvelder J. S., Knezev A., Boon N. and Hammes F. (2017). Flow cytometric bacterial cell counts challenge conventional heterotrophic plate counts for routine microbiological drinking water monitoring. *Water Research*, **113**, 191–206.

Zourob M., Elwary S. and Turner A. (2008). Principles of Bacterial Detection: Biosensors, Recognition Receptors and Microsystems. Springer, New York, USA.

Section 2

Industry needs

Chapter 4

Water quality monitoring at Danish utilities – current state and needs for the future

Sarah C. B. Christensen and Ann-Katrin Pedersen

4.1 INTRODUCTION

Utilities find themselves in the gap between drinking water legislation based on grab sampling and growth based analyses on one side (Miljøstyrelsen, 2017) and on the other side an increasing number of online rapid sensors emerging on the market (Bridle *et al.*, 2014; Højris *et al.*, 2016; Tatari *et al.*, 2016; Besmer & Hammes, 2018).

Traditional approaches to microbiological drinking water quality monitoring are often insufficient to detect small contaminations and too belated to prevent contaminated water from reaching consumers. Major limitations of current approaches outlined by regulation are that monitoring is conducted as infrequent grab sampling at the consumer's tap and only small volumes of water are analysed. Thereby monitoring only provides a snapshot of the quality of already distributed water.

Real-time or close to real-time online sensors, which enable fast responses to changes in water quality are highly desired. In order to initiate the appropriate action based on sensor output, information such as content and concentration of contaminants, place of origin and geographical spread of a contamination is required. Currently available online sensors all have different strengths and limitations with regard to e.g. response time and specificity. Furthermore, costs may be an obstacle for thorough coverage of drinking water systems, required for identification of origin and spread of contaminations.

The aim of this chapter is to present utility needs and discuss how they correspond to currently available monitoring technologies. Different objectives of

Editors: Torben Lund Skovhus and Bo Højris
doi: 10.2166/9781780408699_0057

microbiological water quality monitoring are outlined and the ability of monitoring methods, applied at Danish utilities, to meet those objectives are evaluated. Finally, these points are compiled to present a utility vision of the sensor of the future.

4.2 OBJECTIVES OF MONITORING WATER QUALITY

Drinking water utilities monitor drinking water quality to ensure and document that drinking water production and distribution function and consequently that no consumers receive contaminated water.

The objective of European drinking water regulation is to "protect human health from the adverse effects of any contamination of water intended for human consumption by ensuring that it is wholesome and clean" (The Council of the European Union, 1998). The directive sets minimum requirements for water quality parameters and frequency of sampling, which is ratified in the Danish Water Supply Act (Miljø- og Fødevareministeriet, 2017) and associated Statutory order (Miljøstyrelsen, 2017). Danish drinking water is based solely on groundwater and distributed without disinfection residual (DANVA, 2017). In general, raw water is treated only by aeration and biological rapid sand filtration and in few cases UV treatment. Surveillance of microbiological water quality in Denmark reflects these conditions since systems without disinfection during treatment as well as distribution are vulnerable to intrusion of microorganisms in all steps from abstraction to consumer's tap. High background levels of naturally occurring bacteria in drinking water without disinfection (Besmer & Hammes, 2016) pose a challenge of detecting unwanted changes in microbiological water quality. This is both due to the fact that indicator bacteria typically constitute less than one in a million bacteria and that detection of increase of total bacteria is challenging due to the high bacterial background levels.

4.2.1 Legal compliance

4.2.1.1 Grab samples

Compliance with Danish regulation on microbiological drinking water quality is met by grab sampling at consumer's tap (Table 4.1).

The degree of monitoring varies among utilities (DANVA, 2017). Some small Danish utilities comply with minimum legal requirements, meaning one or two grab samples per year (Miljøstyrelsen, 2017) while other utilities take additional samples.

However, grab sampling at the consumer's tap does not fulfil the aim of assuring that no consumers receive contaminated water. A minimum of 18 hours incubation is required for detection of coliform bacteria, which means that water is consumed before a potential contamination is detected when sampling at consumer's tap. End-point grab sampling at consumer's tap is therefore often described as "too little, too late". In order to detect a contamination before it reaches consumers or detecting an emerging issue in due time to prevent a contamination, many utilities perform surveillance beyond legal requirements (DANVA, 2017). This includes monitoring of microbiological water quality during production and distribution. Some utilities supplement traditional grab sampling with either physical/chemical or microbiological sensors or filtration of large water volumes prior to analysis.

Table 4.1 Microbiological drinking water quality parameters given by the Danish Act on drinking water quality and surveillance (Miljøstyrelsen, 2017). The Act only defines parameter values for water sampled at consumer's tap.

Parameter	Unit	Parameter Value at Consumer's Tap
Coliform bacteria	Per 100 mL	Not detectable by the assigned method
Escherichia coli (*E. coli*)	Per 100 mL	Not detectable by the assigned method
Heterotrophic Plate Counts at 22°C	Per 1 mL	200
Enterococci	Per 100 mL	Not detectable by the assigned method
Clostridium perfringens (if affected by surface water)	Per 100 mL	Not detectable by the assigned method

4.2.1.2 Online microbiological sensors

Danish legislation on drinking water quality (Act on drinking water quality and surveillance) was revised in 2017 (Miljøstyrelsen, 2017). In the former version of the Act, monitoring by online sensors was not included. The current Act opens up possibilities for the use of online sensors during contamination events: "during contamination events, in which instances accredited water sampling cannot be awaited, or in instances where online monitoring or its like is suitable".

Use of online sensors during contamination events makes it possible to obtain results within a short time span. In order to act fast on online sensor results it is important that the sensor is capable of supplying data, which give sufficient information on the contamination to decide on the required action. Optimally, a sensor detects specific microorganisms or groups of microorganisms specific to intruding water and preferably comply with legal requirement to act as decision support for authorities. When applying new methods during contamination events it is crucial that the technology is implemented before occurrence of an event in order to know baseline microbiological levels obtained by the technology and get acquainted with possible false negative and false positive results. A limitation is that currently no real-time online microbiological sensors target indicator microorganisms for drinking water contaminations.

4.2.2 Detection of contamination

One objective of water quality monitoring is to detect contaminations. Detection of contaminations typically relies on indicator parameters such as indicator microorganisms, smell and taste, discoloration or change in chemical/physical parameters. To distinguish a contamination event from other events such as hydraulic disturbances, baseline levels must be established. Baseline levels such as absence of faecal indicators are broadly applicable while others are system specific and require series of measurements to distinguish normal fluctuations from contamination events.

Besides application of suitable monitoring parameters, identification of monitoring locations is important. Monitoring at points of abstraction and production aims at hindering contaminations from reaching distribution systems. Since distribution systems are vulnerable to contaminations, especially during maintenance work or leakages, surveillance of distribution networks is also critical. Identification of monitoring sites in distribution systems is complex and is a compromise between placing a monitoring site close to the end user versus in a central part of the distribution system. Several risk-based decision support tools have been developed (Larsen *et al.*, 2017) to aid identification of optimal monitoring sites.

In the event of contamination, three essential questions are addressed:

(1) How severe is the contamination?
(2) What is the source of contamination?
(3) Which actions must be initiated?

4.2.2.1 Severity of contamination

Concentration of contaminants and identification of microorganisms or groups of microorganisms in the contamination are key factors to identify the severity of the contamination. Another important aspect is coverage of the geospatial spread of contamination, which is dependent on the number and placement of monitoring sites and the available hydrological models to obtain knowledge on the transport pattern of the contamination. Finally, coverage of contamination duration depends on the frequency of sampling with regard to grab sampling or in case of use of online sensors, frequency of measurements.

4.2.2.2 Source tracking

Source tracking is crucial in order to stop the contamination. Data should reveal whether the contamination originates from faeces, soil/surface water or regrowth of bacteria within the system. To identify the point of origin of contamination, frequent monitoring of water sources and main junctions, such as water works outlets and entrances to sections in distribution systems, narrows down the search. Production may also be monitored closely at all steps of the treatment processes in order to enable rapid identification of whether a contamination emerges from e.g. raw water abstraction, raw water pipes, aeration, filtration or storage.

4.2.2.3 Decision support

Finally, monitoring data should be sufficient to decide, which actions are required. This includes where to flush and where to take additional water samples, whether to stop production from a specific source or shut off transmission lines or sections in the distribution system. Drinking water authorities decide on actions with regard to the public in agreement with public health inspectors. Monitoring data obtained by the utility must provide sufficient information to form a basis for decision on whether to issue a boiling advice, estimate how large an area is affected and what the duration of a boiling advice must be.

4.2.3 Prevention of contamination

Monitoring may not only be used for detection of contaminations but as a preventive means. A monitoring strategy focuses on preventing contamination from occurring or reaching the distribution system by detecting emerging issues such as wells in risk of collapse, leaky filter beds, small cracks in clean water tanks or stagnant water due to e.g. faulty closure of valves in the system. Monitoring at various points in the system such as wellfields and throughout production identifies place of concern. However, a low detection limit and rapid results are important in order to discover an emerging contamination.

4.2.4 Process control and optimisation

Changes in water quality may imply that a process is malfunctioning or not optimised. Process control should reveal issues such as insufficient aeration, low velocities during distribution, old and stagnant water, and lack of biostability causing regrowth of bacteria. Process control monitoring may be different from monitoring, which is directed towards detection of contamination and process control does not need to be within the frame of regulatory demands.

4.3 CURRENTLY APPLIED WATER QUALITY MONITORING METHODS

The field of drinking water quality monitoring is at an intersection between traditionally applied grab sampling combined with growth-based analysis methods and a rapid development of new online sensors. Legislation still only acknowledge grab sampling and growth-based methods for regular monitoring but with the revision of the Danish drinking water Act in 2017 (Miljøstyrelsen, 2017) use of online sensors during contamination events may increase. The use of additional methods has always been optional for utilities in addition to the regular monitoring program but whether the change in legislation will open up for authority decision-making based on new methods is still unexplored. The gap between methods approved by legislation and the rapid development of new methods is large. In this section, monitoring of Danish drinking water is covered in two parts reflecting different approaches:

(1) Standardised methods generally applied by Danish utilities
(2) Other commercial methods applied by some Danish utilities

4.3.1 Standardised methods generally applied by Danish utilities

To comply with legal requirements, Danish utilities perform grab sampling and analyse for:

- Total coliform bacteria
- *E. coli* (*Escherichia coli*)
- Heterotrophic Plate Counts at 22°C
- Enterococci (mandatory from October 2017)
- Heterotrophic Plate Counts at 37°C (until October 2017)

These growth-based analyses provide good information on the source and concentration of contamination, but have high detection limits. The microorganisms and groups of microorganisms are chosen as indicators for different types of contamination: total coliform bacteria without presence of *E. coli* typically occur due to intruding water or soil from cracks in tanks or due to insufficient procedures during repair work or new installations. Presence of *E. coli* indicates presence of faeces in the contamination but does not reveal whether it is due to animal faeces or wastewater. Elevated heterotrophic plate counts may be caused either by migration of nutrients from materials inside the supply system or inflow of impurities or regrowth in stagnant or slowly flowing water. Elevated heterotrophic plate counts at 37°C are typically registered when in-house installations are not mounted properly. Enterococci are, as *E. coli*, indicators of faecal contamination but typically Enterococci occur in lower concentrations than *E. coli* but persist longer in drinking water systems (WHO, 2017) thereby being a good indicator for assuring termination of a contamination.

Incubation times of the growth-based methods vary from 18 to 72 hours, which is not sufficient to react in a timely manner to contaminations once in the distribution system.

4.3.2 Other commercial methods applied by some Danish utilities

To meet the need for rapid responses and specific source tracking some utilities apply additional microbiological methods. The extent of usage varies greatly in terms of both prevalence at utilities and frequency of usage.

4.3.2.1 Commercial sensors applied by utilities

Several online and at-line sensors are available for additional monitoring besides required grab sampling. An overview of commercially available microbiological sensors currently applied by Danish utilities and sensors under development is provided by Tatari *et al.* (2016). Experiences from utilities were described for Grundfos BACMON (Grundfos), TECTA B16 (Veolia), BACTControl (MicroLAN) and ALARM (Colifast).

- *Grundfos BACMON*: Optical quantification of total bacteria and inorganic particles. At-line and rapid (10 minutes processing, samples every 10 minutes). Results presented as cells/mL and non-bacteria/mL.
- *TECTA B16*: Enzymatic activity of total coliform bacteria and *E. coli*. Manual sample loading. Semi-rapid (2–20 hours processing). Results presented as CFU/100 mL.
- *BACTControl*: Enzymatic activity of total coliform bacteria or *E. coli*. At-line and semi-rapid (2–4 hours processing, samples up to four times per day). Results presented as pmol/minute/100 mL.
- *ALARM*: Enzymatic activity of total coliform bacteria or *E. coli*. At-line and semi-rapid (6–15 hours processing, samples once per day). Results presented as presence or absence.

4.3.2.2 Additional commercial methods applied by utilities

Rapid methods:

- *Bactiquant-water (Mycometer)*: Based on enzyme activity thereby targeting a broad spectrum of bacteria. Manual but rapid (<30 minutes) semi-quantitative method. Requires establishment of background level.

Continuous sampling:

- *Volume sampler (HOFOR)*: Continuous filtration of water (adjustable flow and time) and subsequent manual sampling and analysis of the concentrated sample. Currently utilities apply analyses of coliform bacteria/*E. coli* in order to identify type of contamination (soil/surface water or faeces) but other analyses can be applied.
- *Pansi 1000/Alonda 1000 (Amphi-Bac)*: Continuous filtration of water (adjustable flow and time) and subsequent manual sampling and analysis of the concentrated sample. Alonda 1000 may be activated by sensor alarms, which trigger automatic sampling.

Analysis of specific organisms:

- MALDI–TOF MS (Matrix-Assisted Laser Desorption/Ionization – Time Of Flight Mass Spectrometry). In monocultures or samples with only few bacteria species present, MALDI–TOF Mass Spectrometry provides a rapid (few hours) identification to genus level. For inhomogeneous environmental samples, an overnight growth step must be included.

- PCR (Polymerase Chain Reaction) and sequencing of 16S and 18S. Current technologies enable identification of all organisms in a sample if previously identified and present in databases. Long analysis time when purchased at commercial laboratories (approximately one week).

Quantitative PCR (qPCR) is not discussed in this chapter since no application for water quality monitoring at Danish utilities has been reported. Quantitative PCR is a faster approach than PCR and sequencing and may target traditional indicator bacteria (Aw & Rose, 2012; Krapf *et al.*, 2016).

Besides microbiological methods, many utilities apply online sensors mainly at water works to monitor physical/chemical parameters such as turbidity, oxygen, conductivity, temperature, pH or ammonium. This is mainly used as process control but significant contaminations might be detected e.g. as rise in conductivity and turbidity or decrease in oxygen concentrations. In-depth knowledge about production and distribution systems may expose errors such as closed valves causing adverse hydraulic conditions alone by temperature measurements. As with all other water quality data this requires that irregularities are noticed and used actively.

4.4 RESEMBLANCE BETWEEN MONITORING OBJECTIVES AND CURRENTLY APPLIED METHODS

None of the methods, currently applied at Danish utilities meets all monitoring objectives listed in Section 4.2 (Table 4.2). This is expected, as statutory requirements are based on culturable methods with incubation times of a minimum 18 hours, which is not desirable in order to initiate fast actions. Since it will take some time before online sensors may be included as a means of fulfilling statutory monitoring requirements, online sensors are add-ons to utilities' regular monitoring programs. In the short run, a sensor may therefore not need to be able to fulfil the same functions as traditional grab sampling and growth-based analyses when serving only as a supplement. In the longer run, a valuable sensor is one that is able to replace currently applied methods by also fulfilling legal requirements and thereby eliminate duplication of effort. However, one sensor may not fulfil all monitoring objectives and the end goal is not necessarily to apply one sensor type for all purposes.

The online/at-line sensors described in Section 4.3 have not been widely implemented by Danish utilities. Most of the sensors (TECTA B16, BACTControl, ALARM) build on detection of total coliform bacteria and *E. coli*, which is desired by utilities in order to identify contamination source and severity. These sensors are able to obtain results in less than 18 hours, which is currently required by traditional methods. In order to obtain shorter analysis time than 18 hours some

Table 4.2 Sensors and analytical methods applied at Danish utilities (Information from Tatari *et al.*, 2016 is included). The table presents, which utility monitoring objectives from Section 4.2 are met by which sensors. Producer and principle of sensor/analytical method is described in Section 4.3.

Commercial Microbiological Sensors and Methods Applied at Danish Utilities and their Compliance with Utility Monitoring Objectives	Compliance with Legal Requirements (Section 4.2.1)	Identification of Group of Microorganisms or Specific Genera (Section 4.2.1, 4.2.2, 4.2.3 and 4.2.4)	Quantification of Microorganisms (Section 4.2.1, 4.2.2, 4.2.3 and 4.2.4)	Installation On-Site (Section 4.5)	Automated Sampling and Reporting (Section 4.4)	Analysis Time (Section 4.2.2, 4.2.3 and 4.2.4)	Continuous Sampling (Section 4.2.2, 4.2.3 and 4.2.4)	Limit of Detection (Section 4.2.1, 4.2.2, 4.2.3 and 4.2.4)
Growth based analysis of indicator microorganisms, assigned by regulation (Total coliform bacteria, *E. coli*, HPC 22C, Enterococci)	Yes	Total coliform bacteria, Enterococci *E. coli*, fractions of heterotrophic bacteria	Yes	No	No	18–72 hours	No	HPC:1 CFU/mL Coliforms/*E. coli*: 1 MPN/100 mL Enterococci: 1 CFU/100 mL
BACMON	No	No – total counts	Yes	Yes	Yes	10 min.	10 min. intervals	167 cells/mL
TECTA B16	No	Total coliform bacteria and *E. coli*	Yes	No	Manual loading, automated reporting	2–20 hours	No	1 viable target organism/100 mL
BACTControl	No	Total coliform bacteria or *E. coli*	No (enzymatic activity)	Yes	Yes	2–4 hours	No	0.1 pmol/min/100 mL
ALARM	No	Total coliform bacteria or *E. coli*	No (presence/absence)	Yes	Yes	6–15 hours	No	1 viable target organism/100 mL

(*Continued*)

Table 4.2 Sensors and analytical methods applied at Danish utilities (Information from Tatari *et al.*, 2016 is included). The table presents, which utility monitoring objectives from Section 4.2 are met by which sensors. Producer and principle of sensor/analytical method is described in Section 4.3. (*Continued*)

Commercial Microbiological Sensors and Methods Applied at Danish Utilities and their Compliance with Utility Monitoring Objectives	Compliance with Legal Requirements (Section 4.2.1)	Identification of Group of Microorganisms or Specific Genera (Section 4.2.1, 4.2.2, 4.2.3 and 4.2.4)	Quantification of Microorganisms (Section 4.2.1, 4.2.2, 4.2.3 and 4.2.4)	Installation On-Site (Section 4.5)	Automated Sampling and Reporting (Section 4.4)	Analysis Time (Section 4.2.2, 4.2.3 and 4.2.4)	Continuous Sampling (Section 4.2.2, 4.2.3 and 4.2.4)	Limit of Detection (Section 4.2.1, 4.2.2, 4.2.3 and 4.2.4)
Bactiquant-water	No	No - fraction of total bacteria	No (enzymatic activity)	No	No	<30 min.	No	Not available
Volume sampler + Colilert	No	Total coliform bacteria and *E. coli*	Yes	Yes	No	Optional filtration time + 18 hours	Yes	0.001 MPN/100 mL
Pansi 1000/ Alonda 1000 + Colilert	No	Total coliform bacteria and *E. coli*	Yes	Yes	Possibility of automated sampling with Alonda 1000	Optional filtration time + 18 hours	Yes	0.001 MPN/100 mL
MALDI–TOF MS	No	Identification of specific genera and often species	No	No	No	Few hours for monocultures and few species. Otherwise overnight growth	No	Concentrated samples, grab samples or incubated samples
PCR and sequencing	No	Identification of specific genera or species	No	No	No	1 week for total water analysis at commercial laboratories	No	Currently only applied on concentrated or incubated samples at Danish utilities

HPC = heterotrophic plate counts, CFU = colony forming units, MPN = most probable number.

sensors compromise on other parameters such as presence/absence output instead of quantification, a limited number of samples per day or detection of only total coliform bacteria or *E. coli* rather than detection of both concurrently. The sensor with close to real-time response (Grundfos BACMON) is not specific to genera or groups of bacteria and therefore does not provide results, which are easily applied by authorities as decision support in case of high bacteria counts.

Many utilities await two to three days incubation of heterotrophic bacteria in order to approve repair work or new installations in the drinking water system. A rapid response would facilitate onset of operation with minimum delay and thereby be beneficial to both consumers and utilities. Bactiquant-water is sometimes applied for this purpose but works only as a rule of thumb since it does not comply with legal requirements nor holds a simple correlation with heterotrophic plate counts.

Methods for detection of specific microorganisms (MALDI–TOF MS and PCR) are included in the overview of currently applied methods since they provide a valuable tool in contamination cases where source tracking is central. However, currently applied specific methods other than detection of *E. coli* are laboratory based and require sophisticated and expensive analysis equipment. A future sensor providing a rapid, specific, quantitative response on genus level, or even species or strain level, would make additional source tracking dispensable in most instances.

Methods of concentrating water samples (Volume sampler, Pansi 1000, Alonda 1000) in order to obtain low detection limits are included in the overview since they fulfil many of the utility monitoring objectives in combination with analysis of specific bacteria or groups of bacteria. Objectives of main importance in this regard are prevention of contaminations and source tracking. The currently applied filtering method (Volume sampler) at HOFOR utility requires manual sample collection and is limited by the incubation time of the applied analytical method. However, due to the low limit of detection, like Pansi 1000 and Alonda 1000, it allows for timely reaction and source tracking. Volume samplers filter large volumes of water (up to 100 litres) continuously from a split flow stream during an optional period of time. This enables continuous sampling and detection of low concentrations (0.001/100 mL if 100 litres are concentrated) of indicator microorganisms of microbiological contaminations. Any type of analysis may be performed on the concentrate but typically, samples are analysed for concentrations of total coliform bacteria and *E. coli*. When combined with analysis of total coliform bacteria and *E. coli* volume sampling comply with several of the utility monitoring objectives listed in Section 4.2, e.g. detection of contamination, prevention of contamination and source tracking. Two examples are given:

Case 1: Volume sampling as a preventive mean. Analysis of small volumes required by regulation does typically not reveal emerging contaminations, even when performed daily. Emerging issues are detected by volume sampling due to both large sample volume and continuous sampling. Low concentrations detected by volume sampling at water works often correlate with rain event in case of cracks in clean water tanks. Detection allows for repair while damage is still minor and thereby prevents emerging contaminations.

Case 2: Volume sampling for source tracking. Volume sampling may be applied as process control and source tracking at several steps during drinking water production. During investigations of a water works with reoccurring events of coliform bacteria, volume samplers were installed at the following points: raw water, after pre-filtration, after biological sand filtration and after storage in clean water tank before entry into the distribution system (Figures 4.1 and 4.2). Results from volume samples in autumn 2016 revealed that no coliform bacteria were present in raw water samples or samples collected after pre-filtration while volume samples collected after sand filtration and after storage contained coliform bacteria (Figure 4.1). Whereas the usual suspects are clean water tanks, the results revealed that the entry point of coliform bacteria was at the point of biological sand filtration, allowing for timely intervention at the right point in the treatment facility.

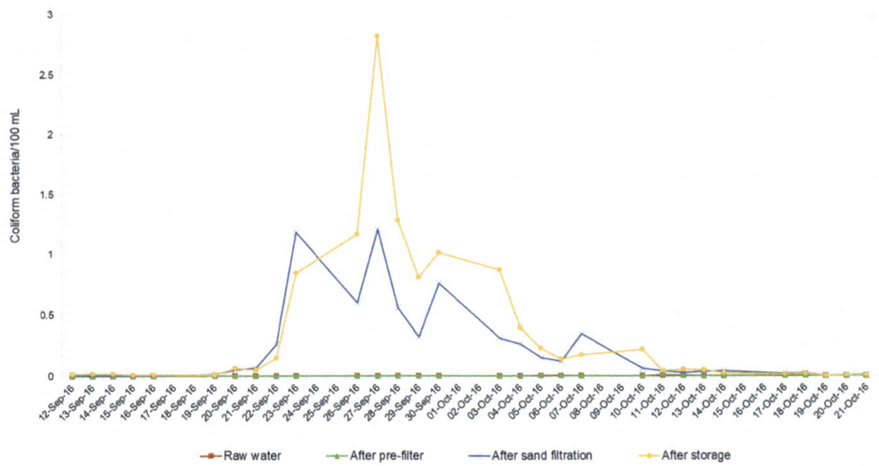

Figure 4.1 Coliform bacteria in water from a Danish water works. Samples were collected by volume sampling (filtration of 100 litres of water) at four different stages in the water works (raw water, after pre-filtration, after sand filtration and after storage) from 12.09.16 to 21.10.16. The x-axis shows date of sample collection.

A subsequent incident in autumn 2017 revealed presence of coliform bacteria in water sampled after the pre-filter, identifying the pre-filter as point of entry for surface water into the system (Figure 4.2).

Volume sampling is time-consuming with regard to sample collection and preparation as well as analysis time but currently serves as the most cost-effective method for timely targeted interventions at HOFOR utility.

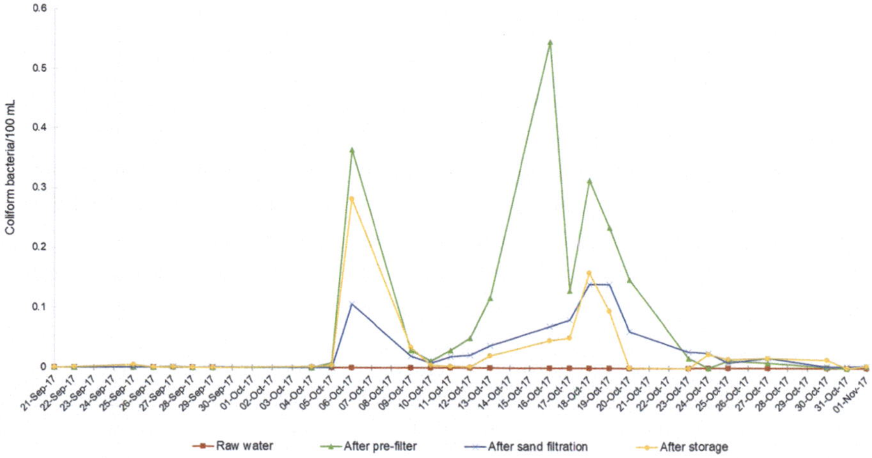

Figure 4.2 Coliform bacteria in water from a Danish water works. Samples were collected by volume sampling (filtration of 100 litres of water) at four different stages in the water works (raw water, after pre-filtration, after sand filtration and after storage) from 21.09.17 to 01.11.17. The x-axis shows date of sample collection.

4.5 FUTURE NEEDS FOR WATER QUALITY MONITORING

A list of key points about the "ideal sensor" from a utility point of view was formulated by a group of large Danish utilities (Tatari *et al.*, 2016). Subsequently HOFOR utility has extended the list according to regulation revisions and utility experiences and the list now contains seven main features:

- online or at- line installation
- rapid response
- detection of indicator microorganisms (e.g. *E. coli* and total coliform bacteria, Enterococci, *Bacteroides*) or various pathogens
- high sensitivity to detect very low concentrations of indicator microorganisms or pathogens in drinking water
- low maintenance requirements
- low false-alarm occurrence
- continuous sampling or high frequency of analyses

Most of the points relate to early detection of contaminations in order to execute timely implementation of appropriate intervention. A challenge for all sensors is to detect low numbers of contaminants within the high background concentration of naturally occurring drinking water bacteria. Currently a low limit of detection may be obtained by application of a pre-concentration step, which also extends the sampling period from snap shots to continuous sampling. To detect specific microorganisms within a short time span, promising methods are based on molecular or immuno-based techniques. One main obstacle is the low occurrence of indicator bacteria and pathogenic bacteria in drinking water sample compared to e.g. samples from the medical field where sensors with those features are currently applied (Sapsford *et al.*, 2008). In drinking water, pre-concentration and specific analyses seem to be an approach, which could be achieved within a foreseeable future.

Low false-alarm rate is crucial in order to rely on sensor results to act instantly. Low maintenance requirements are sought as well, which to a great extent is connected to the point that sensors currently are supplements to statutory monitoring and therefore typically not a main priority. Cost of methodologies is of major importance for the number of monitoring points a utility can apply throughout production facilities and distribution networks. Exhaustive coverage of a drinking water system enhances the ability to detect a contamination and locate origin. As with maintenance requirements, an economic restraint on implementation of sensors is currently that expenses to comply with legal requirements must be prioritized first and foremost.

Coverage of all parts of a drinking water system at all times is not possible and consequently contaminations and smaller irregularities may reach consumers. If a contamination is detected e.g. as consumer complaints, mobile sensor units could be installed on site and at key points upstream and downstream of the contamination. Furthermore, a laboratory version of any online or at-line sensor is desired, enabling control of water samples, which have been sampled manually at locations without sensors or where it is not feasible or possible to temporarily install a mobile sensor.

In a distant future, the number one utility tool is a sensor, which provides rapid analyses, continuous responses around the clock, quantitative results with low limit of detection, no false alarms, low level of maintenance, and in a price range that allows for thorough coverage of drinking water systems. Analysis parameters are several optional and changeable specific microorganisms. Instead of relying solely on indicators of contamination, pathogens are included not only as bacterial pathogens but also as viruses and higher organisms such as single-celled eukaryotic pathogens. The vast amounts of data are analysed to create alarms, and data procession with artificial intelligence detects irregularities and potential for process optimisation. Analysis parameters may even change from several specific microorganisms to a full microbiological community description. Such a sensor may not focus on unwanted bacteria alone but also serve as process control by monitoring functional drinking water bacteria. To best belief, *E. coli* and probably also total coliform bacteria will be maintained in order to uphold the long timeline of continuous use of these indicator microorganisms, which allows for comparison of long-established global monitoring results.

4.6 REFERENCES

Aw T. G. and Rose J. B. (2012). Detection of pathogens in water: from phylochips to qPCR to pyrosequencing. *Current Opinion in Biotechnology*, **23**, 422–430.

Besmer M. D. and Hammes F. (2016). Short-term microbial dynamics in a drinking water plant treating groundwater with occasional high microbial loads. *Water Research*, **107**, 11–18.

Bridle H., Miller B. and Desmulliez M. P. Y. (2014). Application of microfluidics in waterborne pathogen monitoring: a review. *Water Research*, **55**, 256–271.

DANVA (2017). Vand i tal 2017. *DANVA Statistik og Benchmarking* (Statistics and Benchmarking).

Højris B., Christensen S. C. B., Albrechtsen H.-J., Smith C. and Dahlqvist M. (2016). A novel, optical, on-line bacteria sensor for monitoring drinking water quality. *Science Reports*, **6**, 23935.

Krapf T., Kuhn R. M., Kauf P., Gantenbein-Demarchi C. H. and Fieseler L. (2016). Quantitative real-time PCR does not reliably detect single fecal indicator bacteria in drinking water. *Water Science & Technology: Water Supply*, **16**(6), 1674–1682.

Larsen S. L., Christensen S. C. B., Albrechtsen H.-J. and Rygaard M. (2017). GISMOWA: Geospatial risk-based analysis identifying water quality monitoring sites in distribution systems. *Journal of Water Resources Planning and Management*, **143**(6), 04017018.

Miljø- og Fødevareministeriet (2017). Lov om Vandforsyning. LBK nr. 125 af 26/01/2017. Miljø- og Fødevareministeriet, Styrelsen for Vand- og Naturforvaltning, j.nr. 029–00510 (Legal Act in Danish).

Miljøstyrelsen (2017). Bekendtgørelse om vandkvalitet og tilsyn med vandforsyningsanlæg BEK nr. 1147 af 24/10/2017. *Miljø- og Fødevaremin., Miljøstyrelsen, j.nr. SVANA-400–00054* (Statutory order in Danish).

Sapsford K. E., Bradburne C., Delehanty J. B. and Medintz I. L. (2008). Sensors for detecting biological agents. *Materials Today*, **11**(3), 38–49.

The Council of the European Union (1998). Council Directive 98/83/EC of 3 November 1998 on the quality of water intended for human consumption. *Official Journal L*, **330**, 0032–0054.

Tatari K., Corfitzen C. B., Albrechtsen H.-J. and Christensen S. C. B. (2016). Sensors for microbial water quality. Report. DTU Environment, Technical University of Denmark, Kgs. Lyngby, Denmark. Available at: http://orbit.dtu.dk/files/124881515/Sensors_for_microbial_drinking_water_quality_2016_w_logo.pdf (accessed 4 May 2018).

WHO (2017). Guidelines for Drinking-Water Quality: Fourth edition incorporating the first addendum. World Health Organization, Geneva, Switzerland.

Chapter 5

On-line monitoring of bacteria in drinking water systems: Expected benefits and new challenges for utility operators

Sophie Courtois, Guillaume Cussonneau, Gilles Fay and Zdravka Do-Quang

5.1 INTRODUCTION

Consumers expect water supply companies to deliver safe Drinking Water (DW) that meets health quality standards, and aesthetic aspects such as colour, turbidity, taste and odour. Meeting these expectations requires robust (multi-barrier) treatment design, proper operation of the treatment and distribution system, and adequate risk management. To this end water supply companies have developed quality management systems and monitoring strategies, using well established laboratory techniques.

Water quality and public health risks need to be analysed and minimized, but to do so utilities must have knowledge of their raw water (quality and catchment) and their treatment processes. Ensuring minimal risks in the distribution and domestic plumbing systems is the final step towards safe drinking water.

While conventional water quality monitoring programs are not always able to detect a microbial contamination of drinking water due to the limitations of grab sampling, on-line monitoring in DW systems offers the operators the ability to provide safety and respond to incidents (accidental or intentional). In past years there has been a growing interest in the use of on-line monitoring systems, mainly for reasons of their response time and recent security concerns.

On-line monitoring equipment can be installed as early warning systems for water intake, treatment process monitoring and main entry points to the

 Microbiological Sensors for the Drinking Water Industry
Editors: Torben Lund Skovhus and Bo Højris
doi: 10.2166/9781780408699_0073

distribution system. All monitoring systems should be able to distinguish abnormal changes from normal variations. This issue will be enhanced by the ability to detect, identify, and quantify contaminants in the shortest time-lapse possible. The general framework would be based on: (1) continuous on-line monitoring (Early Warning System); (2) rapid sample analysis (in the field when possible) and (3) definitive sample analysis (in the lab).

Currently, on-line sensors are frequently deployed for monitoring physical parameters such as flow rate, turbidity, pH and water temperature. Multi-parameter water quality monitors, or sensor panels, are also mainly used in finished water. Typical parameters and techniques used in these monitors include water quality parameters such as electric conductance, turbidity, dissolved oxygen, oxidation–reduction potential and free chlorine. The technical flowchart illustrated in Figure 5.1 highlights the fact that on-line monitoring should not be seen as an issue only related to sensor technology by itself but should cover all the path.

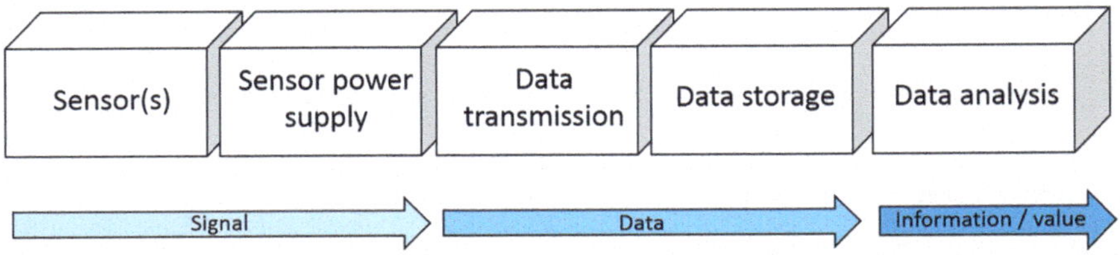

Figure 5.1 Flowchart for on-line water quality monitoring.

Missing so far were reliable on-line systems that measure bacterial concentration within short intervals, bridging the gap between the use of non-specific on-line sensors and the time-consuming lab methods (Lopez-Roldan *et al.*, 2013). The current development of new approaches for microbiological quality on-line monitoring is opening new perspectives in operator's practices for enhancing safety issues and operations improvements by integrating monitoring into vulnerability assessments and Water Safety Plans. We present here the expected benefits but also the multiple challenges raised for water operators in order to ensure the success of new monitoring strategies.

5.2 EXPECTED BENEFITS

There is currently a fast development in various sensors and on-line monitoring (analyser) systems that may have clear and multiple benefits for water utilities such as lower costs and more information in time and space. Data produced by on-line monitoring systems are expected to provide real-time information on water quality. This differs from the established practice of laboratory analysis which can only confirm that the water is of satisfactory quality, in many cases, after the water has been distributed.

Moreover, routine grab sampling followed by laboratory analysis collects only a relative too small dataset to describe sample quality variation. Potentially important events that occur between the less frequent grab sampling were lost or unusual results may be dismissed.

5.2.1 On-line monitoring in raw water at Drinking Water Treatment Plant (DWTP)

5.2.1.1 Fecal contamination monitoring

Microbial monitoring in resource water provides a good indication of the level of fecal contamination in the raw water and of seasonal and annual variation. The quality of source water varies with point and nonpoint discharges upstream from the intake or as a consequence of weather, biological activity and hydrologic cycle. Lakes and surface water quality present the highest variability in space and time although groundwater can also be affected by fecal contamination. Such information, combined with microbial risk assessment, is important for determining sufficient water treatment and disinfection to ensure safe drinking water. However regular monitoring based on weekly (or less) sampling does not necessarily reflect the deterioration of the raw water quality during acute pollution episodes, e.g. caused by heavy rainfall (Tryland *et al.*, 2011). Therefore, characterizing such events may be important for the optimization and daily control of the drinking water treatment plant.

To address this risk, microbial contamination monitoring is usually assessed by turbidity measurements performed at drinking water plants. Despite its lack of sensitivity (there is no direct relation between pathogen density in water and turbidity), the choice of this parameter is relevant because: (i) the suspended particles are a safe and supportive environment for pathogens and a place of protection against disinfectants; (ii) turbidity lowers the efficiency of disinfection processes by the presence of oxidisable associated organic matter. Some studies have shown significant correlations of microbial contamination with the risk of endemic gastro-enteritis (Beaudeau *et al.*, 2011). However, turbidity itself is a bulk parameter without any information about origin or nature. Moreover, the relevance of turbidimetry may be limited since the presence of colloids in water creates interference with the nephelometric response.

In the last two decades, there has been an increasing interest in determining direct GLUC (β-D-glucuronidase) activity, specific to 97% of *E. coli* strains in order to evaluate microbiological water quality (Fiksdal *et al.*, 1994; Servais *et al.*, 2005) and recently automated devices have been developed to measure the enzymatic activity in water by fluorescence photometry on-site (see Chapters 7 and 8). In a first field test study of instruments for on-site enzymatic activity detection for stream water with high suspended sediment loads, results showed that although GLUC activity was not proven as a quantitative proxy for culture-based *E. coli* concentrations, this parameter could be of great value for understanding contaminant transport processes in such habitats (Stadler *et al.*, 2016).

5.2.1.2 Cyanobacteria-associated bloom monitoring

The effects of global climate change appear to enhance the development of potentially toxic cyanobacterial blooms in surface water sources worldwide (Paerl & Paul, 2012). Some cyanobacteria species are producers of a variety of potent toxins and/or taste and odour (T&O) compounds such as geosmin and 2-methylisoborneol (MIB), while microcystins are the most commonly detected cyanotoxins. The ability to promptly and accurately monitor cyanobacteria, as well as associated toxins and T&O compounds, in order to correctly identify the exceeding of an alert threshold, is then a key factor in the implementation of a successful risk management strategy for recreational activities, drinking water production and water reuse.

Currently, the water industry relies on laboratory analysis of cyanobacteria that can take two to five days; there is therefore a need to improve response time. On-line fluorometric probes (also called "fluorescence probes" in some publications) are available for the rapid detection of cyanobacteria cells via measurement of specific pigmentation; although water quality interference with probe measurements in natural environments hinders their wider application as reviewed by Zamyadi *et al.* (2016). However, fluorometric probes and especially those targeting phycocyanin (the phycobilisome pigment of blue-green cyanobacteria) can help with decision-making during plant operation and have the potential to be applied as a management technique (Izydorczyk *et al.*, 2005). Thanks to the combination of on-line fluorescence monitoring implemented in the treatment plant water intake and predictive tools , innovative solutions can also be applied by DWTP operators to define probability distributions of cyanobacterial concentrations and associated algal metabolite concentration in raw water (Steinmann *et al.*, 2016).

5.2.2 On-line monitoring in Drinking Water Treatment Processes (DWTP)

As recommended by World Health Organization (WHO, 2005), Water Safety Plans are an improved risk management tool designed to ensure the safety of drinking water through the use of a comprehensive risk assessment and risk management approach that encompasses all steps in water supply from catchment to consumer.

A critical part of any treatment Water Safety Plan is operational monitoring to ensure that treatment processes are operated as planned and within expected limits. Wherever practical, continuous on-line monitors should be used to measure performance against critical control limits. The selection of specific equipment and methodologies must take into account the costs involved, the time and expertise needed for their operation and maintenance and the relevance of the data obtained. The equipment must be validated and regularly calibrated to ensure that measurement errors are within acceptable limits and that data obtained are reliable. The use of equipment such as pH, turbidity or chlorine sensors by operators has also been shown to improve their understanding of the water treatment processes. Microbial monitoring at the first stages of treatment would provide the most direct, site-specific information, however this may not be feasible at later stages (Smeets *et al.*, 2010) owing to the lack of sensitivity and time for response of current monitors.

5.2.3 On-line monitoring in Drinking Water Distribution System (DWDS)

The microbiological quality in water distribution systems results from a complex series of physical, chemical, and biological reactions that occur after the treatment process. Living organisms and nutrients may enter the distribution system with the treated water or from sources such as leaks, cross-connections or back-siphonages. Bacterial growth may occur at the pipe surface (biofilm formation), the interface of suspended particulates, inside living organisms (amoeba) and within the water itself (Berry *et al.*, 2006). Opportunistic pathogens like *Legionella* spp., Mycobacteria, and *Aeromonas* spp. are able to grow in water distribution systems and are recognized as major waterborne pathogens (Szewzyk *et al.*, 2000).

As revealed by recent studies based on the use of advanced and sensitive methods such as flow cytometry or next-generation sequencing, spatial and temporal changes in microbiological water quality in both chlorinated and non-chlorinated DWDS can be observed while no change was detected with conventional cultivation-based methods (Douterelo *et al.*, 2014; Prest *et al.*, 2016). However, such investigations are still scarce and the dynamics of the highly heterogeneous microbial communities living in DWDS should be further investigated with the help of on-line monitoring means.

The main factors influencing hygienic water quality in DWDS are numerous and can be divided into three main categories:

- *The biological stability of drinking water*: different parameters can promote the growth of bacterial communities within DWDS, such as water quality (residual chlorine, assimilable organic carbon, minerals), water age and temperature, as well as pipe characteristics (Henne *et al.*, 2012).
- *The sources of pressure transients*: caused by sudden change in demand (e.g. flushing operations; fire hydrant opening); service interruptions (power failure, main breaks) and all distribution system operation (pump start-up and shut-down; valve operation; reservoir tank draining, pipe system repairing).

- *The potential for accidental intrusion*: aging and/or cross-connected distribution system infrastructure; backpressure (from a customer's pressurized system through an unprotected cross-connection back into the potable water supply) and back-siphonage (backflow resulting from a negative or reduced pressure in the water distribution supply).

Therefore, monitoring water quality dynamics in the distribution network is of great importance. Biostability is generally desired in drinking water guidelines but may be difficult to achieve in large-scale complex distribution systems that are inherently dynamic (El-Chakhtoura *et al.*, 2015). Today, routine quality monitoring is unable to diagnose the general hygienic quality of drinking water, mainly due to the low frequency of the sampling. Likewise, an effective response to a water contamination incident is based on minimizing the time between the detection of a contamination incident and the implementation of effective response actions that mitigate further consequences.

To conclude about the expected benefits for water utility operators, different drivers for setting up the on-line monitoring strategies must be clearly distinguished between regulatory concerns, safety issues/event detection (concerning mainly DWTP outlet and DWDS) and operation control and improvements (concerning raw water and DW processes). The success of the monitoring strategy implementation – starting with the choice of the sensor(s) – will depend on how well the specified objectives and aims are addressed.

It can be meaningful to design monitoring strategies for specific sections of the system or for specific scenarios. This demands that each utility performs a system analysis to identify potential risks for different sections of the system, and identifies what kind of information would be most beneficial to acquire in each case. This is a process closely linked to the utilities' Water Safety Plans work, and scenario-based monitoring planning can be a tool to optimize monitoring strategy and prepare for acute situations.

5.3 CHALLENGES

Although sensors are increasingly appearing on the market, effective implementation in water utilities has not succeeded to date, particularly for sensors based on microbiological detection processes. There have been several reasons suggested for the lack of implementation:

- sensors do not meet practical utility needs
- verification schemes do not sufficiently match utility practices or regulation drivers
- purchase and maintenance costs
- the challenge of managing large data quantities and translating them into meaningful information for operational processes.

In order to overcome these barriers, different challenges for sensor manufacturers and water utilities detailed in the following section have been raised.

5.3.1 On the track of the ideal sensor

The top three concerns of utilities about any kind of on-line monitoring equipment are its reliability, cost and ease of installation and use (including maintenance requirements). Implicitly, the method is fully automated, and the measured data are continuously available (preferably in "real time" or within a timeframe that allows corrective actions to be taken) in a centralized information system for monitoring and managing operation performances.

Experience has shown that the following aspects are critical: accessibility, power supply and data communications, and absence of waste streams. If the primary objective of on-line monitoring is to provide timely notification of unexpected changes in water quality, then sensor(s) would be ideally located at each service connection where response actions can be performed (e.g. increased chlorine injection, substitution or diversion of water supply). Such Early Warning Systems will require the best spatial distribution as possible to cover customer density or vulnerable sites. Consequently, an end-user could expect to be able to place the sensor in manholes, if the issues related to the size, power supply, data communication and waste streams of the sensor are solved.

Above the technical specifications of the ideal sensor by themselves, all the issues related to the quality of data produced by the sensor are of concern. Full validation of the measurement technology in terms of specificity and sensitivity of the result must be performed to ensure low rates of false-positive/false-negative alarms. This validation must be done first in laboratory conditions but also in the field, based on a wide range of water qualities in order to document the analytical performance that will confirm the reliability of the sensors for the end-user and its confidence to engage corrective actions.

5.3.2 Monitoring parameters

Typically, water systems (from freshwater to drinking water) contain 10^4 to 10^6 harmless microbes per 100 mL; among them fecal indicators and pathogens are a minor part in raw water or absent as expected in water intended for human consumption. Among the number of different methods available for on-line bacteria monitoring, some are highly specific to single species or groups of bacteria, while others simply measure the total microbiological activity or bacteria counts.

These different levels of specificity for microbial monitoring instrument have a direct relationship with the time-to-result offered by current sensors. Typically, slower on-line methods (e.g. enzyme-activity-based assays providing results in at least 3 hours) provide a much greater deal of specificity than faster methods (based on optical or fluorescence measurement) that have the potential to provide results within few minutes. Consequently, the choice of the sensor will be directly driven by the type of specificity required according the monitoring purpose, as illustrated in Figure 5.2.

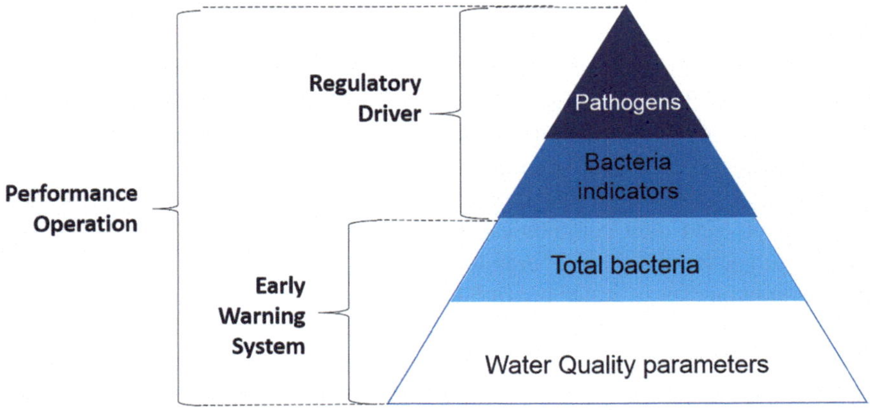

Figure 5.2 What type of targets to measure according the bacteria on-line monitoring purpose?

First, sensors for monitoring physicochemical parameters (e.g. turbidity, oxygen, conductivity, pH, temperature and chlorine) can provide some information about a potential microbial contamination. In particular, a combination of several parameters for detecting and identifying intentional contamination in distribution networks can be used. As example, on-line detection of *Escherichia coli* intrusion in a pilot-scale drinking water distribution system study showed that electric conductivity (EC) measurement gave the strongest signal compared with the measured baseline during the *E. coli* intrusion. Integral calculations showed that the peaks in the EC, pH, temperature, turbidity and UV-absorbance at 254 nm data were detected corresponding to the time predicted (Ikonen *et al.*, 2017).

In case of chlorinated drinking water distribution systems, deploying free chlorine sensors has been shown as surrogate monitors for bacterial contamination events on a laboratory scale (Helbling & VanBriesen, 2008). Monitoring free available chlorine in conjunction with turbidity and temperature can also be used to determine if the conditions within the distribution system may be conducive to microbiological risk according to biological instability or potential intrusion. In that context, sensors for total bacteria providing near real-time monitoring can also be used to identify changes from a background level.

Currently, some methods for total bacterial counts/activity using different approaches (optical, ATP or flow cytometry) are now available (Hammes *et al.*,

2008; van der Wielen & van der Kooij, 2010; Højris *et al.*, 2016) and have been shown to accurately assess the drinking water biological quality and regrowth, giving a great advantage by their rapidity, easiness and sensitivity (Liu *et al.*, 2013). However, this type of parameter is still not included in any regulation or drinking water quality standards and the establishment of background levels and variations under normal operating conditions for the specific system is necessary, requiring a good understanding of and experience with the water system.

If the monitoring purpose is restricted to the regulatory compliance, the measure of specific parameters such as *E. coli* or coliforms is then required and the on-line instrument should provide high sensitivity to detect very low concentrations of indicator microorganisms in drinking water (<1 cfu/100 mL), which is quite challenging to date. Some instruments proposed for the on-line measurement of *E. coli* and total coliforms are based on fluorescence/colour detection of enzymatic (β-galactosidase and β-glucuronidase) activity. However, reagents and additional steps for either incubation or concentration of the water sample in case of direct enzymatic activity assay are necessary, increasing the time to result, the cost of the equipment and the frequency of maintenance.

To conclude this section about the type of parameter to monitor, it is important to observe that many of the current on-line technologies will be complementary to existing microbiological standard parameters and will not (or with very high difficulty) be designed to replace any of the proven surveillance practices based on culture methods.

5.3.3 Validation of measurements

Whatever the main goal driving the use of on-line monitoring, the reliability ascertained by a validation procedure is a key parameter before considering the implementation of on-line sensors through drinking water treatment or distribution processes. Although enough sensitivity and accuracy towards detection limits according quality guidelines or regulation are required, ensuring the good specificity of the measurement is also of great importance for the confidence of the operators towards the results. In particular, the risk of false-positive or false-negative results due to non-target substances (including non-viable cells) or interference from mineral or organic compounds in the water sample, respectively must be documented.

Ideally, there are three stages of testing during the validation of a monitoring system. The first tests are carried out by the vendors. Secondly pilot testing is performed at research institutes or Environmental Technology Verification (ETV) Programs. Finally, water utilities test the overall performance of a system. All stages require harmonized validation procedures that are currently missing. One general feature is the difficulty in comparing results from continuous monitoring (field measurement) with results from grab sampling and laboratory analyses, especially for biological samples.

Moreover, all validation strategies rely on the comparison between results from on-line monitoring systems and standard reference methods that can be missed for parameters that are not considered by the regulation. In particular, the use of standard methods based on culture (such as Heterotrophic Plate Counts or fecal indicator culture methods) should be *de facto* excluded from the reference methods for the validation of on-line monitoring systems measuring total bacteria counts/activity, since no agreement can be expected between the total and culturable bacteria fractions. Therefore, new standard procedures are required for validating such new sensors.

5.3.4 Data management

The ultimate challenge raised by the development of on-line monitoring is the large amount of data that needs to be transmitted, stored and interpreted for acting as Early Warning System.

First, experience has shown that a system like supervisory control and data acquisition (SCADA) is mandatory to manage the huge amount of data produced by on-line measurement systems. The use of sensors in combination and in more locations in the drinking water production and distribution chain leads to high quantities of data in space and time.

With increasing use of chemical and microbiological sensor technology, the complexity of the data produced will also increase. For example, in order to be able to differentiate variations in total bacteria counts due to normal operating conditions (e.g. hydraulic conditions, resource combination, pump regimes) from variation due to contamination entering the system, it is of primary importance to perform more statistical analysis and interpretation.

Since in-depth knowledge of the system is crucial when navigating through increased data logs to ensure meaningful interpretation of variations, algorithms specifically fitted for each system are necessary tools to identify an unexpected variation that requires further action. With that objective, SUEZ has developed the Aquadvanced Quality module addressing one central task: the monitoring of water quality in drinking water distribution networks (Campan & Do-Quang, 2016). The main functionalities of this new tool (Figure 5.3) enable users to:

- *Display*: measurement (from sensors), virtual sensors (from a model) and lab analyses (from grab sampling)
- *Detect*: abnormality, pollutant intrusion, risk anticipation
- *Analyse*: geographical data access, data combination, simulation

Using such new tools, water utilities will get a real-time view of the potable system by detecting water quality changes in real time and reviewing multiple data sets at one go to make timely decisions.

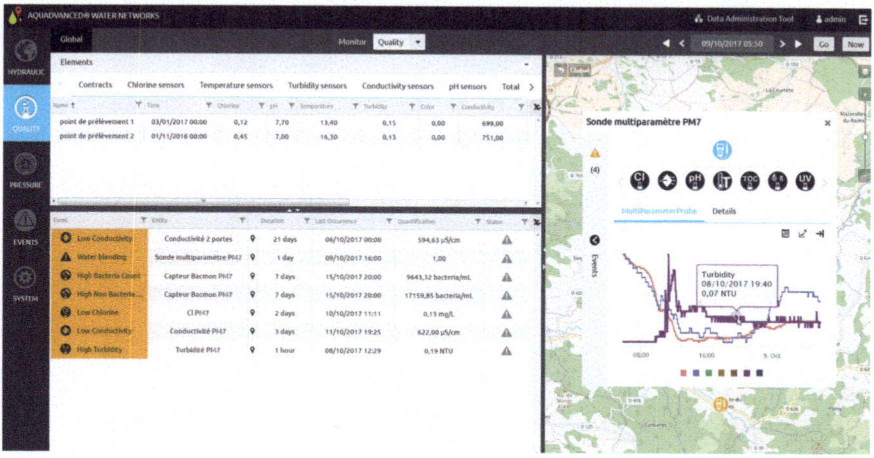

Figure 5.3 Screenshot of the Aquadvanced suite user interface.

5.4 CASE STUDY

In order to illustrate the above points about the benefits and challenges, we propose one example of feedback from the field related to the use of on-line bacteria monitoring in a chlorinated drinking water distribution network. In this case study, one multi-parameters sensor station (i::Scan Nanostation monitoring 7 parameters: chlorine, pH, conductivity, temperature, turbidity, UV254 and TOC) and one BACMON (Grundfos) analyser were installed near a public tap for 4 weeks. During the monitoring period, all water parameters measured by sensors exhibited a quite stable level, with smooth variations, in the range shown in Table 5.1.

Table 5.1 Water quality parameters in the study case during the monitoring period.

	5th - 95th Percentiles
pH	7.3–7.6
Temperature (°C)	20–24
TOC (mg/L)	0.51–0.65
Conductivity (µS/cm)	698–784
Turbidity (NTU)	0.02–0.04
UV254 (ABS/m)	1.1–1.41
Free chlorine (mg/L)	0.08–0.19
Bacteria (number/mL)	395–2295
Particles (number/mL)	1650–5875

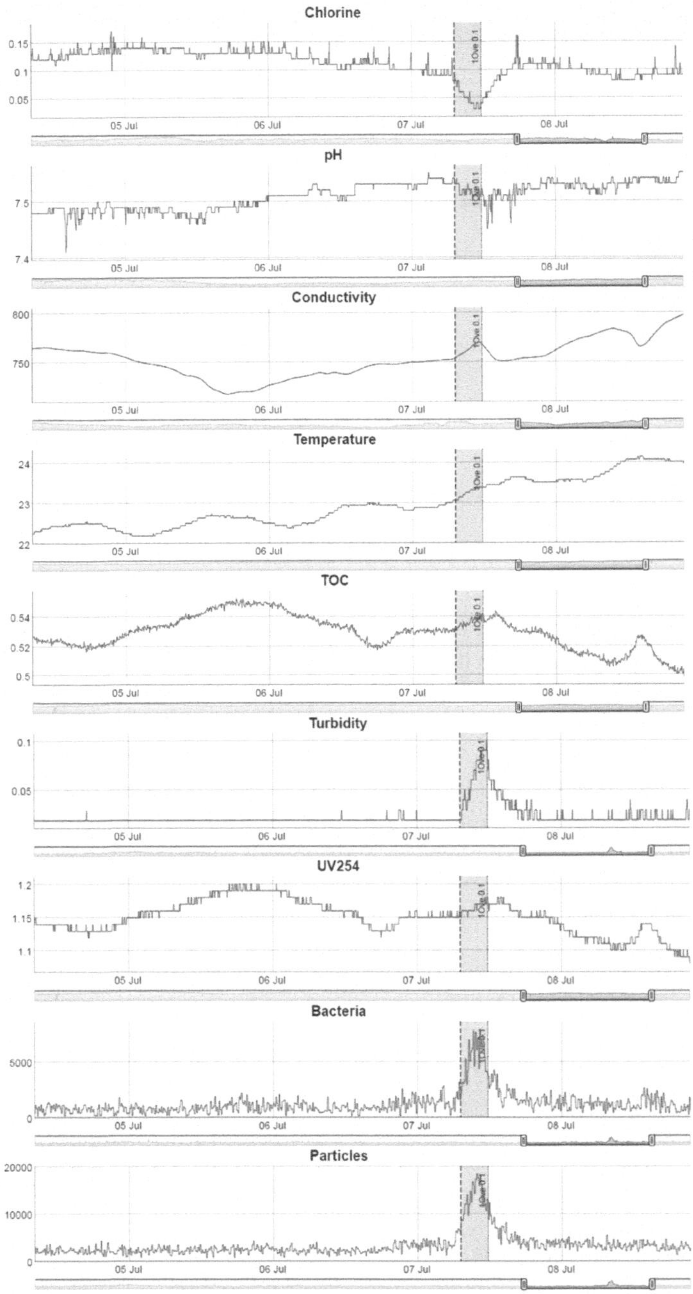

Figure 5.4 Event detection characterized by multi-parameters and bacteria on-line monitoring in a drinking water distribution network.

The extensive data sets derived from the on-line monitoring measurements (Figure 5.4) allowed for the establishment of baseline data as well as for the recognition of daily variations and specific events that would most likely be missed (or mis-characterized) by regular sampling. Statistical learning therefore brings strong added value in integrating the cumulated past data on all the measured parameters to detect future anomalous events. Combined with appropriate transformations of the raw signals, it enables determining if a situation is significantly new and far from previously observed patterns. The sensitivity of such a detector can be adapted to operators' expectations.

Based on water quality history and expertise, complementary rules are defined to pre-characterize events raised by the statistical detector. The simultaneous variations, depending on parameters types, enables users to extract evidence leading to identification of a potentially occurring phenomenon. A criticality indicator is computed to reflect the intensity of variations in parameters for this specific event as compared with observed variability in available historical data. This calculated indicator is a number between 0 and 1, 1 being the highest level of criticality.

Using such tools, events with pre-characterization and criticality can be automatically detected and ranked in real-time, helping drinking water system operators to implement proactive actions. The example illustrated in Figure 5.4 shows an event characterized by a simultaneous increase of turbidity, total particles and bacteria and chlorine consumption, in a short period (around 4h30). The calculated criticality level was low (0.1) and the combined variation of parameters suggest that such short event is likely due to an over-speed of the water flow in the drinking water network.

5.5 CONCLUSION

On-line, real-time monitoring is an upcoming approach that will have implications for the operational framework for assessing water quality. Data produced by on-line monitoring systems can provide real-time information on water quality. This differs from the established practice of laboratory analysis which can only confirm that the water is of satisfactory quality, in many cases, after the water has been distributed. However, to be successful the monitoring purpose must be well defined to select the most relevant parameters to be monitored and how this information will be used. Data management is required to ensure that the data is consistent, of high quality, easy to collect and record, and suitable for analysis. Currently there is no available technology that will specifically identify a microorganism at the genus or species level, with a sufficient sampling rate as to be recognized as "real-time". Nevertheless, thanks to some innovative technologies for on-line bacteria counting, the water professionals are now able not only to constantly monitor bacteria levels, but also study the water quality dynamics in the distribution network. In addition, these advances have been accompanied with new improvements in rapid and automated detection methods for microbial indicators or pathogens, recognized as a critical need in order to be combined with early warning system.

5.6 REFERENCES

Beaudeau P., Pascal M., Mouly D., Galey C. and Thomas O. (2011). Health risks associated with drinking water in a context of climate change in France: a review of surveillance requirements. *Journal of Water and Climate Change*, **2**, 230–246.

Berry D., Xi C. and Raskin L. (2006). Microbial ecology of drinking water distribution systems. *Current Opinion in Biotechnology*, **17**, 297–302.

Campan F. and Do-Quang Z. (2016). AQUADVANCED Quality: a smart real-time monitoring platform for water quality in drinking water distribution networks. Singapore International Water Week, July 2016.

Douterelo I., Boxall J. B., Deines P., Sekar R., Fish K. E. and Biggs C. A. (2014). Methodological approaches for studying the microbial ecology of drinking water distribution systems. *Water Research*, **65**, 134–156.

El-Chakhtoura J., Prest E., Saikaly P., van Loosdrecht M., Hammes F. and Vrouwenvelder H. (2015). Dynamics of bacterial communities before and after distribution in a full-scale drinking water network. *Water Research*, **74**, 180–190.

Fiksdal L., Pommepuy M., Caprais M. P. and Midttun I. (1994). Monitoring of fecal pollution in coastal waters by use of rapid enzymatic techniques. *Applied and Environmental Microbiology*, **60**, 1581–1584.

Hammes F., Berney M., Wang Y., Vital M., Koster O. and Egli T. (2008). Flow-cytometric total bacterial cell counts as a descriptive microbiological parameter for drinking water treatment processes. *Water Research*, **42**, 269–277.

Helbling D. E. and VanBriesen J. M. (2008). Continuous monitoring of residual chlorine concentrations in response to controlled microbial intrusions in a laboratory-scale distribution system. *Water Research*, **42**, 3162–3172.

Henne K., Kahlisch L., Brettar I. and Höfle M. G. (2012). Analysis of structure and composition of bacterial core communities in mature drinking water biofilms and bulk water of a citywide network in Germany. *Applied and Environmental Microbiology*, **78**, 3530–3538.

Højris B., Christensen S. C. B., Albrechtsen H.-J., Smith C. and Dahlqvist M. (2016). A novel, optical, on-line bacteria sensor for monitoring drinking water quality. *Science Reports*, **6**, 23935.

Ikonen J., Pitkänen T., Kosse P., Ciszek R., Kolehmainen M. and Miettinen I. T. (2017). On-line detection of *Escherichia coli* intrusion in a pilot-scale drinking water distribution system. *Journal of Environmental Management*, **198**, 384–392.

Izydorczyk K., Tarczynska M., Jurczak T., Mrowczynski J. and Zalewski M. (2005). Measurement of phycocyanin fluorescence as an online early warning system for cyanobacteria in reservoir intake water. *Environmental Toxicology*, **20**, 425–430.

Liu G., Van der Mark E. J., Verberk J. Q. J. C. and Van Dijk J. C. (2013). Flow cytometry total cell counts: a field study assessing microbiological water quality and growth in unchlorinated drinking water distribution systems. *BioMed Research International*, **2013**, 595872.

Lopez-Roldan R., Tusell P., Cortina J. L., Courtois S. and Cortina J. L. (2013). On-line bacteriological detection in water. *TrAC Trends in Analytical Chemistry*, **44**, 46–57.

Paerl H. W. and Paul V. J. (2012). Climate change: Links to global expansion of harmful cyanobacteria. *Water Research*, **46**, 1349–1363.

Prest E. I., Weissbrodt D. G., Hammes F., van Loosdrecht M. C. M. and Vrouwenvelder J. S. (2016). Long-term bacterial dynamics in a full-scale drinking water distribution system. *PLoS One*, **11**, e0164445.

Servais P., Garcia-Armisen T., Lepeuple A. S. and Lebaron P. (2005). An early warning method to detect fecal contamination of river waters. *Annals of Microbiology*, **55**, 151–156.

Smeets P. W., Rietveld L. C., van Dijk J. C. and Medema G. J. (2010). Practical applications of quantitative microbial risk assessment (QMRA) for water safety plans. *Water Science & Technology*, **61**, 1561–8.

Stadler P., Blöschl G., Vogl W., Koschelnik J., Epp M., Lackner M., Oismüller M., Kumpan M., Nemeth L., Strauss P., Sommer R., Ryzinska-Paier G., Farnleitner A. H. and Zessner M. (2016). Real-time monitoring of beta-d-glucuronidase activity in sediment laden streams: a comparison of prototypes. *Water Research*, **101**, 252–261.

Steinmann D., Humpage A., Cook D., van der Linden L., Courtois S., Bonnard R., Kaeding U., Alba Fernandez S. and Newcombe G. (2016). Management of cyanobacterial metabolite risk in DWTPs: a packaged solution for prediction & operating control through water monitoring and operational tools. *Water Quality Technical Conference*, Indianapolis, Indiana, USA.

Szewzyk U., Szewzyk R., Manz W. and Schleifer K.-H. (2000). Microbiological safety of drinking water. *Annual Review of Microbiology*, **54**, 81–127.

Tryland I., Robertson L., Blankenberg A. G., Lindholm M., Rohrlack T. and Liltved H. (2011). Impact of rainfall on microbial contamination of surface water. *International Journal of Climate Change Strategies and Management*, **3**, 361–373.

van der Wielen P. W. and van der Kooij D. (2010). Effect of water composition, distance and season on the adenosine triphosphate concentration in unchlorinated drinking water in the Netherlands. *Water Research*, **44**, 4860–4867.

WHO (2005). Water Safety Plans. Managing Drinking-Water Quality from Catchment to Consumer. WHO Geneva, Switzerland. WHO/SDE/WSH/05.06.

Zamyadi Z., Choo F., Newcombe G., Stuetz R. and Henderson R. K. (2016). A review of in situ multi-wavelength fluorometers for real-time management of cyanobacteria: recent advances and future direction. *Trends in Analytical Chemistry (TrAC)*, **85**, 83–96.

Section 3

Sensor technologies

Chapter 6

Use of fully automated bacterial monitors for enhanced water safety

Henrik Braathen, Ida M. Oeverleir and Goran Khalaf

6.1 HISTORICAL DEVELOPMENT OF AUTOMATED BACTERIAL MONITORS BY COLIFAST AS

The Norwegian company Colifast AS was established in 1992 as a spin off from the Norwegian University of Science and Technology in Trondheim, Norway. The idea behind the new company was to utilize new knowledge and available substrate technology for the development of new and rapid microbial methods for water and food. The traditional laboratory methods were both time and labour intensive, often including several preparation steps, with results available after 2–4 days. Waterborne diseases are known to occur frequently and many outbreaks are undetected until disease is reported by the local health care system. Faeces from humans, and occasionally other animals, are the main sources of pathogenic microorganisms, and water quality control is usually based on analysis of faecal indicator organisms such as *E. coli* (WHO, 2017; EU, 2006). The emerging new methods were simplified techniques for the monitoring of specific indicator groups (Berg & Fiksdal, 1988), where the most promising substrates were chromogenic or fluorogenic substances that utilize the detection of β-D-galactosidase and β-D-glucuronidase enzyme activities to indicate the presence of total coliforms or *E. coli*, respectively. Figure 6.1 shows hydrolysis by a common coliform enzyme.

Editors: Torben Lund Skovhus and Bo Højris
doi: 10.2166/9781780408699_0091

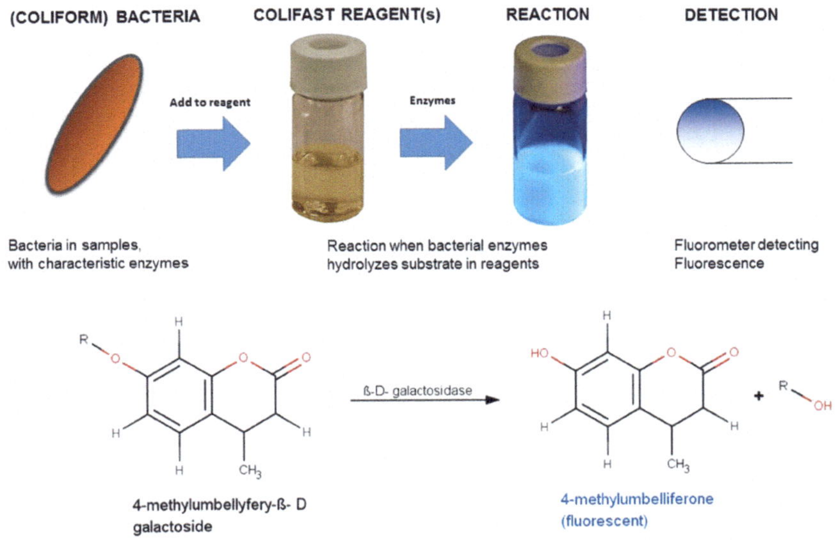

Figure 6.1 The enzymatic reaction in which a fluorescent signal used for detection is generated.

6.1.1 Analysis reagents

The focus of the company was rapid detection of the common indicator bacteria analysed in water and food. The first phase of development was production and testing of different compositions of growth media, both solid agar based and liquid formulas.

After the first 3–4 years, the company had established a product base of approximately six different test reagents, all specific for target groups of bacteria and more rapid than the traditional methods used in the laboratories. Introducing new methods for the laboratory market turned out to be challenging and time-consuming. In addition, there was competition from other vendors also trying to get new and similar products into the laboratory market.

6.1.2 First automated analyser

Colifast continued with further improvement of the methods by designing associated instrumentation, which enabled automatic detection, decreasing the analysis time further. In 1995–1996 the first semi-automated analyser was assembled: the prototype of Colifast Analyser (CA) 100 (see Figure 6.2). This system was comprised of a two-rack incubator module, a robot with a septum-piercing needle for subsampling from incubator positions, and an analyser module with pumps, valves and an integrated detector unit. A basic flow injection technique was used for preparation of subsamples prior to reading in the detector unit. The system was controlled by custom software on a laptop computer. The operator would prepare and fill water samples into vials with Colifast medium, place these into the incubator module and start analysis.

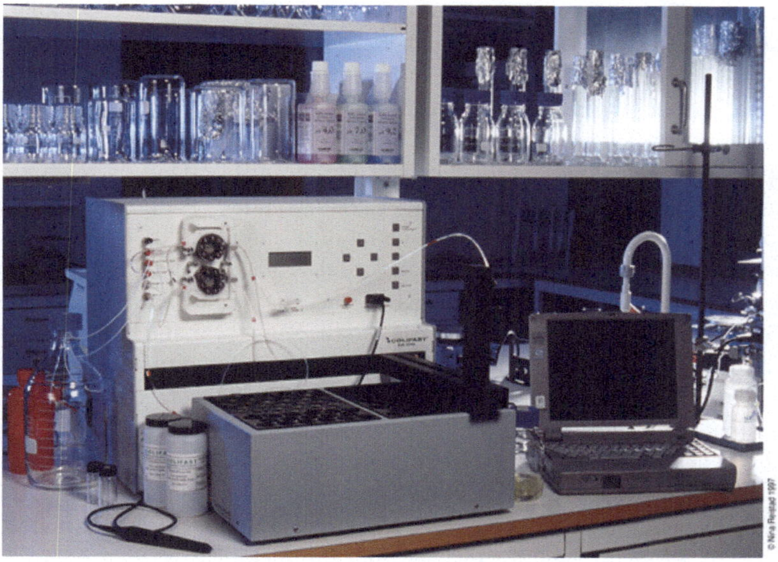

Figure 6.2 The first semi-automated Colifast laboratory analyser: CA 100.

A couple of years later several of these analysers were included in different projects in Scandinavia and Europe, all for rapid analysis of coliforms in water. Several methods were tested, both rapid screening generating results in <1 hour and growth methods generating results in 6–11 hours (Tryland *et al.*, 2001). The promising rapid screening methods generated results close to real-time but also showed some limitations when analysing low levels of target bacteria and environmental samples with shifting water quality and in different seasons (Tryland *et al.*, 2002). The company's experience was that rapid methods would not fit all applications and the more robust methods would have to be based on selective growth of the target group of bacteria. As the laboratory reference methods count number of viable bacteria able to form colonies, methods based on similar bacterial growth showed best correlation. In general, the technology proved very useful and some users asked for further development.

6.1.3 Flexible on-line analyser

There was a need for a fully automated system, which could be installed at key sampling sites and with minimal maintenance requirements. This led to the development of the Colifast At-Line Monitor (CALM), which was launched in 2001. CALM had some similarities with the CA but all modules were redesigned with new components and functions, and the system was controlled using a new software. It utilized a flow injection technique for subsample reading and a robot that would both dispense samples to incubator positions and aspirate subsamples. The system could be connected directly to a water sample flow and suck in a software-specified sample volume at pre-programmed intervals, directly followed by analysis. The maximum capacity of this system was 76 analyses and the maintenance interval was typically once every second week. This system was included in a water quality project and sold to customers in Europe. CALM generated valuable data for the utilities but as a standalone system with continuous operation, it needed frequent technical support. To improve the quality, the company used a lot of resource in the first years, which included several technical modifications and replacement of components. In 2003 a new improved version of CALM was ready and verified as a robust high-quality system. This unit was installed the same year at the river raw water intake for the water treatment plants (WTPs) in a larger city in Sweden. Right from the start the automated results from this system were of high value for the customer which used the data directly to switch between water sources, thereby reducing the level of faecal contamination in the raw water. The CALM system was designed to be very flexible and able to run four different methods with numerous setups, which made it adaptable to many different applications and waters. With the new CALM version, the technical service interval was now once or twice a year, depending on the analysis frequency. In Sweden, they selected a most probable number method which generates a statistical estimate of the *E. coli* number per 100 ml water sample. This is a bacterial growth method which rely on prolonged incubation of samples, but it turned out to meet the requirements and is robust and accurate for this application. The analysis time is 10 hours for early warning and 12 hours for final reading and result. The system was connected by digital and analogue signalling to the local control system and results and alarms were thereby presented in the control room at the WTP.

This application was clearly a success and for Colifast this changed the focus more towards automated on-line analysers and away from the laboratory market. However, even with good references and technology verifications, it was not a straightforward process to introduce the technology. The market was used to the traditional microbial analysis at laboratories and often sceptical of moving this to an automated system. Also, as these were unique products with no competition in the market, there was no former information or sales activity related to this type of technology. The first customers were typically larger water utilities with highly educated personnel with an interest in new innovative solutions, like a large water utility in Rogaland, Norway (see Figure 6.3).

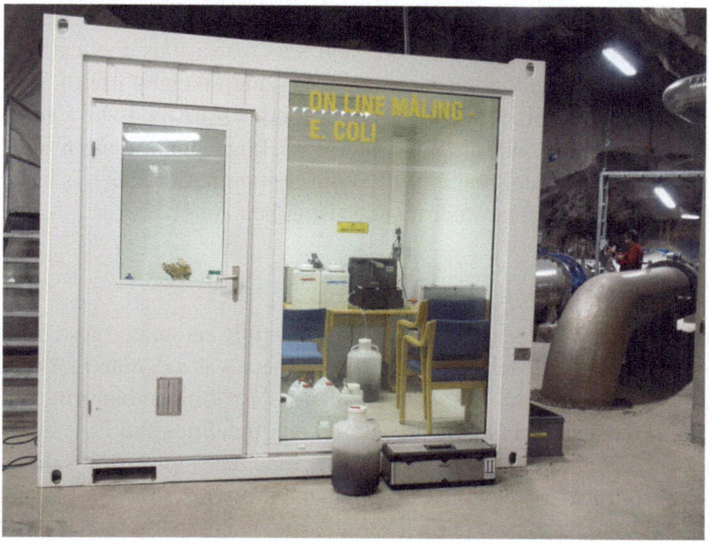

Figure 6.3 Remote CALM installation at a raw water intake. Lake Stoelsvann in Rogaland, Norway.

Through 2004 to 2007, several additional CALM systems were installed, mainly in Scandinavia at remote locations for raw water monitoring. From 2007 until today, the system has also sold to several applications in Europe and Asia. The CALM system underwent several adjustments and modifications to meet specific application and technical needs, resulting in new model numbers. Larger updates saw the system renamed as CALM II and CALM III.

By this technical development, the company product line was altered, replacing the former semi-automated laboratory systems with new field analysers.

6.1.4 Manual Field Kit

The second field product was the Colifast Field kit, launched in 2005. This kit was based on feedback from several smaller municipalities looking for a system which could rapidly track faecal contamination sources in smaller rivers and lakes. These municipalities had problems tracking the source of contamination when using the traditional manual grab samples, transport of these to a laboratory and then waiting for the results. Usually this could take 2–3 days and by the time they received the results it was not possible to confirm contamination by new samples as the weather conditions often had changed and the contamination source was inactive. The Field Kit was based on a sensitive handheld, battery-driven fluorometer, and analysis required some manual work by the operator. In addition, the field kit metal box also contained a small incubator with cables and power converter, various disposables, and prefilled test tubes which combined everything needed for field analysis. Colifast has by 2017 sold more than 60 Field Kits for various applications, mostly used for tracking contamination in urban rivers and lakes by use of a 45 minute faecal coliforms rapid screening test. This method has proved to be a good tool for monitoring the level of contamination and for tracking contamination sources. The method works well in water bodies with some contamination, but it is less accurate for low faecal levels and sea water testing (Braathen *et al.*, 2006).

6.1.5 On-line drinking water analyser

In 2008 the company started development of a new fully automated system. This time it was also based on feedback from customers. There was a demand for a simplified automated system for drinking water analysis, providing monitoring of critical points out on the distribution network. Based on guidelines and legislation, *E. coli* should not be present in 100 ml samples. This time development was done entirely in-house, in contrast to previous development projects, which usually included some support from external experts, institutes and manufacturers. Also, prototypes and later commercial units where now entirely assembled at Colifast, whereas the other systems where often pre-assembled by different suppliers. This improved the technical knowhow in the company and decreased the development time. The new system was designed to perform a fully automated presence/absence test of 100 ml sample volumes and requiring a minimum of maintenance (Figure 6.4).

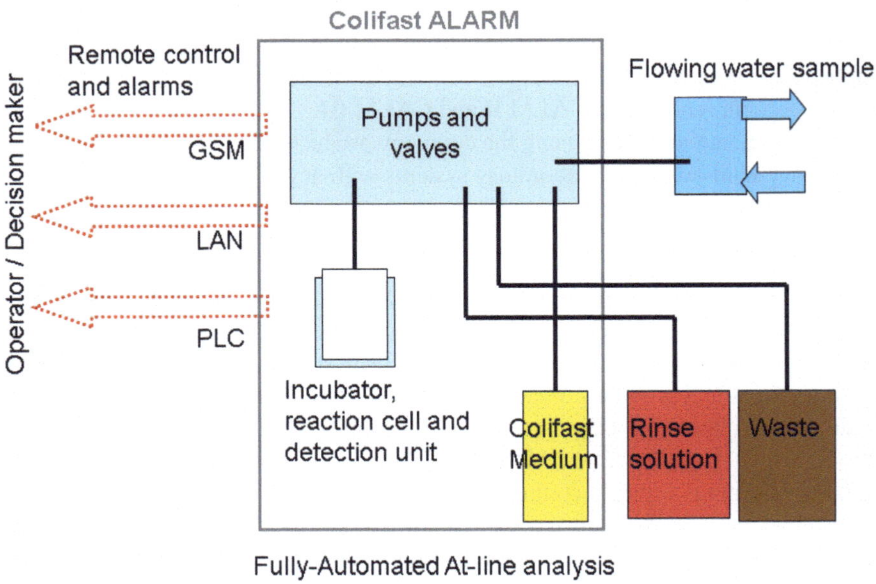

Figure 6.4 Overview of the technical functions of the Colifast ALARM system.

Technically, the new system was less complex and flexible compared to the CALM system. It was specialized mainly for drinking water analysis. To simplify the system there was no robot and multi-position incubator, and with direct sample reading without using flow injection technology. A new and user-friendly software was designed and the analyser was assembled into a cabinet with a touchscreen-panel computer. To avoid any technical problems and to ensure high-quality performance, the system was tested for one year before release. In 2010 the new Colifast At-Line Automated Remote Monitor (ALARM) was launched. Two units were sold and installed directly after product launch, one at a drinking water reservoir in Copenhagen, Denmark (Bjerrgaarde & Hansson, 2012) and the other at a small elevated reservoir on an island outside of Stavanger, Norway (Figure 6.5). The ALARM typically analyses 21 samples and is visited by the operator every third week for replacing medium bottle, empty waste and restart system. Both units performed well from the start and generated valuable data for the customers. The first unit installed is still in operation and has been analysing with minimal technical problems during the past seven years. Since then, many more units have been sold and installed, in Europe, the Middle East and in Asia. Based on experience with good technical stability over time, the technical service interval of Colifast ALARM is limited to one service per year.

Figure 6.5 One of the first Colifast ALARM installations at a reservoir far out on the water distribution network. Mosteroy Island, Stavanger, Norway, 2010.

In combination with the number of CALM units, Colifast has by 2017 delivered and installed close to 40 fully automated systems.

6.2 METHODS

During the past 25 years of technology development, production and installations for various applications, Colifast has developed and utilized several different formats for analysis. At the beginning the customers prepared Colifast reagents for the instruments in a laboratory. Later, Colifast developed pre-filled disposable trays, tubes and flasks for use on the systems, and today all systems can be operated without the need of a laboratory.

The Presence/Absence (P/A) method is usually for low numbers of target bacteria in large volumes. Typically, a 100 ml sample is analysed for presence or absence of *E. coli* or coliform bacteria. Colifast growth medium is mixed into the sample and after a defined incubation time, adequate for bacterial growth, the colour/fluorescence of the liquid mix is measured. Colour/fluorescence above a selected threshold value indicates presence, and no or low colour/fluorescence indicates an absence. This is a robust method, based on both selective growth and specific enzyme activity of the target group. The P/A analysis of 100 ml sample corresponds to international guidelines and legislation which define sample volumes and the absence of *E. coli* in drinking water.

The Most Probable Number (MPN) method is based on the P/A format but quantifies the number of target bacteria growing in the medium based on distribution of the sample to several duplicates, for example 5 × 10 ml. Based on volume and the number of positive and negative duplicates, the MPN-value can be found using a statistical calculation. The calculated result will typically be presented as number of bacteria per 100 ml and is the result of using Thomas' simple formula:

$$MPN/100ml = \frac{\text{no. of positive vials} \times 100}{\sqrt{(\text{ml sample in negative vials} \times \text{ml sample in all vials})}}$$

The MPN method is robust, normally generating accurate results and this format is commonly used in several laboratory methods.

The Time to Detect (TTD) method generates a semi-quantitative bacterial number based on difference in growth time, according to the number of bacteria in the sample. More bacteria will give a positive detection quicker compared to a lower number. The number of hours to get a positive fluorescent signal after an exponential increase in fluorescence (growth) can be used to indicate a bacterial level by correlating detection time to a table created with local empirical data. The TTD method is quite rough as variation in the stress level, different bacteria strains and local environmental factors can influence the result. This method is normally only suitable for water samples containing stable and homogeneous strains of target bacteria (Figure 6.6).

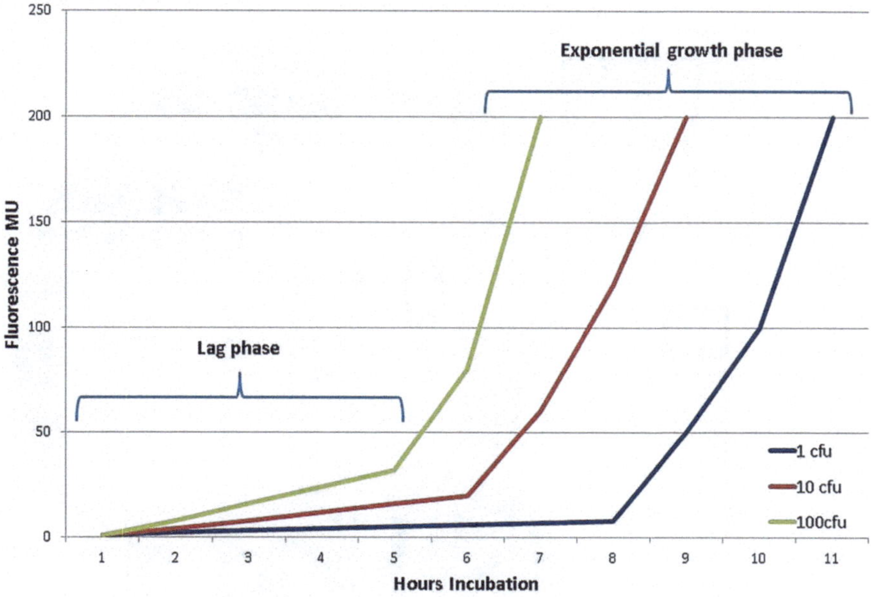

Figure 6.6 Illustration of time difference in bacterial growth based on the number of target bacteria in sample which can be used in a TTD format.

The rapid screening method utilizes the release of a fluorescent end-product during the first minutes after mixing sample and medium. The analysis time would often be less than 1 hour. The end-product released in the reaction between the bacterial enzymes and the substrate is typically 4-methylumbelliferone (MU), and this technique is often referred to as the MU-production (MUP) method. A sample with high number of bacteria will contain more enzymes and generate more fluorescence (MU) compared with a sample with less bacteria. The increase of fluorescence over time (slope) can thereby give an indication of the number in the sample when correlated to a table with corresponding bacterial numbers. This method relies on change in fluorescence prior to any growth and is suitable for rapid screening of sewage and surface water containing medium to high levels of faecal contamination (e.g. contaminated rivers). For samples containing low levels of target bacteria, this method is less accurate compared to growth methods.

The Colifast water analysers (Figure 6.7) can run different formats. All the described methods can be performed on the Field Kit and results calculated by the operator. The most common format for the Field Kit is the rapid screening (MUP) method. CALM can also run all formats by fully automated analysis, and the system software directly calculates and presents the results. The ALARM system is dedicated to drinking water analysis and is limited to the P/A method only.

Figure 6.7 2017 Colifast products for water analysis. The manually operated Field Kit and the two stand-alone online systems for automated analysis.

6.3 APPLICATION EXAMPLES, OPERATION AND INTEGRATION

6.3.1 Monitoring raw water quality – Gothenburg, Sweden

The City of Gothenburg is, with a population close to 600 000, the second largest city in Sweden after Stockholm. The Göta river, originating from Lake Vänern and mouthing out in the North Sea, is the raw water source for the city's two WTPs, Aleyckan and Lackarebäck. The latter treats river water indirectly through the lake-reservoir Delsjön. The reservoir is supplied from the river through a 9 km tunnel.

The Göta river is long, passing through large areas of farmland, industrial areas, smaller cities and villages, which affect the water quality and the level of faecal pollution (Figure 6.8).

The WTPs both facilitate adequate disinfection, but by risk assessment, there is a theoretical chance of incidents where pathogens can pass through treatment and out into the water distribution system. In periods with elevated levels of faecal pollution in the river, the water utility has the ability to run both WTPs on reservoir water.

Figure 6.8 Göta river, a challenging raw water source.

In the facilities of the Lackarebäck WTP there is an accredited laboratory and in the 1990s, rapid methods by use of the CA 100 lab analyser were tested in the laboratory. This system required some manual sample preparation prior to analysis and it performed both a rapid screening method and a time-to-detect method for detection of faecal coliforms. The results were used to support decisions regarding closing the river water inlet. This helped the water works to switch to the water source with the best available quality and thereby effectively reduce the level of faecal contamination in the raw water, which increased the safety of the drinking water production.

In periods, the 1 hour rapid screening results showed some inaccuracy compared with results from standardized laboratory methods, and was not considered satisfactorily reliable for decisions regarding which raw water source to use. The growth methods were more robust and accurate, but slower both in analysis and time spent on bringing in samples and preparation. Later, when the CALM system was developed, analysis by CA 100 was discontinued.

Since summer 2003 Gothenburg Water Works has used CALM for fully automated *E. coli* analysis of raw water. After a period of quality assurance testing the system was approved, purchased and permanently installed (Figure 6.9).

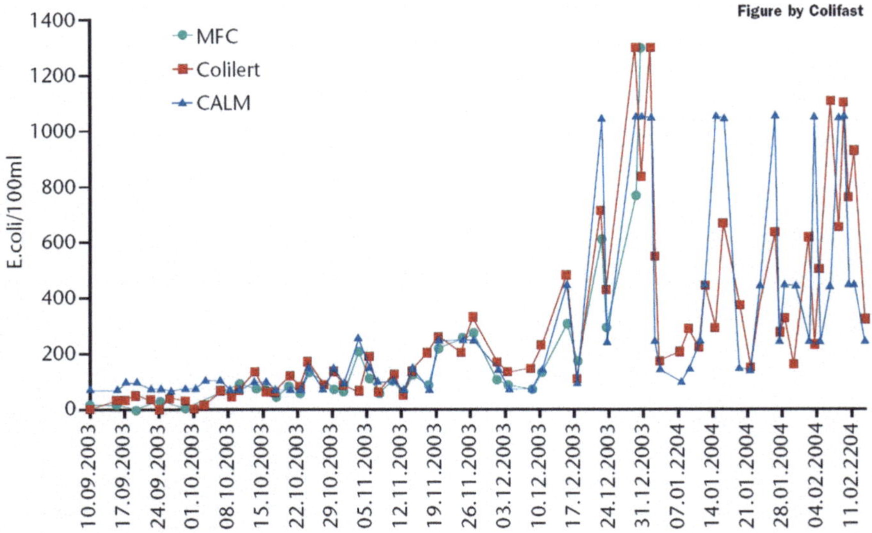

Figure 6.9 Early correlation data from the CALM installation at the water intake. CALM MPN results correlated with laboratory results by two different standard methods, membrane filtration (m-FC method) and Colilert MPN.

The system automatically collects water samples for *E. coli* analysis, followed by electronical reporting of results and alarms to the waterworks control room (Braathen *et al.*, 2005). After two years of successful operation, a second system was purchased and some years later a third system. Altogether, these three CALM systems monitor water at the primary raw water intake from the River Göta älv, upstream for early warning and at a lake reservoir. In addition, the river is used for recreation in some areas, and ideas have been raised for the establishment of an urban aquatic centre. In this regard, the faecal pollution level measured by the CALM systems is relevant, showing if the water is safe for recreation and bathing.

At the raw water intake at Lärjeholm, the raw water quality is constantly monitored by two analyses per day and by 2017 this system had been in continuous operation for 14 years (Figure 6.10). CALM measures *E. coli* by an MPN technique and the analysis is completed in 12 hours. The results are directly transferred to the control room at Alelyckan WTP and the reported bacterial numbers and alarms are directly used for regulation of the raw water intake. Normally, the river intake is closed numerous times each year due to increased *E. coli* levels reported by CALM. For instance, in 2007 it was closed 52 times due to high levels of *E. coli* (Svenskt Vatten, 2009).

The monitoring station in Garn, Lilla Edet, serves as an upstream early warning of pollution. A CALM system performs *E. coli* analysis twice every 24 hours. This system is the third generation of CALM, model 3. The results are transmitted and displayed at the waterworks control room.

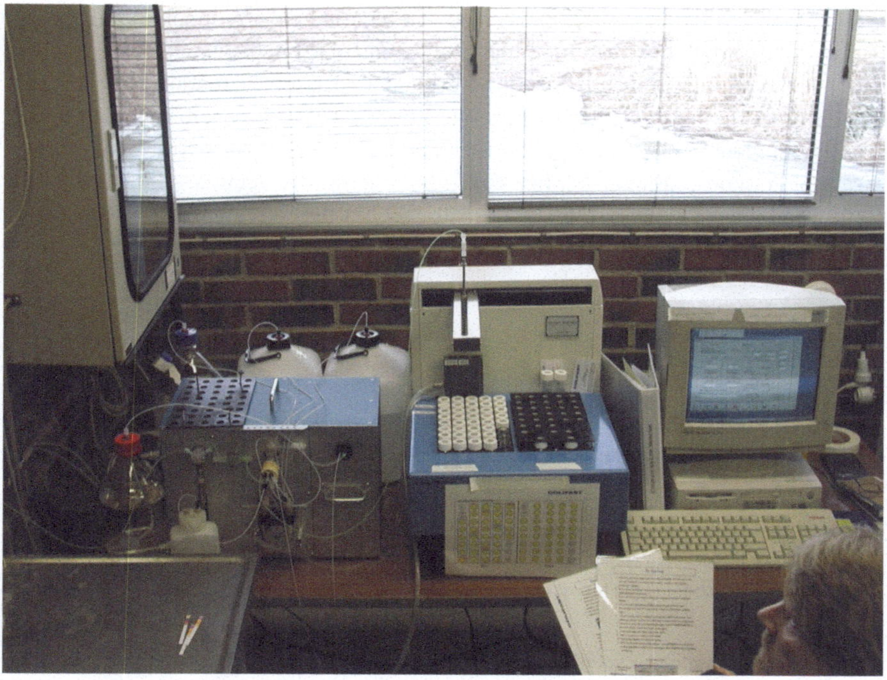

Figure 6.10 CALM model 1 installed in 2003 at the Göta river raw water intake building.

The CALM at Lake Rådasjön analyses for *E. coli* once every day. This unit is an upgraded CALM model 2. The results provide valuable information about the quality of this backup raw water supply that can be used when the river raw water intake is closed. Water is either transferred directly from Lake Rådasjön to a tunnel between Stora Delsjön and Lackarebäcks WTP or to Delsjöarna. Lake Rådasjön thereby functions as a raw water reserve for both WTPs. The results are transmitted to the control room and displayed together with the results from the intake and Garn. In addition, Gothenburg Water Works performs frequent laboratory analysis on samples brought in from several sample sites. By the use of the CALM systems to monitor three key sampling sites, they are able to reduce the number of manual laboratory analysis. Over the years, these systems have generated more than 20 000 items of water quality data (see Figure 6.11).

As the CALM results are integrated in the WTP main control system and used for closing and opening the raw water intake, the results and alarms are handled directly, following detailed routines. The data is processed by the control system, which activate level alarms and display a basic overview of the current *E. coli* level. These results are easily interpreted and handled by personnel with different background and education. Operators at the control room will switch to reservoir and back without the need of any additional expert confirmation. Every year, results from the past 12 months generated by the CALM systems are compared with available corresponding laboratory results, as a part of the internal quality assessment. These annual quality controls validate the accuracy of the CALM results and approve the systems as tools for regulating the raw water.

CALM Most Probable Number (MPN) Method

Stable, Robust and High Quality Results

3 selective steps ensure detection of viable organisms with high sensitivity and specificity:

- ✓ **Selective growth of target bacteria in growth medium containing inhibitors and specific target activators**
- ✓ **Accurate detection of target bacteria based on group-specific enzyme activity**
- ✓ **Selective threshold value for accurate detection of viable target bacteria**

The quality of the method is verified by the good correlation between CALM results and numerous laboratory reference analyses (past 14 years)

CALM Unit	Installed	Site and Application	Mode	Analysis results*
1 (CALM I R1)	2003 -	Lärje. River intake quality control	Continous , 2 analysis per day (MPN)	10 200
2 (CALM II R1)	2005 -2009	Surte. Upstream river monitoring	Continous, 2 analysis per day (MPN)	2 900
2 (CALM II R1)	2009 -	Råda. Lake quality control	Continous, 1-2 analysis per day (MPN)	4300
3 (CALM II R3)	2012 -	Garn. Upstream river monitoring	Continous, 2 analysis per day (MPN)	3700

Figure 6.11 The selective steps of the CALM MPN method that ensure accurate results and an overview of the >20000 results generated by the systems during the past 14 years. *Approximate number by 2017.

6.3.2 Process control – treatment step in small WTP

Many of the Colifast on-line installations are at smaller WTPs, elevated reservoirs and pumping stations. One example is the Colifast ALARM installation in Jevnaker Norway (Figure 6.12). The municipal WTP supplies drinking water for approximately 5000 inhabitants. The water is treated by ultrafiltration followed by chlorination and the installation of a Colifast on-line system was based on technical concerns related to the filtration process. In this situation, additional sampling was desired, either by sending more samples to a laboratory or use an alternative system to measure the hygienic quality. Based on recommendations, verifications and user-friendliness, they selected a fully automated system from Colifast. One Colifast ALARM was installed at the filtration rig in 2012, at the water works situated by the south end of Lake Randsfjorden. The raw water from Randsfjorden usually contains low levels of coliforms and *E. coli* and the system monitors the processed water directly after filtration. The operators adjust sampling frequency based on current risk calculation and the instrument analyses one sample every day or one every second day. When analysing one sample every second day the Colifast ALARM is able to run unattended for 6 weeks before it needs supervision and restart. It works as a hygienic security system for the water treatment as filter damage and breakthrough might affect the drinking water quality. The system is interfaced and integrated by digital outputs into the plant control system and will alarm operators if it detects 1 or more total coliforms in 100 ml sample. These alarms are displayed at the plant monitors together with SMS notifications to operator's cell phones. In a typical small facility, there is a limited number of staff and there is no specialization in water microbiology. No special skills are required for the ALARM operation, the automated results are easily handled by standardized procedures, and the operators do not have to rely on expert advice. The system has proved to be a good quality control tool and thereby the WTP do not need additional laboratory analysis.

Figure 6.12 The Colifast ALARM installed at the filtration rig, Jevnaker WTP, by Lake Randsfjorden, Norway.

6.3.3 Oslo municipality water and wastewater

6.3.3.1 Monitoring of drinking water bacterial quality with Colifast ALARM

Oslo Municipality Water and Wastewater (Oslo VAV) is a publicly owned water and wastewater company responsible for providing clean water to the inhabitants of the capital of Norway. The main drinking water source is Lake Maridalsvannet and the water extracted from the lake is treated at the Oset WTP. As the lake is located in a forest area and with residential areas close by, horses and dogs as well as wild birds or other animals may cause faecal pollution. The sewage generated in the residential area is however treated under strict control.

Oslo VAV installed a Colifast ALARM in 2012 at their raw water intake as a part of a research project coordinated by the Norwegian Institute for Water Research (NIVA). The water utility was interested in how the levels of faecal indicator bacteria in their surface raw water source could be affected by different weather conditions. From December 2012 to November 2014 the instrument analysed a 100 ml water sample every day and provided a presence/absence result after 6–15 hours. With this sampling frequency, the normal maintenance interval was 21 days. Maintenance was limited to replacing reagents and restart analysis. The instrument detected *E. coli* at as small a concentration as 1 colony forming unit (cfu) per 100 ml water sample. Weekly sampling and laboratory testing was also done for comparison of results (Tryland *et al.*, 2015).

Daily measurements indicated that most faecal pollution events happened in autumn turnover periods (Figure 6.13). The data also pointed to that winters without ice cover or temperature stratification as physical barriers can give less protection to the raw water source. As Figure 6.13 shows, during the winter of 2013 ("Winter 1") not one *E. coli* was detected in the water samples. In the winter of 2014 ("Winter 2") several samples were analysed as positive. This was a much milder winter with more rainfall and a shorter period with ice cover compared to the winter of 2013. The results indicate that milder winters and more rainfall, possibly caused by climate change, could have a negative effect on the natural hygiene barrier of drinking water sources such as Maridalsvannet.

During the summer of 2013, the weekly sampling and lab testing detected *E. coli* only once. This would normally give an indication that the raw water source is subject to few and isolated *E. coli* events. The quality monitoring by the Colifast ALARM however detected *E. coli* seven times during July, typically in hot and warm days (Tryland *et al.*, 2015). Sampling performed only once every week might lead the water utility to think that the temperature stratification that forms in summer is a potent hygienic barrier. The raw water is extracted below the thermocline but it would seem from the daily monitoring that faecal pollution, presumably from birds located near the water intake pose a hygienic risk after all.

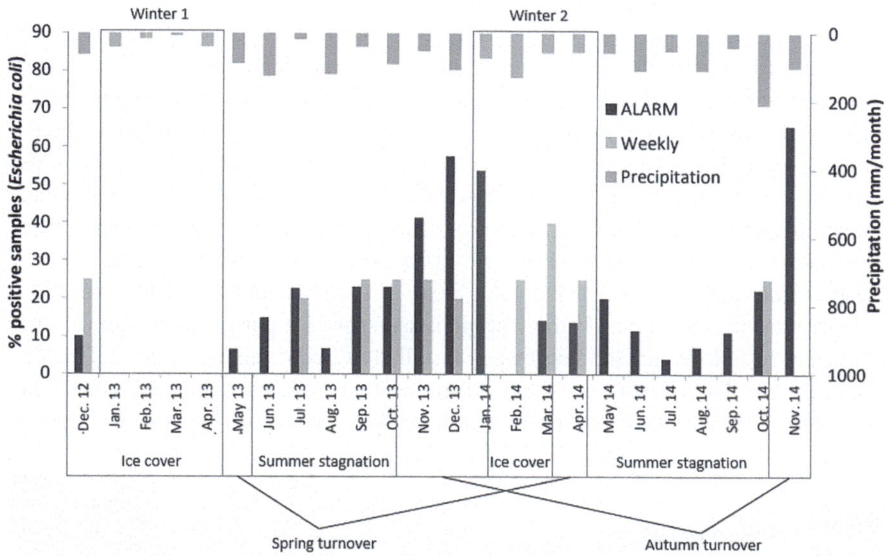

Figure 6.13 Comparison of Colifast ALARM reports of *E. coli* positive samples ("ALARM"), weekly laboratory samples ("Weekly") and precipitation in the period December 2012 to November 2014 at Oset WTP. Modified from Tryland *et al.* (2015).

In 2016 the same Colifast ALARM was moved to a pump station in the middle of the city for more frequent monitoring of the bacterial quality on the distribution network. Especially in urban areas, the water utilities have growing concerns regarding combined sewers, aging pipes and more extreme weather. Having daily updates on the bacterial quality of the water on a particular part of the network being sent directly to the WTP control room provided the utility with an extra security.

6.3.3.2 Raw water monitoring with CALM

As Lake Maridalsvannet is a surface water body it is subject to rivers and creeks that can carry pollution with them. Oslo VAV discovered that Movassbekken, one of the creeks that end up in their raw water source, occasionally carried faecal pollution with it. They installed a CALM system on the river to monitor the level of faecal coliform bacteria 2–3 times a day in spring, summer and autumn. Movassbekken has a remote location which makes it time and labour consuming to collect water samples manually and at the desired frequency. The CALM system uses the MPN method to statistically calculate the number of faecal coliforms in a water sample based on sample volume and number of vials with growth. Results are then automatically sent via a programmable logic controller (PLC) to Oset WTP control room.

6.3.3.3 Environmental monitoring with CALM

Originating in Oslo's drinking water source Lake Maridalsvannet, the urban river Akerselva runs down through the city until it reaches the Oslo Fjord. An industrial area was for over a hundred years centred on the river. Along with industrial spills, sewage pipe leaks and an urban location lead to a generally poor water quality in the river. In newer times, the quality improved due to shut down of factories and better sewage treatment and handling, and introduction of regulations involving the creeks surrounding the river. Combined sewer overflow is however a more modern threat to the bacterial quality of the recreational areas of the river and the Oslo Fjord close to where the river runs out. In the bathing season this is a particular concern for the municipality. Oslo Water and Wastewater installed a CALM unit on Akerselva in 2011 (Figure 6.14). The instrument measured the content of faecal coliform bacteria (as requested by the customer) on a daily basis in the bathing season (Tryland *et al.*, 2012).

Figure 6.14 CALM system at Myraløkka monitoring the levels of faecal coliforms in the urban river Akerselva.

The CALM system measures bacteria in the river water using one out of several possible analysis principles. The MPN format was used for the majority of the measurements, which takes 11 hours from sampling to result, and gives a statistical calculation of the number of bacteria present.

Occasionally the instrument has performed a rapid test on the water sample. This analysis is non-growth dependent and therefore only takes 2.5 hours from sampling to when a quantitative result is obtained (Tryland *et al.*, 2016).

Figure 6.15 shows the results from the summer of 2011, where the daily analyses performed by the CALM system were compared to the weekly standard analysis performed in the laboratory as well as precipitation at the same time. The municipality wanted more frequent motorization of that particular localization, Myraløkka, prior to their plans of establishing a public bathing zone on the riverbank. The European Guidelines for Recreational Water states that recreational water where bacterial levels exceed 1000 cfu of faecal coliforms per 100 ml water sample is unfit for bathing.

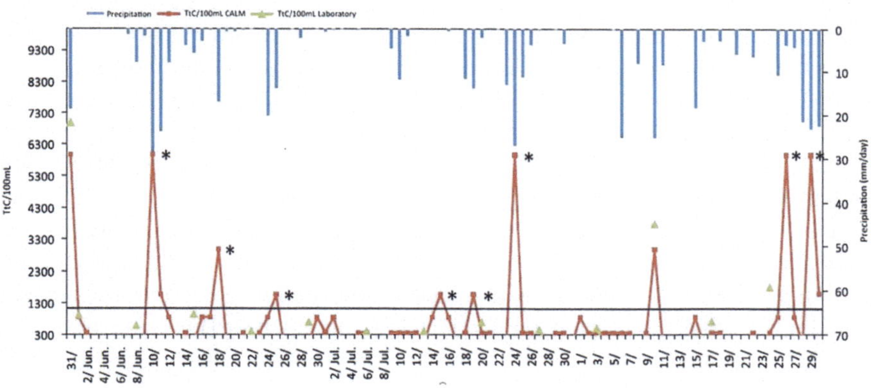

Figure 6.15 Daily results from CALM (lines with squares) compared with weekly lab results (triangles) and precipitation (top bars) in Akerselva summer 2011. *Y* axis shows number of thermotolerant coliform bacteria (TKB) e.g. faecal coliform bacteria. Horizontal line marks the European Guidelines for Recreational Water: 1000 cfu faecal coliforms (FC) per 100 ml. Stars indicate events not recorded by weekly sampling and lab testing. Modified from Tryland *et al.* (2012).

After monitoring the Myraløkka area for around four years Oslo Water and Wastewater moved the instrument downstream. Now the results from the CALM instrument are used in their internal control of the city's sewage system. If the results indicate levels of faecal coliforms above 1000 per 100 ml over a longer time period and in dry weather the department starts source tracking to see where the contamination may originate from.

6.4 BENEFITS AND LIMITATIONS

During the past 25 years several automated systems for monitoring bacteria in water have been developed. Colifast AS has been central in this development, launching the first semi-automated system in the mid-90s, followed by fully automated on-line systems from 2001. During this period there has been some sporadic competition in the market but most of these products disappeared after a short time while some were more successful and are still available (Movig *et al.*, 2017). Other analysis principles have been tested, such as genetic markers in polymerase chain reaction, optical particle counting and analysis by use of target specific antibodies, but these formats where found to be both difficult to automate and less sensitive and accurate. Today, the majority of these systems utilize quite similar methodological principles, using the specific enzymatic activity of the target bacteria for detection. There are different detection formats, some rely on bacterial growth in combination with enzyme activity, and other rely on measurements of the initial enzyme activity prior to bacterial growth (Fiksdal & Tryland, 2008). The rapid methods with results generated in 0.5 to 2 hours are less accurate compared with the growth methods with results available after typically 6 to 14 hours. The latter show good correlation with the standard laboratory reference methods.

The growth methods are more accurate and more robust due to three selective steps, involving use of a medium with inhibitors, which facilitate growth of the target group, the group specific enzyme activity and a detection threshold based on high level of end-product after growth. Most Colifast customers use one of the growth formats for the automated analysis. Growth methods require prolonged incubation and analysis time, but the automatic at-site sample handling, processing and direct reporting would be much quicker compared to the traditional manual sampling, transportation, preparation and laboratory analysis (Figure 6.16).

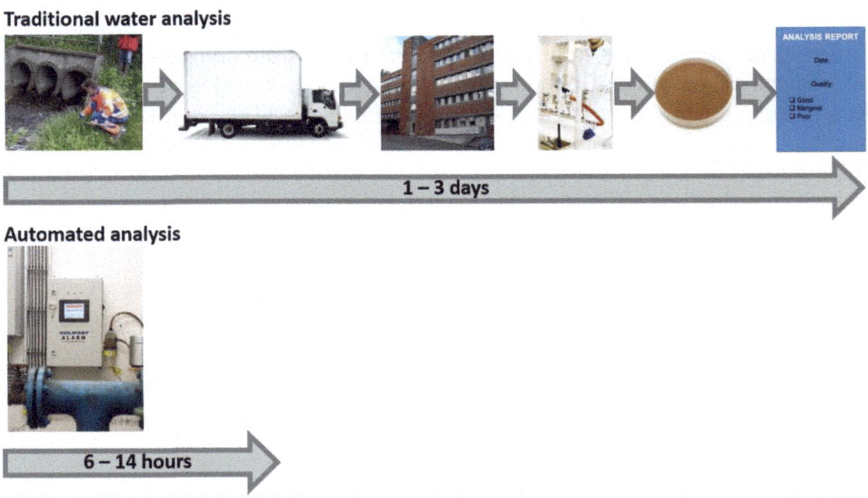

Figure 6.16 Traditional microbial laboratory analysis including the typical steps involved in the analysis, compared to Colifast ALARM automated at-site analysis. Analysis time showed in grey arrows.

The rapid methods rely only on the specific enzyme activity for detection, not including pre-growth and a threshold level to discriminate environmental interference. Sunlight, holding time in water, per-treated pollution and temperature can alter the enzymatic activity and bacteria population in water bodies which can affect the accuracy of the rapid methods (Davies *et al.*, 2005; Fiksdal & Tryland, 1999). The Colifast systems for rapid screening are typically used for monitoring urban rivers with elevated bacterial levels, where there can be big shifts in the faecal pollution level and where an exact bacterial number is of less importance for the customer.

Sufficient documentation includes test reports, customer reviews and verification statements, and are important for any customer considering implementing this type of technology. The first sales and installations were therefore quite difficult. When there is an established customer base and documentation, it is easier. For technical documentation, scientific papers describing the technology can be of good help. The company has created and published many scientific papers during the past 25 years. Some of these are: Anglès d'Auriac *et al.* (2000), Braathen *et al.* (2005), Eckner *et al.* (1999), Movig *et al.* (2017), Tryland *et al.* (2001, 2002, 2012, 2015, 2016). In addition, third-party verifications and pre-validations are valuable and Colifast has gone through both an EU project with technology pre-validation of test formats and instrumentation, and a separate US EPA technology verification (ETV) of the ALARM system (Demowatercoli, 2003; US EPA ETV, 2011). Still, the technical stability and practical experience information available through operators and references seem to be most important for any potential customer, and for Colifast. Examples like the first CALM and ALARM units, with reported continuous operation since 2003 and 2010, are very good as references. However, the lack of a full validation for use as alternative method is a problem for many prospective buyers. If these automated systems could replace the mandatory laboratory analysis it would also be a greater economic benefit for the utility in having this technology installed. International and local legislation and guidelines describe the methods to be used, minimum number of analyses and parameters. The methods are to be performed at laboratories accredited for specific methods. Colifast has several times looked for ways to get the technology approved as a standard reference method but it has proved to be difficult. As all the approved microbial methods and results are solely related to laboratory preparation and laboratory accreditation including certification of the personnel, there are not any protocols or standardized ways to compliance test and get accreditation for a fully automated on-line system. As the ALARM system has the US ETV statement, Colifast has been in contact with US EPA to discuss ways to get it approved as a standard method. So far, it is not possible. If solely used in an accredited lab, operated by certified personnel, it would be possible, but not as standalone on-line analysis out in the field. As the world steadily becomes more automated, it might be reconsidered and possible to get automated bacterial monitors approved in the future.

6.5 FUTURE PERSPECTIVES

From the year 2001 until today the number of on-line analysers in use has steadily been growing which indicates an increasing interest in the market for automated bacterial monitors. The water utilities are getting more used to on-line detectors for various parameters and implementation of these has shown to be beneficial. This makes the integration of a fully automated bacterial monitor more relevant and interesting.

As these systems prove to successfully monitor relevant faecal indicators, decrease the time to results, increase the test frequency and replace costly and time-consuming manual analysis, the demand for this automated technology is likely to further increase, especially for utilities which have some concerns regarding the hygienic water quality. Even if these systems today cannot replace the mandatory analyses, they are very good supplements to the low required number of laboratory analyses (Figure 6.17).

Figure 6.17 A CALM system generating rapid results and increasing the test frequency on water out from a pumping station in Poland.

The rapid and more frequent automated results can directly improve the safety in the water supply and for recreational waters, when integrated in local control systems, which displays the measurements and generate alarms at set threshold levels. In this way, the results are directly available and valuable for the operators. Another important benefit is the high analysis frequency provided by on-line measurements. The high number of results generates a better overview of the water quality and also provides data during periods with higher risk of pollution and in weekends and holidays. The traditional laboratory analysis might thereby miss important results from periods with higher risk of contamination. The third important benefit is the ease of use and high technical quality, which requires a minimum of supervision. Colifast has worked constantly over the years to provide stable and robust systems and today these systems have excellent uptime and would typically only require one annual technical maintenance service. This is very important for many utilities, as the benefit of automated systems and the results also relies on uptime and manual work required for operation. In addition, as these systems performs well, all the bigger water utilities, which originally purchased one Colifast system, have today several units.

On-line analysis by the Colifast systems has proved to be powerful in water monitoring and it is hoped that these systems can be accepted as official alternative analysis methods in the near future.

6.6 REFERENCES

Anglès d'Auriac M. B., Roberts H., Shaw T., Sirevåg R., Hermansen L. F. and Berg J. D. (2000). Field evaluation of a semiautomated method for rapid and simple analysis of recreational water microbiological quality. *Applied and Environmental Microbiology*, **66**, 4401–4407.

Berg J. D. and Fiksdal L. (1988). Rapid detection of total and fecal coliforms in water by enzymatic-hydrolysis of 4-methylumbelliferone-beta-d-galactoside. *Applied and Environmental Microbiology*, **54**, 2118–2122.

Bjergaarde N. E. and Hansson E. (2012). New methods for on-line detection of coliform bacteria in drinking water. *Proceedings from the Eighth Nordic Drinking Water Conference*, Stockholm, 18–20 June 2012, pp. 105–114.

Braathen H., Berg J. D. and Hermansen L. (2006). Technical Note No. 8. Evaluation of a MU-production (2 h 15 min) Method for Detection of Thermotolerant Coliforms in Water using Colifast Microdetector. http://www.foodsafety.gr/brochures/FIDS_FieldKit_Technote.pdf (accessed 22 March 2018).

Braathen H., Ranneklev S., Sagstad H. and Rydberg H. (2005). Fully automated monitor reduces intake of *E. coli*-contaminated raw water. *Water and Wastewater International*, **30**(2), 35.

Davies C. M., Apte S. C. and Peterson S. M. (1995). Beta-d-galactosidase activity of viable, non-culturable coliform bacteria in marine waters. *Letters in Applied Microbiology*, **21**(2), 99–102.

Demowatercoli Final Report. EC Framework 5 (2003). http://www.colifast.no/wp-content/uploads/pdf/EU%20Demoatercoli.pdf (accessed 22 March 2018).

Eckner K. F., Jullien S., Samset I. D. and Berg J. D. (1999). Rapid, enzyme-based, fluorometric detection of total and thermotolerant coliform bacteria in water samples. *Water Supply*, **17**(2), 73–80.

European Union (2006). Directive 2006/7/EC of the European parliament and of the council of 15 February 2006 concerning the management of bathing water quality and repealing directive 76/160/EEC.

Fiksdal L. and Tryland I. (1999). Effect of UV light irradiation, starvation and heat on *Escherichia coli* beta-d-galactosidase activity and other potential viability parameters. *Journal of Applied Microbiology*, **87**, 62–71.

Fiksdal L. and Tryland I. (2008). Application of rapid enzyme assay techniques for monitoring of microbial water quality. *Current Opinion in Biotechnology*, **19**, 289–294.

Movig T., Braathen H. and Stenersen H. (2017). Handbook of Online and Near-real-time Methods in Microbiology. CRC Press, Boca Raton, Florida, USA, Chapter 1, pp. 1–22.

Svenskt Vatten (2009). Risk assessment case study Göteborg Sweden. Svenskt Vatten Rapport Nr 2009-03 Tidiga Förvarningssystem – Är det någonting för våra kommuner? http://vav.griffel.net/filer/2009-03.pdf (accessed 22 March 2018).

Tryland I., Braathen H., Beschorner A.-L. and Muthanna T. (2012). Vurdering av metoder for overvåkning av hygienisk badevannskvalitet. Monitoring of hygienic bathing water quality. *VANN*, **47**(2), 194–206.

Tryland I., Braathen H., Wennberg A. C., Eregno F. E. and Beschorner A. L. (2016). Monitoring of β-d-galactosidase activity as a surrogate parameter for rapid detection of sewage contamination in urban recreational water. *Water*, **8**(2), 65.

Tryland I., Eregno F. E., Braathen H., Khalaf G., Sjolander I. and Fossum M. (2015). On-line monitoring of *Escherichia coli* in raw water at Oset Drinking Water Treatment Plant, Oslo (Norway). *International Journal of Environmental Research and Public Health*, **12**(2), 1788–1802.

Tryland I., Samset I. D., Hermansen L., Berg J. D. and Rydberg H. (2001). Early warning of faecal contamination of water: a dual mode, automated system for high (<1 hour) and low levels (6–11 hours). *Water Science and Technology*, **43**(12), 217–222.

Tryland I., Surman S. and Berg J. D. (2002). Monitoring faecal contamination of the Thames estuary using a semiautomated early warning system. *Water Science and Technology*, **46**(3), 25–31.

US EPA ETV (2011). Environmental Technology Verification Report: Colifast Alarm at-line automated remote monitor. http://nepis.epa.gov/Adobe/PDF/P100BSVR.pdf (accessed 22 March 2018).

World Health Organisation (WHO) (2017). Guidelines for Drinking Water Treatment, 4th ed. World Health Organisation, Geneva, Switzerland. Chapter 7: Microbial aspects.

Chapter 7

Online flow cytometry: Towards a rapid, robust, and reliable microbial sensor

Frederik Hammes and Michael D. Besmer

7.1 INTRODUCTION: MICROBIAL DYNAMICS MATTER IN AQUATIC ECOSYSTEMS

Indigenous bacterial communities inhabit nearly every natural and engineered aquatic ecosystem and often play a crucial role in ecosystem functioning and stability (Fenchel & Jørgensen, 1977; Paerl & Pinckney, 1996; Girvan *et al.*, 2005; Verstraete *et al.*, 2007). Typical concentrations of suspended cells are in the order of 10^2–10^4 cells/mL (groundwater), 10^3–10^5 cells/mL (drinking water), 10^5–10^6 cells/mL (surface waters), and 10^6–10^7 cells/mL (wastewater effluent) (Wang *et al.*, 2007; Berney *et al.*, 2008). Since microorganisms are an inherent part of treatment and distribution systems, bacterial concentrations and composition can often be reflective of the performance of treatment steps, the overall ecosystem stability, as well as the effects of a variety of periodic and aperiodic system changes or perturbations (Gunther *et al.*, 2012; Nescerecka *et al.*, 2014; Van Nevel *et al.*, 2017; Props *et al.*, 2016b).

Natural and engineered ecosystems are subject to a myriad of perturbations (i.e., deviations from their perceived normal state) over considerably different time scales (Figure 7.1). Some changes occur over extended time scales from months to years (e.g., seasonal changes, global warming impacts). For example, Prest and colleagues (2016) described a periodic 6-fold increase in flow cytometric (FCM) total cell concentration (TCC) during warm summer months in the effluent of a drinking water treatment plant. This was ascribed to seasonally increased bacterial growth and detachment in a final biofiltration treatment step (Figure 7.1). However, many

Editors: Torben Lund Skovhus and Bo Højris
doi: 10.2166/9781780408699_0115

perturbations occur at considerably shorter time scales, ranging from seconds to weeks. For example, natural aquatic ecosystems such as groundwater, rivers, lakes, and oceans are influenced by short-term periodic perturbations such as photosynthesis, temperature, anthropogenic activities (Kirchner *et al.*, 2004; Hayashi *et al.*, 2012; Zeng *et al.*, 2012) and aperiodic perturbations such as precipitation (rainfall) events, snow melt, and acute pollution (Kistemann *et al.*, 2002; Geyer *et al.*, 2008; Guadayol *et al.*, 2009; Stadler *et al.*, 2010). Engineered aquatic ecosystems such as drinking water and wastewater treatment plants are subject to such aperiodic fluctuations in their raw/source water (Astrom *et al.*, 2007; Plummer & Long, 2007; Foladori *et al.*, 2010; Gunther *et al.*, 2012), but often also experience periodic operational changes (e.g., backwashing of filters, varying water production rates, hydraulic changes (Stevenson, 1997; Bakker *et al.*, 2003; Props *et al.*, 2016b)). Figure 7.1b shows an example of daily perturbations in a drinking water distribution system occurring at minute-to-hour resolution. In this example, the two-fold increase in suspended bacterial cell concentration was ascribed to diurnal hydraulic changes in the system, resulting in increased bacterial detachment/suspension (Nescerecka *et al.*, 2014).

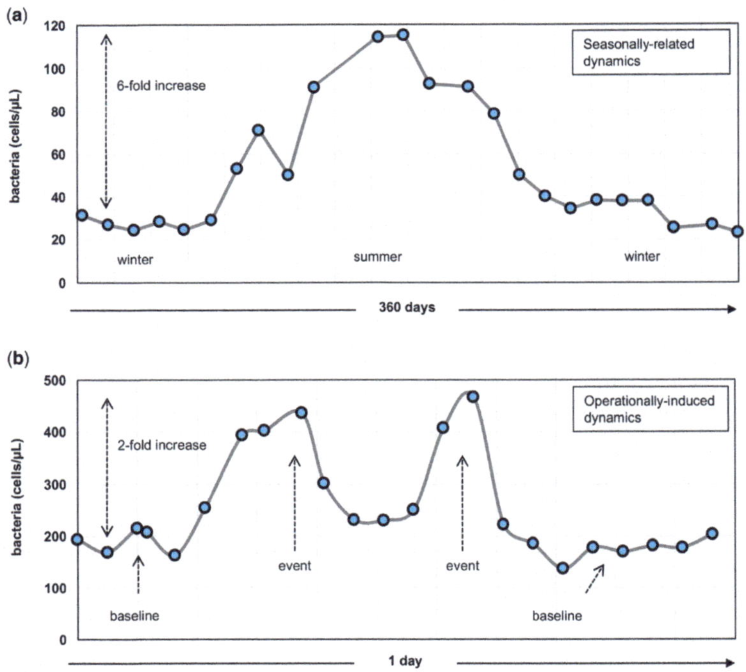

Figure 7.1 Different microbial dynamics occurring in different drinking water systems and at different time scales. (a) Periodic seasonal dynamics in the effluent of a drinking water treatment plant in the Netherlands (adapted from Prest *et al.*, 2016). (b) Operationally induced fluctuations in distributed drinking water in Latvia (adapted from Nescerecka *et al.*, 2014).

A basic requirement for good management of any natural or engineered ecosystem is proper understanding of the system itself (Krewski *et al.*, 2002; Verstraete *et al.*, 2007; Petterson & Ashbolt, 2016). Understanding ecosystem dynamics allows for (1) judgment of the relevance of a given dynamic, (2) optimized monitoring of the dynamics, and (3) tailored corrective measures against problematic dynamics where required and when feasible. A clear example is distributed drinking water, which can be subject to both periodic (seasonal) fluctuations (Figure 7.1a) as well as occasional aperiodic hydraulic perturbations (Figure 7.1b). It is evident that in such a case, the system's baseline for cell concentrations cannot simply be characterized by fixed average value of data collected over an extended time period. Rather, the baseline becomes time-dependent relative to the periodic fluctuation, and deviations due to aperiodic changes need to be assessed against this "dynamic baseline" (Murray *et al.*, 2010; Arad *et al.*, 2013; Prest *et al.*, 2016). Moreover, dynamic periods of high microbial loads can be of short duration but are often the predominant drivers of overall risk (Signor & Ashbolt, 2006).

Aquatic ecosystem dynamics are described by their frequency, duration, and magnitude, and hence tools are needed to characterize these aspects accurately. It is evident that microbial system perturbations in aquatic environments can neither be fully addressed by the use of abiotic online monitoring tools (e.g., turbidity, conductivity) (Page *et al.*, 2017), nor by the use of conventional cultivation-based microbial approaches (Van Nevel *et al.*, 2017). Fast, accurate, robust and automatable microbial monitoring technologies are needed. Methods that emerged during the last few years include automated FCM for auto-fluorescent and green fluorescent protein (GFP)-labelled organisms (Thyssen *et al.*, 2008; Broger *et al.*, 2011; Arnoldini *et al.*, 2013; Brognaux *et al.*, 2013; Pomati *et al.*, 2013; Heck *et al.*, 2014), adenosine triphosphate (ATP) (Vang *et al.*, 2014), indicator organism detection after targeted sampling (Stadler *et al.*, 2008), and automated optical detection methods (Hojris *et al.*, 2016), several of which are discussed in other chapters of this book. Here we will focus exclusively on the development and use of automated online FCM.

7.2 FCM IS AN AUTOMATABLE MULTIVARIATE MICROBIAL SENSOR

FCM has been applied during the last three decades for the quantification and characterization of microorganisms from a variety of samples from natural and engineered aquatic ecosystems (Muirhead *et al.*, 1985; Czechowska *et al.*, 2008; Van Nevel *et al.*, 2017). The two main factors driving increasing interest in the technology in recent years were (1) the emergence of small, user friendly instruments, and (2) increased awareness of, and general knowledge on, indigenous microbial communities in aquatic ecosystems. More recently, developments in FCM data analysis (Props *et al.*, 2016a; Koch *et al.*, 2014), automated systems (Hammes *et al.*, 2012; Van Nevel *et al.*, 2013) and linking FCM data with microbial community analysis (Prest *et al.*, 2014; Props *et al.*, 2016b) have opened new and exciting opportunities in this field.

The basic principle of FCM is that a water sample is pumped through a capillary in such a manner that particles (e.g., bacteria) in the sample are passing one after another (Figure 7.2). A laser beam is directed at the capillary and both the scattering of the light and excited fluorescence signals are collected for every single particle by detectors and recorded (for more details, see Shapiro, 2003). A broad variety of instruments is commercially available from multiple suppliers, which allows for individual selection of lasers and filters to detect the desired fluorescent signals.

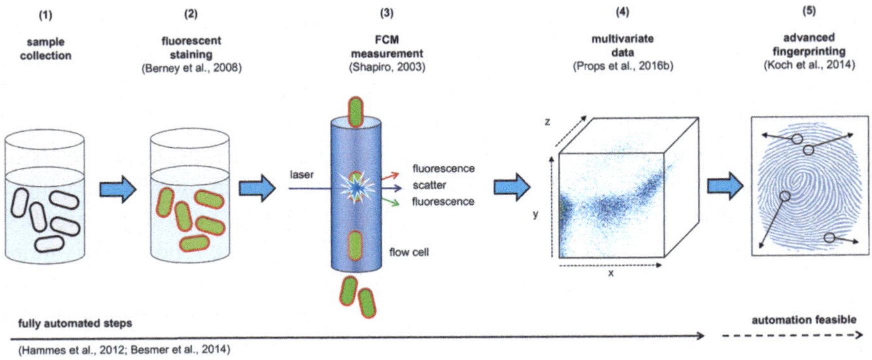

Figure 7.2 Overview of the different steps in a flow cytometric measurement of bacteria in a water sample.

The fluorescence signals can be used to separate microbial cells from abiotic background particles, either by utilizing auto-fluorescence from photo-pigments (e.g., chlorophyll in algae, cyanobacteria) or by labelling cells with fluorescent molecules (Figure 7.2). For the latter, commonly referred to as staining, fluorescent dyes binding specifically to cellular compounds such as nucleic acids (e.g., SYBR Green I) are used (Prest *et al.*, 2013). Dyes can also be used for more complex characterizations of bacteria in samples. In this regard, a multitude of dyes and labelled molecules targeting different aspects of cell viability (e.g., propidium iodide), activity (e.g., cFDA) and identity (e.g., specific antibodies) exist, and have been reviewed in detail elsewhere (Hammes *et al.*, 2011; Gasol & del Giorgio, 2000; Czechowska *et al.*, 2008). As a consequence, the instrumentation enables the detection of different biotic particles including viruses, bacteria, yeasts, algae and protozoa. Moreover, each particle/cell that is analysed yields multivariate information that can be used to identify, characterize and classify the particle. The multivariate information gained on single-particle level can in turn be combined into a FCM fingerprint of the analysed community, for which several new approaches have been described and tested in detail in recent years (Koch *et al.*, 2014; Props *et al.*, 2016b). One relatively commonly used approach differentiates between so-called large high nucleic acid (HNA) content bacteria and small low nucleic acid (LNA) content bacteria as a straightforward fingerprint compatible with different instruments (Prest *et al.*, 2013). The HNA/LNA fingerprinting approach broadly differentiates between groups with remarkably different phylogeny (Vila-Costa *et al.*, 2012), and is used complimentary to cell concentration measurements to track changes in the microbial composition of a water sample (Prest *et al.*, 2014).

During the last decade, considerable effort was made towards standardizing FCM protocols for analysing bacteria in aquatic samples (SLMB, 2012; Prest *et al.*, 2013; Nescerecka *et al.*, 2016a) leading to further acceptance and application of the method in water-related industries (reviewed in Van Nevel *et al.*, 2017). The advantages, challenges, and overall application value of FCM in comparison to conventional cultivation-based methods are discussed extensively by Van Nevel *et al.* (2017). For additional information, readers are referred to the review papers dealing with (a) FCM methodology (Shapiro *et al.*, 2003; Hammes *et al.*, 2011) and (b) specific stains and applications (Czechowska *et al.*, 2008; Van Nevel *et al.*, 2017).

Importantly, all the steps involved in an FCM measurement can be automated (Figure 7.2). The potential for FCM automation is specifically based on (1) the rapid preparation and measurement of samples (<15 minutes time-to-result), (2) low requirements for sample volume (usually 50–200 μL), (3) the relative simplicity of sample preparation (addition of 1–2 reagents, mixing, incubation) and (4) affordable reagents enabling cost-effective measurements (Hammes *et al.*, 2012; Besmer *et al.*, 2014). In fact, many commercial instruments are equipped with multi-well plate auto-loader systems, thus providing a built-in degree of automation already (Van Nevel *et al.*, 2013). Several automated online FCM systems were previously developed and applied in the fields of medicine and biotechnology (Omann *et al.*, 1985; Lindberg *et al.*, 1993; Zhao *et al.*, 1999; Abu-Absi *et al.*, 2003; Broger *et al.*, 2011; Brognaux *et al.*, 2013) and for phytoplankton enumeration and characterization (Dubelaar *et al.*, 1999; Thyssen *et al.*, 2008; Pomati *et al.*, 2013). Cell detection in many of these early instrument developments were based on either size/scatter properties (Arnoldini *et al.*, 2013), auto-fluorescence (Dubelaar *et al.*, 1999) or expression of fluorescent proteins (Broger *et al.*, 2011; Arnoldini *et al.*, 2013), thus not requiring the addition of fluorescent dyes. However, in many aquatic systems, indigenous bacteria are small and usually not auto-fluorescent, thus rendering them indistinguishable from background particles without prior fluorescent staining. In the present chapter, we focus specifically on the development and application of automated online FCM systems targeting bacterial samples from aquatic ecosystems that require staining for detection/characterisation.

A first prototype of an online FCM system specific for aquatic ecosystems was proposed previously by Hammes and colleagues (Hammes *et al.*, 2012). This system in essence combined an automated sampling and sample preparation device, including a temperature controlled incubation chamber (enabling staining of bacteria under controlled conditions), and a hard- and software interface, with a commercially available flow cytometer. This approach was improved and optimised, resulting in more reliable instruments that were ready for use in practice (Besmer *et al.*, 2014). These systems were originally developed to collect and process discrete samples online at approximately 15 min intervals, using the nucleic acid targeting stain SYBR Green I to distinguish bacteria from background noise and abiotic particles (Hammes *et al.*, 2012; Prest *et al.*, 2013; Besmer *et al.*, 2014).

7.3 EXPERIENCE GAINED WITH DATA SETS GENERATED DURING THE LAST FIVE YEARS

Discrete online FCM has been implemented for the past five years in a variety of full-scale case studies and laboratory-scale experiments representing a range of different aquatic ecosystems, predominantly related to drinking water treatment. Here we briefly discuss four examples where different temporal microbial dynamics were studied on varying time scales.

7.3.1 Periodic and aperiodic fluctuations in river water

Rainwater runoff contaminates river water through stagnant storm water drain overflows and direct surface wash-off. Particularly in peri-urban areas, river water is exposed to considerable aperiodic perturbations resulting from localised precipitation events. This is relevant because precipitation-induced river water contamination often results in increased pathogen concentrations in water used for drinking water or recreational purposes (Shehane *et al.*, 2005). Moreover, previous studies described the occurrence of periodic diurnal fluctuations in abiotic variables in rivers, attributed to the precipitation and dissolution of calcite influenced by microbial photosynthesis and respiration (Vogt *et al.*, 2010), but a concurrent link to fluctuations in microbial variables has not yet been established.

The study of Besmer and colleagues (Besmer *et al.*, 2014) used online FCM to investigate fluctuations in TCC, FCM fingerprints and some abiotic variables in a shallow oligotrophic river for a period of 14 days (1104 measurements), during which two small precipitation events occurred. An extract of the data is shown in Figure 7.3. Here the dominant perturbation was an aperiodic precipitation event (3.1 mm during 8 hours) on day 2. FCM-TCC data demonstrate the impact of the precipitation event, with TCC increasing rapidly and continuously at a rate of about 0.2×10^6 cells/hour to a maximum of 2.3×10^6 cells/mL, and decreasing at a considerable slower rate of about 0.07 × 10^6 cells/h after precipitation ended. Simultaneously the FCM fingerprint changed dramatically, with a clear increase in the percentage HNA content bacteria from about 45% to 60% during the precipitation event. This suggests that the contaminating rainwater runoff was dominated by large HNA content bacterial cells. The data also shows discernible periodic diurnal fluctuations, with the lowest TCC in the morning at about 0.85 × 10^6 cells/mL and the highest TCC in the late afternoon at about 1.06 × 10^6 cells/mL, and small concurrent fluctuations in the HNA/LNA bacterial composition (Figure 7.3).

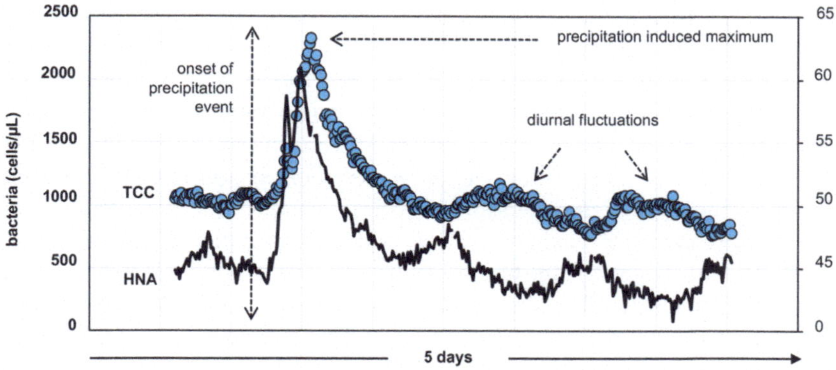

Figure 7.3 Diurnal and precipitation-induced fluctuations in TCC and FCM fingerprints (% HNA) in river water. Samples were collected automatically at 15-minute intervals, stained with SYBR Green I and measured with FCM. A small, localized precipitation event (3.1 mm in 8 h) occurred early on day 2. HNA = high nucleic acid content bacteria. All data adapted from Besmer *et al.* (2014).

The advantages of online FCM in this example are (1) an accurate description of the frequency, duration and magnitude of the precipitation-induced events, and (2) the detailed characterisation of the small diurnal TCC fluctuations. It is unlikely that low-frequency manual grab sampling and laboratory analysis would have captured the latter dynamic accurately. The dataset shows the value of using online microbial sensors to describe natural and/or anthropogenic pollution events, and provides researchers and practitioners with additional tools to characterise and monitor such events. Further application potential lies in the management and control of coastal waters (e.g., localised pollution), lake water (e.g., natural diurnal fluctuations) and groundwater used for drinking water treatment (discussed below). Moreover, this example clearly demonstrates that many aquatic ecosystems have naturally fluctuating bacterial baselines, rather than straightforward average values, which is critical to understand when grab samples are collected for further analysis.

7.3.2 Precipitation-induced contamination of karstic spring water

Karstic groundwater is a critical drinking water source for a considerable number of people worldwide. These aquatic ecosystems are notoriously vulnerable towards surface water infiltration and subsequent contamination following rainfall events, due to their porous geology. Moreover, land use, particularly animal farming, in local areas pose a substantial contamination risk, which has been described in detail before (Pronk *et al.*, 2006; Page *et al.*, 2017). Understanding the relationship between precipitation events and subsequent groundwater contamination would support risk assessment, the development and optimisation of dedicated sampling plans and potential emergency response.

The studies of Besmer and Hammes (2016) and Page *et al.* (2017) investigated different aspects of precipitation-induced TCC fluctuations in various karstic springs during multiple months with online FCM, focusing on the characterisation of dry weather baseline values and contamination breakthrough following localised precipitation. Figure 7.4 shows a typical example of such a spring, with TCC increasing nearly 50-fold from a very stable dry-weather average of about 1.2×10^5 cells/mL (Figure 7.4b) to maximum peaks of about 5.0×10^6 cells/mL after precipitation (about 64 mm in 72 h), and decreasing considerably slower after the precipitation event was over (Figure 7.4a). These increases are particularly relevant, as previous studies have clearly established links between precipitation-induced events and karstic groundwater contamination with faecal indicator organisms (Pronk *et al.*, 2006; Page *et al.*, 2017; Besmer *et al.*, 2017b). The study of Page *et al.* (2017) specifically investigated abiotic variables as predictors of bacteriological changes in springs, and showed that with sufficient data it was possible to link precipitation data and subsequent increases in spring discharge to imminent TCC contamination. Importantly, the data from several studies suggested that no single abiotic sensor (e.g., turbidity) was capable of characterising the microbial dynamics (Figure 7.4) accurately and consistently (Besmer & Hammes, 2016; Page *et al.*, 2017).

The advantages of online FCM in these specific examples are (1) the accurate description of the frequency, duration and magnitude of the precipitation-induced contamination of karstic springs, and (2) having extensive event data to assess abiotic sensors with. The latter is important, as online abiotic sensors are currently much better established with operators as online microbial sensors, but extensive microbial data is needed to link abiotic data with microbial events (Page *et al.*, 2017). The data shows that online microbial sensors can potentially be used as components of early warning systems where dramatic aperiodic fluctuations in source water used for drinking water is expected. Moreover, the data generated from such measurements can be used to predict contamination events and thereby develop tailored monitoring strategies (discussed in detail in Besmer *et al.*, 2017a).

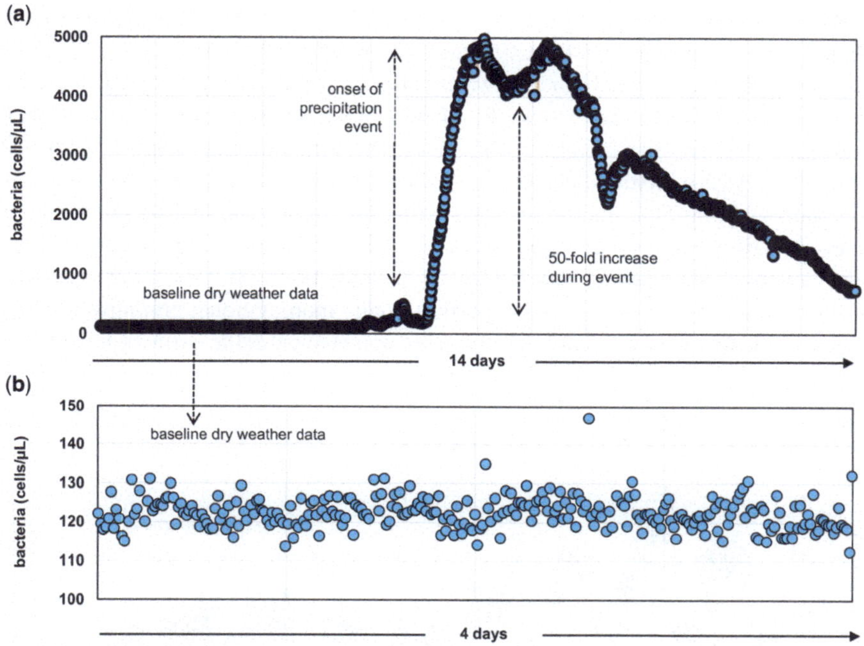

Figure 7.4 (a) Dramatic precipitation-induced increases in TCC in karstic groundwater following a localized precipitation event (64 mm in 72 h). (b) Dry weather data show stable measurements without discernible trends. Samples were collected automatically at 20-minute intervals, stained with SYBR Green I and measured with FCM. All data was adapted from Besmer and Hammes (2016).

7.3.3 Operationally-induced fluctuations in river bank filtered water

River bank filtration is used as a first essential step in many drinking water treatment systems worldwide. Here, the subsoil functions as a particulate filter, removing contaminants such as microorganisms and other particulate matter from the river water. Bacteria found in river bank filtered water can originate from the river as unfiltered contaminants, but can also detach from the soil particles as natural components of the groundwater. Knowing the concentrations, composition and dynamic of bacteria in such engineered ecosystems are important with respect to the type of treatment that is applied further to the water.

In the study of Besmer *et al.* (2016), we tracked TCC for two weeks (1400 samples) in the extraction well of a drinking water treatment plant, which was adjacent to a river, but which was also influenced by regional groundwater. Due to low water demand, the plant was operated in diurnal cycles, with water extraction occurring at night when electricity prices are lower than in the daytime (Figure 7.5). The online

FCM data show clearly a diurnal pattern in TCC caused by the operation of the plant. Cell concentrations in the extraction well ranged between 1.3–1.8 × 10^5 cells/mL during the day in the absence of water extraction, and decreased to 0.9–1.3 × 10^5 cells/mL (ca. 25% decrease) at night when the extraction pumps were activated. Transition between the high and low TCC concentrations was quick (1–2 h) and coincided always with the activation/deactivation of the extraction pumps (Figure 7.5). By comparing the TCC data to regional hydrogeological data, such as groundwater levels in surrounding observation wells, we were furthermore able to explain the mechanisms behind the changes in TCC in the extraction well (Besmer *et al.*, 2016). In short, water extraction resulted in proportionally more regional groundwater with a lower TCC (compared to bank filtered river water) to enter the extraction well.

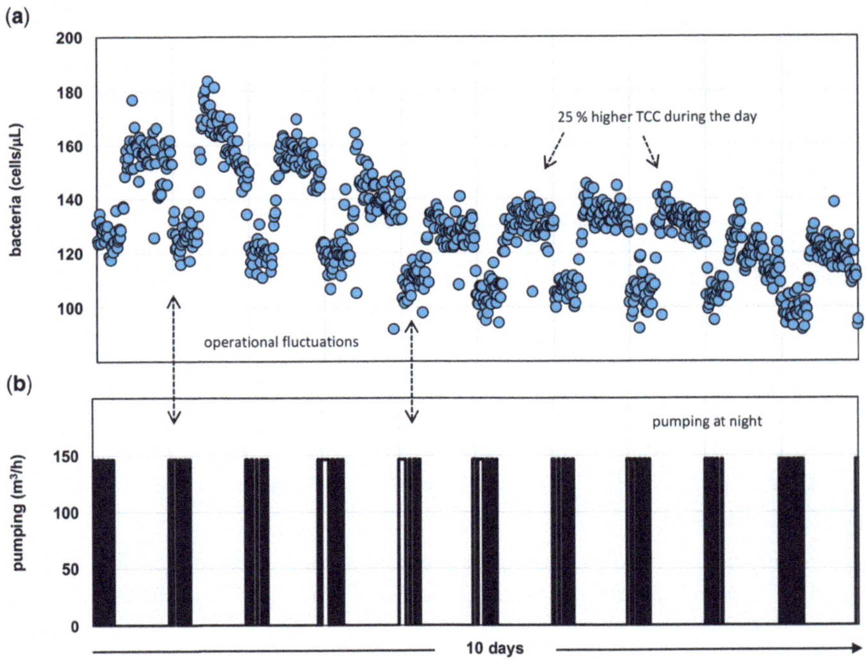

Figure 7.5 (a) Operationally-induced fluctuations in TCC in river bank filtered water during 10 days of measurements, directly linked to (b) diurnal extraction of water from the well. Samples were collected automatically at 15-minute intervals, stained with SYBR Green I and measured with FCM. All data was adapted from Besmer *et al.* (2016).

The advantage of online FCM in this example was the detection and characterisation of dynamic changes in TCC in an engineered system that were not previously known, and which were directly resulting from operation (i.e., extraction pump operation). These diurnal changes are unlikely to be detected by manual low frequency grab sampling, which happens typically only during the day. Such information can empower operators to change the operational practice (if regarded as relevant) or at least to design and execute a sampling plan relative to the microbial dynamic of the system. A further example of the online monitoring of operational-induced fluctuations in water microbial quality is discussed in Besmer and Hammes (2016), where the impact of flow rate through a granular active carbon biofilter on TCC was demonstrated.

7.3.4 Characterising bacterial growth

The examples described above are all related to microbial monitoring of full-scale natural and engineered aquatic ecosystems. However, there is also a considerable need for implementation of these technologies in laboratory-scale research. The most straightforward example is the characterisation of bacterial batch-growth, particularly when uncultivable indigenous bacterial communities are grown on natural organic matter. Bacterial growth is a dynamic event, characterised by changes in growth rate, cell size, and cell concentrations (e.g., Nescerecka *et al.*, 2016a) (Figure 7.6a). Although this is relatively straightforward, conventional microbial methods to characterise bath growth, e.g. plating and/or optical density measurements, are completely insufficient when uncultivable indigenous microbial communities grow at low cell densities (Wang *et al.*, 2009). However, such experiments can be accurately measured with automated FCM (Figure 7.6a; Prest *et al.*, 2013; Nescerecka *et al.*, 2016a).

A laboratory-scale study of Mimoso *et al.* (2015) focused on the quantification and characterisation of indigenous bacterial growth in the effluent of gravity driven membrane filters (GDMF) treating river water, with emphasis on household type filters. Sequential filtration, storage and water usage in such filters would essentially result in sequential batch growth of bacteria in the storage reservoirs. This process was followed with online FCM during multiple operational cycles and with varying experimental conditions, showing rapid adaptation of the indigenous communities to the batch growth environment (Figure 7.6b). The main advantage of the online monitoring here is the high frequency data, which enabled accurate descriptions of the operationally induced dynamic (dilution) and natural dynamic (growth) occurring daily with each operational cycle. Moreover, the online FCM data enabled the detection of subtle differences between different experimental set-ups (Mimoso *et al.*, 2015).

The application potential of online microbial sensors in laboratory scale research is particularly broad, ranging from growth studies as described above and elsewhere (e.g., Wang *et al.*, 2009), to biotechnology applications focussing on chemostats or fed-batch growth under varying conditions (e.g., Broger *et al.*, 2011; Brognaux *et al.*, 2013), to detailed monitoring of microbial response to various stress factors (e.g., Arnoldini *et al.*, 2013).

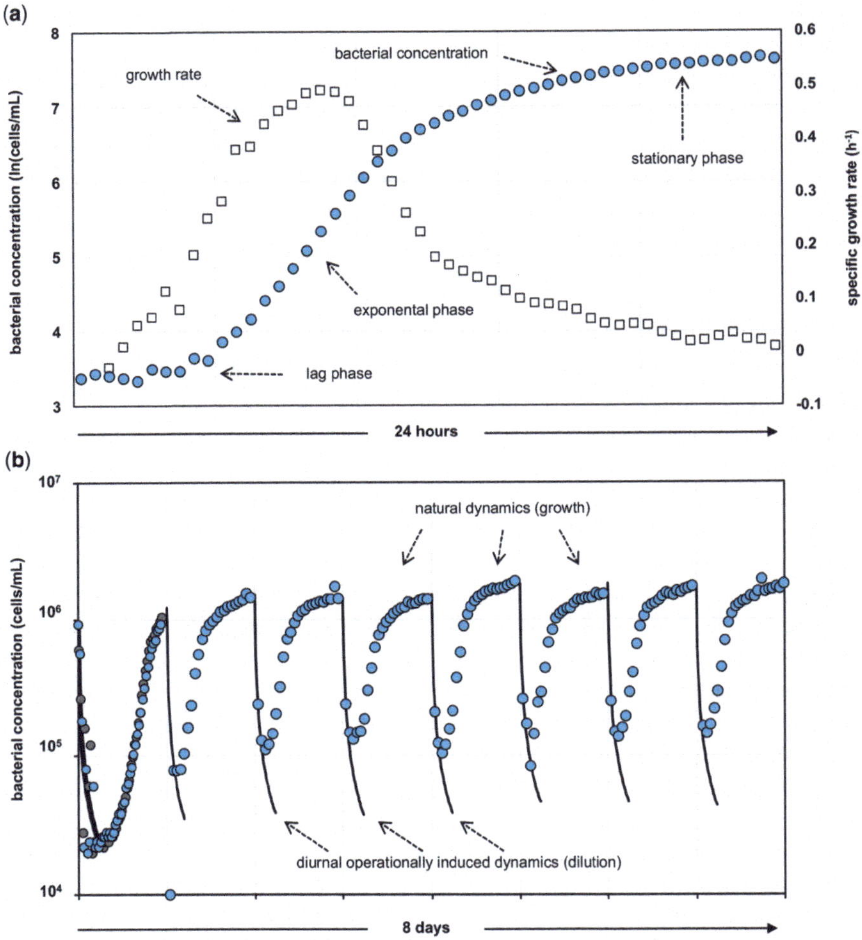

Figure 7.6 (a) Bacterial batch growth of an indigenous community growing in filtered river water and measured with online FCM. (b) Successive cycles of dilution followed by bacterial batch growth in the simulated effluent of a gravity driven membrane filter. Samples were collected automatically at 15-minute intervals, stained with SYBR Green I and measured with FCM. The data in (b) was adapted from Mimoso *et al.* (2015).

7.4 TOWARDS ROUTINE IMPLEMENTATION

Based on the first applications and positive experiences reported in early studies summarised above and elsewhere (Hammes *et al.*, 2012; Besmer *et al.*, 2014; Mimoso *et al.*, 2015; Besmer & Hammes, 2016; Besmer *et al.*, 2016; Page *et al.*, 2016), we argue that microbial monitoring in general, and online FCM in particular, can and should be further developed to enable routine implementation and application in both laboratory-scale research and full-scale monitoring. Such developments include: (1) commercialisation and broad implementation of affordable and easy-to-use instruments, (2) future development of dedicated small instruments such as microflow cytometers (Ateya *et al.*, 2008), (3) routine implementation of more complex staining approaches such as viability staining (Hammes *et al.*, 2012), (4) development and integration of automated advanced FCM data processing (Props *et al.*, 2016; Koch *et al.*, 2014), (5) combination with both abiotic online sensors (Page *et al.*, 2017) and alternative online microbial monitors (other chapters in this book), and (6) integration into broader monitoring schemes by scientists and practitioners alike.

The comparison with existing abiotic online sensors is critical, as such sensors are already well established in current monitoring programs. Such sensors are often cheaper to operate than microbial monitoring tools and are therefore regarded as so-called proxy variables for tracking microbial dynamics. Previous studies highlighted the value of integrating online FCM data with other online measurements (e.g., water throughput, meteorological data, spring discharge, abiotic variables) (Besmer *et al.*, 2014; Page *et al.*, 2017). For example, the study of Page *et al.* (2017) investigated statistical approaches to integrate online FCM data with multiple abiotic variables to describe contamination of karstic aquifers following regional precipitation. However, it is evident that abiotic variables will never fully suffice to accurately capture the myriad of complex microbial dynamics in aquatic ecosystems.

Moreover, there is an inherent need for comparing different online microbial monitors with each other. These systems often target different aspect of bacteria such as specific enzymatic activity (Ender *et al.*, 2017), direct imaging (Hojris *et al.*, 2016), nucleic acid staining (Besmer *et al.*, 2016), ATP content (Vang *et al.*, 2014). Regardless of the goals of commercial enterprises, it is well conceivable that in many circumstances the different technologies would not necessarily be in competition with each other, but would rather be collecting complimentary information (Berney *et al.*, 2008; Hammes *et al.*, 2011). For end-users of these technologies, it would be particularly useful if a detailed comparative overview is made, thus supporting the decision-making with respect to specific technologies.

Apart from direct monitoring applications, the wealth of highly quantitative and reproducible data arising from high-frequency monitoring of bacterial concentrations offers new possibilities for quantitative microbial risk assessment (QMRA). While online FCM does not measure pathogens directly, it can complement specific measurements that suffer from high methodological variability and often low concentrations that make statistical analysis challenging (Krewski *et al.*, 2002; Schmidt & Emelko 2011). Moreover, online FCM allows for proper coverage of microbial peak loads of short duration that are specifically relevant from a QMRA perspective and often missed with current sampling strategies (Krewski *et al.*, 2002; Signor & Ashbolt, 2006; Nilsson *et al.*, 2007). Online

FCM and other high-frequency measurements of bacterial concentrations can thus potentially bridge the gap between pure process variables and specific pathogen detection (Petterson & Ashbolt, 2016).

With the excitement accompanying new technologies, it is also imperative to consider whether online microbial monitoring is really needed. Despite encouraging data, we do not necessarily argue for wholesale installation and implementation of online FCM (or similar tools) in the water industry. The fact is that the depth of data generated with online tools might not always be needed. Targeted investigations for limited time periods with online FCM might be sufficient in many circumstances to gain sufficient understanding of a system and to achieve process optimization and improved monitoring with simpler tools (Krewski *et al.*, 2002).

7.5 THE NEXT FRONTIERS: HIGH FREQUENCY REAL-TIME FLOW CYTOMETRY (RT-FCM) AND ONLINE VIABILITY ANALYSIS

All the data and applications described in Section 7.3 above were based on discrete online FCM, meaning an automated system where individual samples are discretely collected and processed at time intervals (typically >10 minutes). While this was completely meaningful in the given examples, there are a multitude of microbial related events and processes happening on considerably shorter time scales.

An alternative approach is RT-FCM, which collects, process and analyse samples continuously (typically at 10–60 second frequency). This is particularly useful for highly dynamic systems/events, where critical changes occur faster than the sampling frequency of online FCM systems. The basic concept of RT-FCM has been described in detail in Arnoldini *et al.* (2013), with the then unsolved limitation of not being able to stain the targeted cells separate from the experimental/sampling environment. Recent technological developments in this field comprised the development of RT-FCM staining hardware that overcomes the stated limitation. Figure 7.7 demonstrates the high-resolution data that can be obtained with this approach, with an example of river water treated with successive hypochlorite spikes and stained with a combination of SYBR Green I and propidium iodide to detect membrane damage. First applications of the technology are promising, and include monitoring of the dynamic TCC changes in the effluent of a GAC filter during five hours to confirm online FCM observations (Besmer & Hammes, 2016), and a detailed characterisation of a laboratory-scale wastewater contamination and subsequent chlorine remediation event simulation (Besmer *et al.*, 2017).

All the examples in Section 7.3 above showed only TCC data measured after SYBR Green I staining, which is a reliable process variable, but not fully adequate when bacterial viability should be considered. One of the advantages of FCM is the potential to combine the analysis with a broad range of viability dyes (Hammes *et al.*, 2011; Gasol & del Giorgio, 2000; Czechowska *et al.*, 2008). The example above (Figure 7.7) used a combination of SYBR Green I and propidium iodide to detect oxidative membrane damage caused by hypochlorite, and it was previously shown that such FCM measurements of intact cell concentrations correlated particularly well with other viability-based methods such as ATP analysis (Van Nevel *et al.*, 2017). It is envisaged that online microbial sensors based on FCM will rapidly expand towards utilizing the multiple possibilities offered by different viability dyes, with the potential to tailor these sensors individually to each unique ecosystem or experiment under investigation.

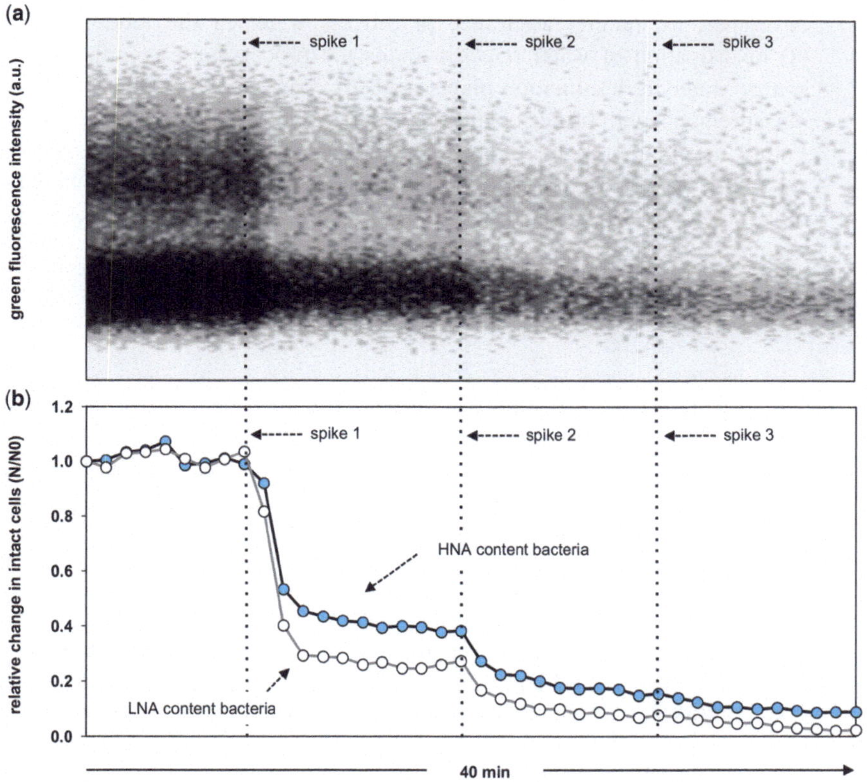

Figure 7.7 (a) Raw data and (b) processed data from real-time flow cytometry (RT-FCM). A river water sample was successively spiked with sodium hypochlorite. Samples were collected continuously, stained with SYBR Green I and propidium iodide and then measured with FCM.

7.6 CONCLUSIONS

- Online FCM has been developed to a technological level where it is regularly used in full-scale case studies and laboratory-scale monitoring of microbial changes in aquatic ecosystems.
- The first datasets describe various dynamic microbial processes/events in detail, which was previously not known or not characterized properly.
- The very short time-to-result of online FCM allows detection of problems (e.g., treatment failure, contamination) early and to assess corrective actions immediately. This is in stark contrast to the current practice at water utilities of grab samples and cultivation-based detection, which strongly limit active monitoring and management of water quality.
- Promising future applications include: (1) investigation of drinking water treatment and infrastructure, (2) process monitoring for bottled water and beverages, (3) quality assurance of process water in the food industry, (4) investigation of water reuse in industrial processes, (5) monitoring of source water in desalination plants, and (6) complementing phytoplankton measurements with bacterioplankton.
- Future developments are likely to include a broad expansion of staining capabilities (e.g., towards detailed viability assessment) and the application of RT-FCM as sensors to track highly dynamic microbial processes and events continuously.

7.7 REFERENCES

Abu-Absi N. R., Zamamiri A., Kacmar J., Balogh S. J. and Srienc F. (2003). Automated flow cytometry for acquisition of time-dependent population data. *Cytometry Part A*, **51A**, 87–96.

Arad J., Housh M., Perelman L. and Ostfeld A. (2013). A dynamic thresholds scheme for contaminant event detection in water distribution systems. *Water Research*, **47**, 1899–1908.

Arnoldini M., Heck T., Blanco-Fernandez A. and Hammes F. (2013). Monitoring of dynamic microbiological processes using real-time flow cytometry. *PLoS One*, **8**(11): e80117. https://doi.org/10.1371/journal.pone.0080117

Astrom J., Petterson S., Bergstedt O., Pettersson T. J. R. and Stenstrom T. A. (2007). Evaluation of the microbial risk reduction due to selective closure of the raw water intake before drinking water treatment. *Journal of Water and Health*, **5**, 81–97.

Ateya D. A., Erickson J. S., Howell P. B. Jr., Hilliard L. R., Golden J. P. and Ligler F. S. (2008). The good, the bad, and the tiny: a review of microflow cytometry. *Analytical and Bioanalytical Chemistry*, **391**, 1485–1498.

Bakker M., van Schagen K. and Timmer J. (2003). Flow control by prediction of water demand. *Journal of Water Supply: Research and Technology – AQUA*, **52**, 417–424.

Berney M., Vital M., Hulshoff I., Weilenmann H. U., Egli T. and Hammes F. (2008). Rapid, cultivation-independent assessment of microbial viability in drinking water. *Water Research*, **42**, 4010–4018.

Besmer M. D. and Hammes F. (2016). Short-term microbial dynamics in a drinking water plant treating groundwater with occasional high microbial loads. *Water Research*, **107**, 11–18.

Besmer M. D., Weissbrodt D. G., Kratochvil B. E., Sigrist J. A., Weyland M. S. and Hammes F. (2014). The feasibility of automated online flow cytometry for in-situ monitoring of microbial dynamics in aquatic ecosystems. *Frontiers in Microbiology*, **5**, https://doi.org/10.3389/fmicb.2014.00265.

Besmer M. D., Epting J., Page R. M., Sigrist J. A., Huggenberger P. and Hammes F. (2016). Online flow cytometry reveals microbial dynamics influenced by concurrent natural and operational events in groundwater used for drinking water treatment. *Science Reports*, **6**, 38462.

Besmer M. D., Sigrist J., Hammes F. and Ort C. (2017a). Evaluating monitoring strategies to detect precipitation-induced microbial contamination events in karstic springs used for drinking water. *Frontiers in Microbiology*, **8**. https://doi.org/10.3389/fmicb.2017.02229.

Besmer M. D., Sigrist J. A., Props R., Buysschaert B., Mao G., Boon N. and Hammes F. (2017b). Laboratory-scale simulation and real-time tracking of a microbial contamination event and subsequent shock-chlorination in drinking water. *Frontiers in Microbiology*, **8**. https://doi.org/10.3389/fmicb.2017.01900.

Broger T., Odermatt R. P., Huber P. and Sonnleitner B. (2011). Real-time on-line flow cytometry for bioprocess monitoring. *Journal of Biotechnology*, **154**, 240–247.

Brognaux A., Han S., Sorensen S., Lebeau F., Thonart P. and Delvigne F. (2013). A low-cost, multiplexable, automated flow cytometry procedure for the characterization of microbial stress dynamics in bioreactors. *Microbial Cell Factories*, **12**(1), 100.

Czechowska K., Johnson D. R. and van der Meer J. R. (2008). Use of flow cytometric methods for single-cell analysis in environmental microbiology. *Current Opinion in Microbiology*, **11**, 205–212.

Dubelaar G. B. J., Gerritzen P. L., Beeker A. E. R., Jonker R. R. and Tangen K. (1999). Design and first results of CytoBuoy: A wireless flow cytometer for in situ analysis of marine and fresh waters. *Cytometry*, **37**, 247–254.

Ender A., Goeppert N., Griimmeisen F. and Goldscheider N. (2017). Evaluation of β-d-glucuronidase and particle-size distribution for microbiological water quality monitoring in Northern Vietnam. *The Science of the Total Environment*, **580**, 996–1006.

Fenchel T. M. and Jørgensen B. B. (1977). Detritus Food Chains of Aquatic Ecosystems: the Role of Bacteria. Advances In Microbial Ecology. M. Alexander (ed.), Springer Boston, Massachusetts, pp. 1–58.

Foladori P., Bruni L., Tamburini S. and Ziglio G. (2010). Direct quantification of bacterial biomass in influent, effluent and activated sludge of wastewater treatment plants by using flow cytometry. *Water Research*, **44**, 3807–3818.

Gasol J. M. and del Giorgio P. A. (2000). Using flow cytometry for counting natural planktonic bacteria and understanding the structure of planktonic bacterial communities. *Scientia Marina*, **64**, 197–224.

Geyer T., Birk S., Liedl R. and Sauter M. (2008). Quantification of temporal distribution of recharge in karst systems from spring hydrographs. *Journal of Hydrology*, **348**, 452–463.

Girvan M. S., Campbell C. D., Killham K., Prosser J. I. and Glover L. A. (2005). Bacterial diversity promotes community stability and functional resilience after perturbation. *Environmental Microbiology*, **7**, 301–313.

Guadayol O., Peters F., Marrase C., Gasol J. M., Roldan C., Berdalet E., Massana R. and Sabata A. (2009). Episodic meteorological and nutrient-load events as drivers of coastal planktonic ecosystem dynamics: a time-series analysis. *Marine Ecology Progress Series*, **381**, 139–155.

Gunther S., Koch C., Hubschmann T., Roske I., Muller R. A., Bley T., Harms H. and Muller S. (2012). Correlation of community dynamics and process parameters as a tool for the prediction of the stability of wastewater treatment. *Environmental Science & Technology*, **46**, 84–92.

Hammes F., Berney M. and Egli T. (2011). Cultivation-independent assessment of bacterial viability. *Advances in Biochemical Engineering/Biotechnology*, **124**, 123–50. doi: 10.1007/10_2010_95.

Hammes F., Broger T., Weilenmann H. U., Vital M., Helbing J., Bosshart U., Huber P., Odermatt R. P. and Sonnleitner B. (2012). Development and laboratory-scale testing of a fully automated online flow cytometer for drinking water analysis. *Cytometry Part A*, **81A**(6), 508–516.

Hayashi M., Vogt T., Machler L. and Schirmer M. (2012). Diurnal fluctuations of electrical conductivity in a pre-alpine river: Effects of photosynthesis and groundwater exchange. *Journal of Hydrology*, **450**, 93–104.

Heck T., Pham P. H., Hammes F., Thony-Meyer L. and Richter M. (2014). Continuous monitoring of enzymatic reactions on surfaces by real-time flow cytometry: sortase a catalyzed protein immobilization as a case study. *Bioconjugate Chemistry*, **25**, 1492–1500.

Hojris B., Christensen S. C. B., Albrechtsen H. J., Smith C. and Dahlqvist M. (2016). A novel, optical, on-line bacteria sensor for monitoring drinking water quality. *Science Reports*, **6**: 23935.

Kirchner J. W., Feng X. H., Neal C. and Robson A. J. (2004). The fine structure of water-quality dynamics: the (high-frequency) wave of the future. *Hydrology Processes*, **18**, 1353–1359.

Kistemann T., Classen T., Koch C., Dangendorf F., Fischeder R., Gebel J., Vacata V. and Exner M. (2002). Microbial load of drinking water reservoir tributaries during extreme rainfall and runoff. *Applied and Environmental Microbiology*, **68**, 2188–2197.

Koch C., Harnisch F., Schröder U. and Müller S. (2014). Cytometric fingerprints: evaluation of new tools for analysing microbial community dynamics. *Frontiers in Microbiology*, **5**, 273. doi: 10.3389/fmicb.2014.00273.

Krewski D., Balbus J., Butler-Jones D., Haas C., Isaac-Renton J., Roberts K. J. and Sinclair M. (2002). Managing health risks from drinking water – a report to the Walkerton inquiry. *Journal of Toxicology and Environmental Health. Part A*, **65**, 1635–1823.

Lindberg W., Ruzicka J. and Christian G. D. (1993). Flow-injection cytometry – a new approach for sample and solution handling in flow-cytometry. *Cytometry*, **14**, 230–236.

Mimoso J., Pronk W., Morgenroth E. and Hammes F. (2015). Bacterial growth in batch-operated membrane filtration systems for drinking water treatment. *Separation and Purification Technology*, **156**, 165–174.

Muirhead K. A., Horan P. K. and Poste G. (1985). Flow cytometry: past, present and future. *Nature Bio/Technology*, **3**, 337–356.

Murray R., Haxton T., McKenna S. A., Hart D. B., Klise K. A., Koch M., Vugrin E. D., Martin S., Wilson M., Cruze V. A. and Cutler L. (2010). Water quality event detection systems for drinking water contamination warning systems: development testing and application of CANARY. U.S. Environmental Protection Agency, Washington, DC, EPA/600/R-10/036, 2010.

Nescerecka A., Rubulis J., Vital M., Juhna T. and Hammes F. (2014). Biological instability in a chlorinated drinking water distribution network. *PLoS One*, **9**(5): e96354. https://doi.org/10.1371/journal.pone.0096354.

Nescerecka A., Hammes F. and Juhna T. (2016a). A pipeline for developing and testing staining protocols for flow cytometry, demonstrated with SYBR Green I and propidium iodide viability staining. *Journal of Microbiological Methods*, **131**, 172–180.

Nescerecka A., Juhna T. and Hammes F. (2016b). Behavior and stability of adenosine triphosphate (ATP) during chlorine disinfection. *Water Research*, **101**, 490–497.

Nilsson P., Roser D., Thorwaldsdotter R., Petterson S., Davies C., Signor R., Bergstedt O. and Ashbolt N. (2007). SCADA data and the quantification of hazardous events for QMRA. *Journal of Water and Health*, **5**, 99–105.

Omann G. M., Coppersmith W., Finney D. A. and Sklar L. A. (1985). A convenient online device for reagent addition, sample mixing, and temperature control of cell-suspensions in flow-cytometry. *Cytometry*, **6**, 69–73.

Paerl H. W. and Pinckney J. L. (1996). A mini-review of microbial consortia: their roles in aquatic production and biogeochemical cycling. *Microbial Ecology*, **31**, 225–247.

Page R. M., Besmer M. D., Epting J., Sigrist J. A., Hammes F. and Huggenberger P. (2017). Online analysis: Deeper insights into water quality dynamics in spring water. *The Science of the Total Environment*, **599–600**, 227–236.

Petterson S. R. and Ashbolt N. J. (2016). QMRA and water safety management: review of application in drinking water systems. *Journal of Water and Health*, **14**, 571–589.

Plummer J. D. and Long S. C. (2007). Monitoring source water for microbial contamination: Evaluation of water quality measures. *Water Research*, **41**, 3716–3728.

Pomati F., Kraft N. J. B., Posch T., Eugster B., Jokela J. and Ibelings B. W. (2013). Individual cell based traits obtained by scanning flow-cytometry show selection by biotic and abiotic environmental factors during a phytoplankton spring bloom. *PLoS One*, **8**(8): e71677. https://doi.org/10.1371/journal.pone.0071677.

Prest E. I., Hammes F., Kotzsch S., van Loosdrecht M. C. M. and Vrouwenvelder J. S. (2013). Monitoring microbiological changes in drinking water systems using a fast and reproducible flow cytometric method. *Water Research*, **47**(19), 7131–7142.

Prest E. I., El-Chakhtoura J., Hammes F., Saikaly P. E., van Loosdrecht M. C. M. and Vrouwenvelder J. S. (2014). Combining flow cytometry and 16S rRNA gene pyrosequencing: A promising approach for drinking water monitoring and characterization. *Water Research*, **63**, 179–189.

Prest E. I., Weissbrodt D. G., Hammes F., van Loosdrecht M. C. M., Vrouwenvelder J. S. (2016). Long-term bacterial dynamics in a full-scale drinking water distribution system. *PLoS One*, **11**(10), e0164445.

Pronk M., Goldscheider N. and Zopfi J. (2006). Dynamics and interaction of organic carbon, turbidity and bacteria in a karst aquifer system. *Hydrogeology Journal*, **14**, 473–484.

Props R., Monsieurs P., Mysara M., Clement L. and Boon N. (2016a). Measuring the biodiversity of microbial communities by flow cytometry. *Methods in Ecology and Evolution*, **7**, 1376–1385. doi:10.1111/2041-210X.12607.

Props R., Kerckhof F.-M., Rubbens P., De Vrieze J., Hernandez Sanabria E., Waegeman W., Monsieurs P., Hammes F. and Boon N. (2016b). Absolute quantification of microbial taxon abundances. *ISME Journal*, **11**(2), 584–587.

Schmidt P. J. and Emelko M. B. (2011). QMRA and decision-making: Are we handling measurement errors associated with pathogen concentration data correctly? *Water Research*, **45**, 427–438.

Shapiro H. M. (2003). Practical Flow Cytometry, 4th edn. Wiley-Liss, New York, USA.

Shehane S. D., Harwood V. J., Whitlock J. E. and Rose J. B. (2005). The influence of rainfall on the incidence of microbial faecal indicators and the dominant sources of faecal pollution in a Florida river. *Journal of Applied Microbiology*, **98**, 1127–1136.

Signor R. S. and Ashbolt N. J. (2006). Pathogen monitoring offers questionable protection against drinking-water risks: a QMRA (Quantitative Microbial Risk Analysis) approach to assess management strategies. *Water Science and Technology*, **54**(3), 261–268.

SLMB (2012). Determining the total cell count and ratios of high and low nucleic acid content cells in freshwater using flow cytometry. Analysis method 333.1, Swiss Food Book. Federal Office of Public Health, Swiss Confederation.

Stadler H., Skritek P., Sommer R., Mach R. L., Zerobin W. and Farnleitner A. H. (2008). Microbiological monitoring and automated event sampling at karst springs using LEO-satellites. *Water Science and Technology*, **58**(4), 899–909.

Stadler H., Klock E., Skritek P., Mach R. L., Zerobin W. and Farnleitner A. H. (2010). The spectral absorption coefficient at 254 nm as a real-time early warning proxy for detecting faecal pollution events at alpine karst water resources. *Water Science and Technology*, **62**(8), 1898–1906.

Stevenson D. G. (1997). Water Treatment Unit Processes. Imperial College Press, London, UK.

Thyssen M., Tarran G. A., Zubkov M. V., Holland R. J., Gregori G., Burkill P. H. and Denis M. (2008). The emergence of automated high-frequency flow cytometry: revealing temporal and spatial phytoplankton variability. *Journal of Plankton Research*, **30**, 333–343.

Van Nevel S., Koetzsch S., Weilenmann H. U., Boon N. and Hammes F. (2013). Routine bacterial analysis with automated flow cytometry. *Journal of Microbiological Methods*, **94**, 73–76.

Van Nevel S., Koetzsch S., Proctor C. R., Besmer M. D., Prest E. I., Vrouwenvelder J. S., Knezev A., Boon N. and Hammes F. (2017). Flow cytometric bacterial cell counts challenge conventional heterotrophic plate counts for routine microbiological drinking water monitoring. *Water Research*, **113**, 191–206.

Vang O. K., Corfitzen C. B., Smith C. and Albrechtsen H. J. (2014). Evaluation of ATP measurements to detect microbial ingress by wastewater and surface water in drinking water. *Water Research*, **64**, 309–320.

Verstraete W., Wittebolle L., Heylen K., Vanparys B., de Vos P., van de Wiele T. and Boon N. (2007). Microbial resource management: the road to go for environmental biotechnology. *Engineering in Life Sciences*, **7**, 117–126. doi:10.1002/elsc.200620176.

Vila-Costa M., Gasol J. M., Sharma S. and Moran M. A. (2012). Community analysis of high- and low-nucleic acid-containing bacteria in NW Mediterranean coastal waters using 16S rDNA pyrosequencing. *Environmental Microbiology*, **14**, 1390–1402.

Vogt T., Hoehn E., Schneider P., Freund A., Schirmer M. and Cirpka O. A. (2010). Fluctuations of electrical conductivity as a natural tracer for bank filtration in a losing stream. *Advances in Water Resources*, **33**, 1296–1308.

Wang Y., Hammes F., Boon N. and Egli T. (2007). Quantification of the filterability of freshwater bacteria through 0.45, 0.22, and 0.1 mu m pore size filters and shape-dependent enrichment of filterable bacterial communities. *Environmental Science & Technology*, **41**, 7080–7086.

Wang Y., Hammes F., Boon N., Chami M. and Egli T. (2009). Isolation and characterization of low nucleic acid (LNA)-content bacteria. *ISME Journal*, **3**, 889–902.

Zeng C., Gremaud V., Zeng H. T., Liu Z. H. and Goldscheider N. (2012). Temperature-driven meltwater production and hydrochemical variations at a glaciated alpine karst aquifer: implication for the atmospheric CO2 sink under global warming. *Environmental Earth Sciences*, **65**, 2285–2297.

Zhao R., Natarajan A. and Srienc F. (1999). A flow injection flow cytometry system for on-line monitoring of bioreactors. *Biotechnology Bioeng*, **62**, 609–617.

Chapter 8

Adenosine Triphosphate (ATP) measurement technology

Patrick A. Whalen, David R. Tracey and Jeremy Duguay

8.1 INTRODUCTION

As water management continues to move to the forefront of worldwide resource concerns, better information is required to facilitate effective decision making. Not only does the information have to be more accurate, it also must be obtained faster and acquired from multiple locations. This challenge is common across all industries where water is a critical commodity.

Heterotrophic Plate Counts and other culture-based tests have traditionally been used to estimate microbiological content, but they must be completed in a laboratory, require two or more days for feedback, and do not truly measure total microorganisms (Bartram *et al.*, 2003). Because of these limitations, municipal water contamination issues can go unnoticed until major deterioration has already occurred. By the time culture test results are realized on contaminated water samples, the potable water has already been distributed to and likely utilized by consumers.

Adenosine Triphosphate (ATP) is the primary energy carrier for all forms of life on Earth. Given its evolutionary importance, its quantification is seen as one of the fundamental tools in understanding biological processes. Since the 1960s,

The chapter is from the book *Microbiological Sensors for the Drinking Water Industry,* Torben Lund Skovhus and Bo Højris (Eds.).
doi: 10.2166/9781780408699_0137

the principles of ATP measurement have been successfully applied to a variety of applications, from environmental (Cairns *et al.*, 1979) to marine monitoring (Holm-Hansen & Booth, 1960) and even in space (Birmele *et al.*, 2010), in addition to food hygiene monitoring. The latter application has been the most successful, in which the vast majority of food processing and distribution companies have adopted ATP methods as a quick test for total contamination under Hazard Analysis Critical Control Point (HACCP) strategies. These First Generation ATP test methods use basic principles and as such provided users with only a semi-quantitative measurement of ATP content. Due to their basic nature, these First Generation ATP kits have limited applicability.

The advantages of ATP testing over traditional culture tests are that the results are obtained in mere minutes (not days), that the results are complete (rather than the small fraction that will grow on culture media), and that the methods are generally field- and user-friendly. However, none of these advantages will provide significant value unless the results produced are reliable and accurate. On that vein, there were two main reasons that the technology did not achieve successful adoption in the field of water: for one, the techniques developed for hygiene monitoring were not designed to be fully quantitative – rather, they were designed to simply provide a clean/not-clean determination for microbiological attachment as well as food residue. Secondly, the industry operated under the assumption that success of such techniques would be based upon achieving direct correlations with incumbent methods (e.g. culture tests), which is quite simply an impossible objective since the tests measure different things.

Second Generation Adenosine Triphosphate (ATP) were introduced early in the 21st century and provided several advancements including enhanced resistance to interferences and high sensitivity, while maintaining the same fast response time. Being the primary energy transfer molecule for all living cells, successful measurement of ATP is a direct indication of the level of total microbiological contamination and serves as the ideal basis for risk assessment and disinfection program optimization. While being a rapid microbiological test makes it valuable, the real power comes from the fact that ATP testing is extremely sensitive to changes by measuring total biomass, which is not easily achieved using traditional methods.

8.2 ATP MEASUREMENT FUNDEMENTALS

ATP is the keystone of metabolic activity. Most of the energy for microbiological processes is stored and transmitted via ATP. ATP is produced as microbiological food is consumed and is subsequently utilized for cell maintenance as well as the synthesis of new cells and biochemicals (Knowles, 1980).

ATP provides energy for cellular operations by donating phosphate groups, resulting in the formation of either Adenosine Diphosphate (ADP) or Adenosine Monophosphate (AMP). ADP and AMP are subsequently recycled and regenerated back into ATP as the organism consumes food. This operation typically occurs many times per second (Goodsell, 1996). Therefore, the measurement of intracellular

ATP (i.e. ATP contained inside living cells) can be considered as the 'potential energy' contained within an active biomass population at any given time. ATP production is directly related to the growth rate of the cell and therefore higher ATP levels are indicative of greater mass and cell volume (Johnston *et al.*, 2012). Additionally, when cells become weak and/or lyse, they release their ATP into their external environment. This is termed 'extra-cellular' ATP (i.e. ATP outside of living cells). The presence of a greater proportion of extra-cellular ATP (relative to intracellular ATP levels or compared to an established "baseline") is indicative of a less healthy population (Maeir *et al.*, 2009).

ATP can be easily measured with high specificity via the firefly luciferase assay. Luciferase is a naturally occurring enzyme that is most commonly found in the tails of fireflies but can now be reliably sourced through several commercial sources thanks to recombinant DNA technology. The reaction between ATP and Luciferase is described in Equation (1):

$$ATP + O_2 + luciferin \xrightarrow{Mg^{++},\, luciferase} AMP + PPi + oxyluciferin + light \quad (1)$$

where,

ATP = Adenosine Triphosphate
AMP = Adenosine Monophosphate
PPi = Pyrophosphate
Mg^{++} = Magnesium ion

The chemical energy produced from the breakdown of ATP is converted into light. Each molecule of ATP consumed in the reaction produces one photon of light. This light output can be quantified using a luminometer within a matter of seconds. A luminometer is a light-detecting instrument like a spectrophotometer; however, luminometers are generally much more sensitive and do not contain a light 'source'. It is the luminescent reaction that acts as the light source in luminometers.

The result produced by luminometers is typically expressed as a Relative Light Unit, or RLU. This value represents the instrument's interpretation of the amount of light detected. As such, RLU results from different instruments will tend to produce significantly different RLUs due to variation in instrument tuning, manufacturer calibration, and other design factors (Berthold *et al.*, 2000). It is therefore important to run ATP standards with the reagent system and luminometer being used to account for these variables.

There is a wide variety of luminometer brands on the market with their designs depending on the intended application. For example, hygiene swab ATP tests often rely on photodiode-equipped luminometers, which, while not overly sensitive, are usually available at a low cost and are acceptable for surface analyses. For more sensitive applications, photomultiplier-tube-equipped luminometers are often used. Photomultiplier tubes, or PMTs, can typically achieve several orders of magnitude worth of enhanced sensitivity compared to photodiodes (Berthold *et al.*, 2000). Luminometers are also available in both single-chamber format as well as 96-well and even 384-well plate instruments (intended for research applications rather than practical field testing), which can sequentially perform many assays in an automated fashion.

8.3 FIRST VERSUS SECOND GENERATION ATP MEASUREMENT

As with any enzymatic reaction, the luciferase assay for ATP is susceptible to interferences that can be caused by various components in a sample. For this reason, ATP testing has not been traditionally applied in water, wastewater and industrial applications. In addition, ATP on its own is highly unstable and readily degrades to ADP (adenosine diphosphate) and AMP (adenosine monophosphate), unless it is complexed with other molecules or cell debris (Cairns *et al.*, 2005).

Developments in the reagents and methodology used have led to substantial improvements in the ATP test method. These new, Second Generation protocols and reagents were specifically designed to mitigate various types of interferences found in water applications including suspended solids, dissolved solids, biocides and organics. This is accomplished using chemical treatment and/or separation to isolate microorganisms from potential interference. These methods also ensure the necessary quantitativeness in ATP recovery, conversion to known levels of ATP concentration, and sufficient sensitivity to meet the needs of drinking water samples.

8.3.1 Overcoming interferences

Second Generation ATP testing methods were originally designed for use in biological wastewater treatment plants. To provide accurate test results in this situation, significant interferences needed to be overcome which were considerably more challenging than what is typically encountered in most other situations where microbiological quantification is carried out. Therefore, the robustness of the reagents developed for wastewater analyses was more than sufficient for application to less contaminated sample types such as potable water, cooling water, oilfield water (e.g. produced water, fracturing fluids, etc.), industrial process streams, petroleum products, and chemical products (e.g. paints, coatings, adhesives, admixtures, emulsions, etc.). The focus was then shifted to establish protocols that provide enhanced sensitivity for applications that deal with smaller levels of bioburden.

The optimum test protocol choice is driven by the properties of the sample to be tested. These tests can be divided into two main classes:

(1) *Filtration-based:* The sensitivity of ATP test protocols can be significantly improved by concentrating microbiological cells on a filter sufficient to capture intact microbiological cells (e.g. 0.2 μm) and directly extracting the ATP from the captured cells. This allows ATP tests to be performed in several "clean" applications for which first generation ATP products are not applicable such as drinking water and purified water.

(2) *Dilution-based:* Many industrial samples such as wastewater or industrial process streams cannot be filtered due to high amounts of suspended solids or high viscosities. The sample is therefore added directly to extraction reagent to release ATP from living cells and stabilize extracellular ATP. Extracellular ATP can be measured separately through an additional dilution-based test that does not harm intact cells; rather, it releases bound extracellular ATP and stabilizes soluble ATP prior to measurement (Cairns *et al.*, 2005).

8.3.2 Extraction and recovery

To get an accurate ATP reading, it is critical that ATP from all cells in a sample is recovered. It is especially difficult to extract ATP from biofilm clumps, microbiological floc, yeasts and moulds. As such, having a weak extraction reagent can lead to incomplete cell lysis and thus only partial ATP recovery yielding lower than actual results. Therefore, it is important to ensure that the extraction reagent used is strong enough to completely lyse all microorganisms present in a sample for maximum ATP recovery, while not being so strong that it inhibits the luciferase enzyme.

Additionally, the high-energy bonds in an ATP molecule make it an unstable molecule and when not stabilized, ATP readily hydrolyses to ADP and phosphate. Most extra-cellular ATP is complexed to various molecules and cell debris and is therefore unavailable to react with the luciferase enzyme. When intra-cellular ATP is extracted, it can readily complex or hydrolyse, making it too unavailable to react with the luciferase enzyme. If left unbuffered, the instability of an ATP extract can lead to lower than actual results since most ATP would have either degraded or complexed before the luciferase assay is performed. It is therefore important to ensure that all complexed extra-cellular ATP and all extracted intra-cellular ATP is stabilized for maximum recovery.

8.3.3 Importance of calibration

As mentioned previously, while many First Generation ATP tests rely on RLU values alone, a low result from an ATP test can be due to low levels of contamination or because the enzyme has lost a significant amount of its activity. Without a way to gauge the activity of the enzyme, it is impossible to tell for what reason a low RLU is observed. Other factors such as the temperature of the reagents, samples, or the surrounding environment as well as the make, model, and condition of the luminometer used can significantly affect the RLU produced. As such, an RLU value on its own is essentially an uncalibrated result.

For this reason, sample RLU's should be converted to ATP concentrations by standardizing the result to a known amount of ATP. The RLU produced by a standard ATP solution can normalize the sample RLU to determine the amount of ATP in the sample. This also provides a direct measure of enzyme activity and thus the ability to determine the lower detection limit and enzyme reagent shelf life. If the sample volume and dilution factor for a given kit are fixed, the sensitivity of the

ATP assay is solely decided by the activity level of the enzyme formulation used. Fresh reagent will produce more light than an older batch when used to analyse the same sample, and for this reason, samples that may have been below the detection limit when using an older batch become within reach of quantification.

8.3.4 Limit of detection

One of the biggest benefits of Second Generation ATP monitoring is the enhanced sensitivity that is made possible through optimized reagents and test protocols. In general, Equation (2) indicates how the ATP concentration is calculated for all test protocols:

$$[\text{ATP}] = \left(\frac{\text{RLU}_{\text{sample}}}{\text{RLU}_{\text{ATP Standard}}} \right) \times \left(\frac{\text{Dilution Factor} \times \text{ATP Standard Conc.}}{\text{Sample Volume}} \right) \quad (2)$$

Aside from the RLU value produced by the sample assay, modifying any of the other three variables in the above equation will affect the dilution factor of the test. The practical limit of detection of the ATP assay is around 0.1 pg ATP/mL, though higher sensitivity may be achieved through filtration of large sample volumes and special care in reagent and test component decontamination.

While high-solids samples such as wastewater must be analysed using a dilution-based method, sample types that can be filtered allow the user to process larger sample volumes, and in doing so, significantly lower the detection limit of the test. When dealing with samples containing very low microbiological concentrations, it is important to keep in mind that sampling error becomes a bigger issue. It also becomes more important to perform replicate analyses to ensure that the results are accurate.

8.4 METHODS FOR DETECTION OF TOTAL MICROORGANISMS

There are several methods presently used to quantify microorganisms in drinking water processes (Ashbolt *et al.*, 2001). Table 8.1 compares each method according to four categories for measuring total microorganisms:

(1) Applicability (i.e. the specificity with which the quantity of total microorganisms is measured);
(2) Speed (i.e. the speed with which the result is obtained);
(3) Degree of Difficulty (i.e. the relative expertise required to complete an analysis);
(4) Relative Cost (i.e. the capital investment required to obtain equipment and recurring consumables required to complete the analysis).

As observed in Table 8.1, successful ATP measurements provide relevant results in a short period of time and at a moderate cost. For quantifying the total microbiological activity in a water sample, ATP provides the greatest cost and labour efficiency, and has the additional benefit of being portable.

Table 8.1 Comparison of microbiological quantification methods.

Method	Measures	Applicability	Speed	Degree of Difficulty	Relative Cost
2nd Generation ATP	Total Active Organisms	High	Fast	Low	Moderate
Culture (i.e. Plate Counts)	Specific Viable Organisms	Low	Slow	Moderate	Low
DNA methods	Specific Organisms	High to Low*	Moderate	High	High
Microscopic observation	Physical Cell Count	High	Moderate	High	High

*Depends on primer selection.

8.4.1 Why measure total microorganisms?

The most widely accepted method for quantifying microorganisms in water systems is the culture test. Of these analyses, the Heterotrophic Plate Count (HPC) is most commonly associated with indication of total microorganisms. This analysis involves addition of a sample to a culture medium that is subsequently incubated until sufficient growth is apparent via visual inspection. Colonies are then physically counted and the concentration of microorganisms in the sample is estimated. More user-friendly adaptations of these analyses have been developed and widely utilized in industry, including the dip slide (Vanderzwaag *et al.*, 2009).

The most widely recognized deficiency of culture tests is slow feedback. Days or even weeks can be required to obtain results due to slow growth rates of certain species. A more recently recognized drawback is that it is impossible to quantify total microorganisms without performing hundreds or thousands of culture tests on a given sample (Laupland & Valiquette, 2013). As such, many researchers view ATP technology as a potential rapid estimator of microorganisms, much in the same way as turbidity is used throughout industry to rapidly estimate total suspended solids. Although ATP test results will strongly correlate with culture test results under most conditions, there are several factors that affect this correlation (Sloan *et al.*, 2008).

8.4.2 Population specificity

ATP tests effectively provide a superior indication of the total amount of microbiological content compared to culture tests since all living cells contain ATP. As such, all living microorganisms in a sample will contribute to the ATP measurement. Conversely, a heterotrophic plate count only recovers a small portion of metabolically active organisms and results will vary a great deal according to the method used. Results have indicated that in drinking water systems, only 0.1–1% of the total microbiological population is detected by HPCs. In fact, when considering all known species of microorganisms, it is estimated that only 0.01% of waterborne microorganisms are heterotrophic bacteria (Bartram *et al.*, 2003).

8.4.3 Particle association and agglomeration

Culture tests tend to underestimate the number of microorganisms because a clump of many organisms produces only one countable colony (Todar, 2008). As with other biochemical and molecular tests, ATP measurements will count all the organisms in a clump separately. It is worth noting however that as with all cellular components, the quantity of ATP will change depending on the size and metabolic state of the cell. All these factors must be taken into consideration by the operator when using and comparing microbiological tests in different circumstances.

8.4.4 Disinfection efficacy

Exclusive use of culture tests can pose disadvantages in the context of disinfection monitoring. In addition to slow feedback, culture tests provide no information about the effectiveness of the biocide treatment on organisms that they do not measure. Furthermore, they can be misleading if a biocide fails to penetrate a clump of microorganisms or, alternatively, disperses the clump. DNA-based methods also struggle to provide relevant information of disinfection efficacy, primarily since there are as of yet no universally accepted mechanisms to distinguish live from dead DNA when using genomic methods (Cangleosi & Meschke, 2014).

Conversely, ATP measurements alone can underestimate the efficiency of a biocide kill since they will detect ATP from cells that are still alive but are rendered unable to reproduce. However, they will quantify the effect of the biocide on the entire population, meaning that it provides a more complete indication of true kill efficacy than do culture tests (Corrin *et al.*, 2009). As such, it would be most effective to use ATP together with conventional culture-based methods, especially when the culture tests used are for nuisance organisms or specific pathogens (e.g. *E. coli* in drinking water).

ATP measurements will therefore only correlate strongly with plate counts if several conditions are met. Many attempts in the past at establishing a correlation between culture-based analyses and ATP analyses have not been routinely successful. Although correlations are typically not as strong as researchers hope, they attribute this to some shortcomings in the plate count methodology since many species will not be detected nor will cells that have been injured to the point where they are incapable of growing in the time permitted (Stopa & Orahovec, 2002).

The interest in conducting comparison studies between culture tests and ATP results is understandable, as culture-based techniques have served as the benchmark to assess the degree of microbiological contamination and to assess disinfectant performance for many years (Bartram *et al.*, 2003). Although plate counts have been the standard to assess the amount of microbiological contamination at a site, any lack of correlation with ATP is not a suitable basis to reject the use of either method. Rather than attempting to completely replace plate counts with ATP tests, ATP monitoring can serve as a screening and routine monitoring tool for detecting the total quantity of active microorganisms where the total population is measured with a routine frequency. Culture-based tests can then be used to troubleshoot revealed issues that become evident through rapid screening.

8.5 CASE STUDIES

8.5.1 Direct comparison of second generation ATP to culture-based methods

One of the reasons why comparisons between ATP and HPC methods have not always shown a strong correlation is because culturing methods tend to vary significantly based on the culture media used. To investigate this, an independent laboratory performed a study using two different types of HPC media alongside the Quench-Gone Aqueous (QGA, LuminUltra Technologies Ltd., Fredericton, NB, Canada) test protocol to examine the difference between the methods for 12 unique sample types including clean samples (e.g. ultra-purified water), dirty samples (e.g. untreated surface water), and a selection of others of varying cleanliness with each test done in duplicate. One low-nutrient media (TGYA) and one high-nutrient media (NWRI) were selected as comparative culture tests. The variation of media was a necessary component of the experiment due to their tendency to encourage the growth of different species of microorganism even though both media are intended to provide a measure of the total microbiological population. The correlations of each method to each other are shown in Table 8.2.

Table 8.2 Correlation (Pearson R^2) between microbiological enumeration methods (L'Abbee & Ritchie, 2008).

Method	Quench-Gone Aqueous	High-Nutrient HPC NWRI	Low-Nutrient HPC TGYA
Quench-Gone Aqueous	1.00	0.88	0.78
High-nutrient HPC (NWRI)	0.88	1.00	0.79
Low-nutrient HPC (TGYA)	0.78	0.79	1.00

The results shown in Table 8.2 indicate a strongly positive correlation between ATP tests and the culture methods chosen. While the R^2 values are strong, the reason that they are not higher lies with the fundamental differences between the methods as it pertains to the fact that ATP measurements detect all species under all different metabolic states as opposed to those that are detected through conventional culturing techniques under very specific conditions.

In addition to comparison tests on discrete samples, analyses using the aforementioned methods were also done before and after chlorination of a selection of samples from the previous sample set, as shown below in Table 8.3.

In addition to illustrating how Second Generation ATP monitoring can provide a good early indication of disinfection efficacy on the entire population; this experiment shows the significant difference that can be encountered when utilizing different HPC media in terms of the magnitude of the microbiological population measured. This comparison highlights the reliance of different culture media on detection of cultivable organisms, which typically represent a small proportion of total microorganisms. Conversely, ATP results revealed the presence of microbiological activity, albeit in small quantities, in several cases where the HPC methods did not. The results of ATP tests were above what would be considered "noise" and therefore indicates the presence of non-culturable species or unhealthy cells which could not be grown in the media, time, or incubation conditions chosen.

The laboratory performing this experiment deemed it to be successful in demonstrating that, while the media chosen significantly affects the outcome of the comparison, HPC and ATP tests generally correlate quite well under the proper circumstances. They stated that method selection should be consistent with the goal of the testing program. That is, if measuring for specific organisms to meet regulatory requirements is the goal of a test, culture-based analyses are required. However, ATP tests may be used as a suitable alternative to HPC tests with certain benefits for operational purposes including total detection and near-real-time feedback.

It is important to point out, though, that while this study showed good agreement between the ATP and HPC tests, this correlation tends to become weak in situations where non-culturable species (i.e. nitrifiers, sulphate-reducers, iron bacteria, etc.) are present. Regrowth within distribution systems due to nitrification is a prime example of when ATP monitoring provides the extra insight required to quickly and properly diagnose problems that conventional methods such as HPC tests would otherwise miss.

8.5.2 Using second generation ATP testing to optimize biologically active filters

The concept of biological filtration has become a common means to remove organics prior to disinfection to minimize formation of disinfection by-products. However, the tools available to assess biological growth tendencies in biological filters are limited. Heterotrophic Plate Counts and other culture-based methods take days to obtain results, only measure very small portions of the true microbiological population, and struggle with the ability of quantifying cells attached to media. By contrast, the ATP measurement has been cited as one of the most reliable tool for monitoring of biological filtration processes (Evans *et al.*, 2013).

Table 8.3 Method comparison in response to shock chlorination (L'Abbee & Ritchie, 2008).

Source	Chlorine Added (mg/L)	Cellular ATP (pg cATP/mL)			HPC – NWRI (CFU/mL)			HPC – TGYA (CFU/mL)		
		Before Chlorination	After Chlorination	% Reduction of ATP	Before Chlorination	After Chlorination	% Reduction of CFU	Before Chlorination	After Chlorination	% Reduction of CFU
River water	10	94.5	0.902	99%	2930	1	100%	1650	3	99.8%
Ground water	10	113	ND	100%	39.0	ND	100%	779	ND	100%
Potable water	10	0.116	ND	100%	ND	ND	100%	3.00	ND	100%
Potable water storage tank	10	1.83	ND	100%	873	ND	100%	778	ND	100%
Reverse osmosis permeate	10	0.223	ND	100%	ND	ND	—	ND	ND	—

ND = not detectable.

The following process monitoring and control opportunities are made possible via analysis of ATP results (LuminUltra, 2017a):

- Assess filter capacity in terms of biomass population size as it relates to organics removal.
- Quantify the biological population at the beginning and end of filter runs to determine the magnitude of growth that occurred during the cycle.
- Investigate the amount of biomass purged from the filter during backwash cycles. Once a baseline has been established for normal operation at the beginning and end of filter runs, targets can be established to optimize the duration of backwash cycles to minimize downtime while avoiding the release of too much or too little biomass.

The above example shown in Figure 8.1 illustrates the following findings via ATP results (as determined using the Deposit & Surface Analysis (DSA) test kit from LuminUltra Technologies):

- Enhanced biological growth capacity is seen in granular activated carbon compared to anthracite.
- The greatest concentration of biomass is found at the top of the filter where there is an abundance of food and nutrients.
- Biomass concentrations decrease proportionally according to bed depth.

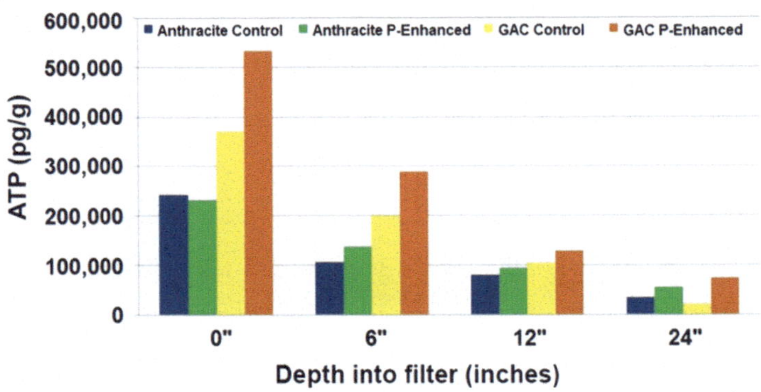

Figure 8.1 Comparison of biomass response to process modifications (Stoddart, 2017).

8.5.3 The Membrane Biofouling Index

Membrane filtration units are widely used in the water treatment industry for applications ranging from groundwater treatment and desalination to industrial water treatment and wastewater processes. While they can produce extremely high-quality water, they require a significant investment to acquire and maintain therefore it is important to protect the membranes from excessive wear to maximize the lifespan of the modules (LuminUltra, 2017c).

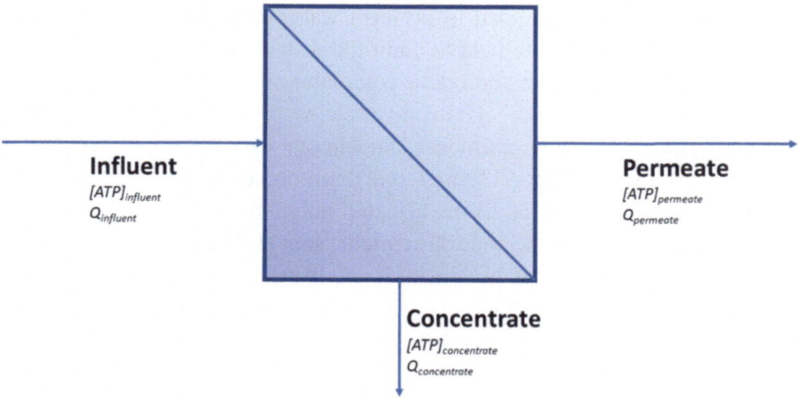

Figure 8.2 A typical 3-flow membrane.

Due to the porous nature of the membranes, they are susceptible to both chemical scaling and biological fouling which can increase energy requirements and operating costs and shorten their useful lifespan. Treatments differ depending on the nature of the fouling and each cleaning cycle imposes extra stress on the membranes, so cleaning chemicals should be targeted to the specific type of fouling that is present and cleaning frequency should be optimized.

Through Second Generation ATP testing, influent water quality can be monitored; permeate water quality can be assessed to reveal contamination and microbiological breakthrough; and membrane fouling can be assessed through a mass balance of total biomass across the membrane (Figure 8.2) according to Equation (3):

$$\text{Membrane Biofouling Index (MBI)} = [(Q_{\text{permeate}} \times ATP_{\text{permeate}}) + (Q_{\text{concentrate}} \times ATP_{\text{concentrate}})]/[(Q_{\text{influent}} \times ATP_{\text{influent}})] \tag{3}$$

By monitoring membrane fouling on a regular basis, the appropriate treatment can be applied, and cleaning cycles can be optimized to reduce excess wear on the membranes (Travis & Tracey, 2015).

8.5.4 Using second generation ATP testing to address biological hotspots

A Southern US municipality experiencing taste and odour issues in a certain neighbourhood was also having difficulty maintaining chlorine residual levels in the area. Biological growth was suspected; however, water leaving the treatment plant met and exceeded all water quality requirements. After several investigations, the source of contamination in the distribution system could still not be identified.

The utility was successfully using Second Generation ATP testing in their water treatment plant to monitor biological fouling in the membrane filtration unit. Since they were already equipped to perform water analyses, it was decided that the kit would be taken into the field to audit the distribution system and determine whether the cause of the taste and odour complaints was biological contamination (LuminUltra, 2017b).

Testing began at the water treatment plant where it was confirmed that biological content was very low (<1pg ATP/mL) and therefore well within the acceptable range. Operators then travelled to the affected neighbourhood and began to audit the system, tracing back to the water treatment plant (Figure 8.3).

The results confirmed that biological growth was significantly higher in the affected neighbourhood with ATP levels in the high-risk range of >10pg ATP/mL. The line was traced back toward the treatment plant and water samples were tested at several major junction nodes. ATP results remained high near the neighbourhood until suddenly, between sample point B and A, there was a significant drop in ATP levels back into the good control range of <1pg ATP/mL (Figure 8.4).

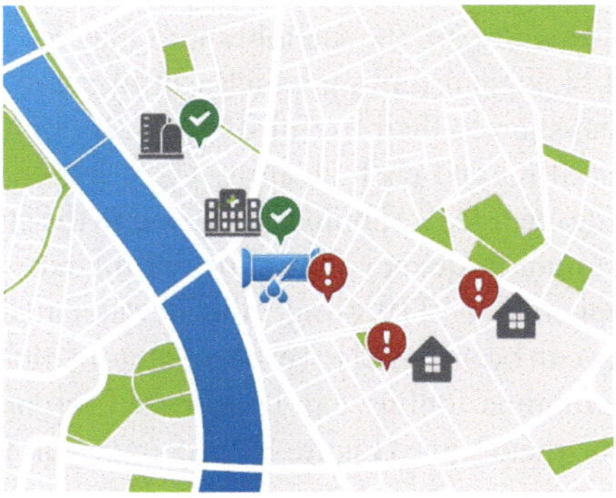

Figure 8.3 Map of distribution system.

Upon further investigation, operators discovered a leaking valve that was not identified on the design drawings. Significant biofilm growth had built up on the valve and was entering the distribution system causing the disinfectant residual loss and water quality issues for consumers. Operators repaired the valve immediately and flushed the lines. Residual chlorine levels were quickly restored, and follow-up ATP testing confirmed that biological growth in the system had dropped back into the good control range.

Having a field-ready rapid microbiological test allowed water distribution system operators to quickly audit their system and pinpoint the source of water quality issues. Routine distribution system monitoring was implemented to proactively monitor for biological regrowth and prevent similar water quality issues from affecting consumers in the future.

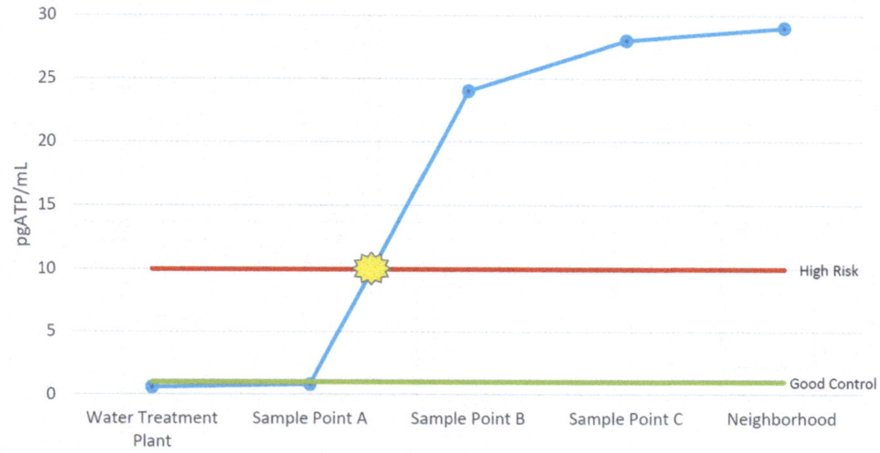

Figure 8.4 Results across distribution system.

8.5.5 Audit of public building water distribution system

A French spa utilized untreated mineral water (from ground wells) in its daily operation (LuminUltra, 2017d). The source water is distributed to 40+ stations for therapeutic purposes. One week prior to the site audit, two stations had tested positive for pathogenic organisms in the form of *Pseudomonas*. Second Generation ATP testing was utilized to locate the source and nature of the contamination using a HACCP-style audit, on the fresh water and various water distribution branches. Samples were taken from the cold and hot water of each bank (Figure 8.5).

Once testing was complete at each location in the network, the results immediately indicated that the feed water was not the source of contamination. Further inspection showed that elevated microbiological content originates in branch #2L and carries over to branch #3 in the hot water system only (Figure 8.6). Therefore, rather than disinfecting the entire water network, additional treatment was necessary in the hot water lines of Bank #2L and #3 to eliminate the biological contamination. This remediation was undertaken and in a matter of hours it was confirmed that clean-up had been achieved.

The results obtained clearly indicated the source (microbiological build-up, or biofilm) and nature (localized) of the water contamination in less than two hours. Using this methodology, any operator of any water system can detect contamination in the same manner and act to correct the problem on the same day. Routine testing of critical control points (for example, feed water, Bank #2L) facilitates preventative maintenance to maintain compliance.

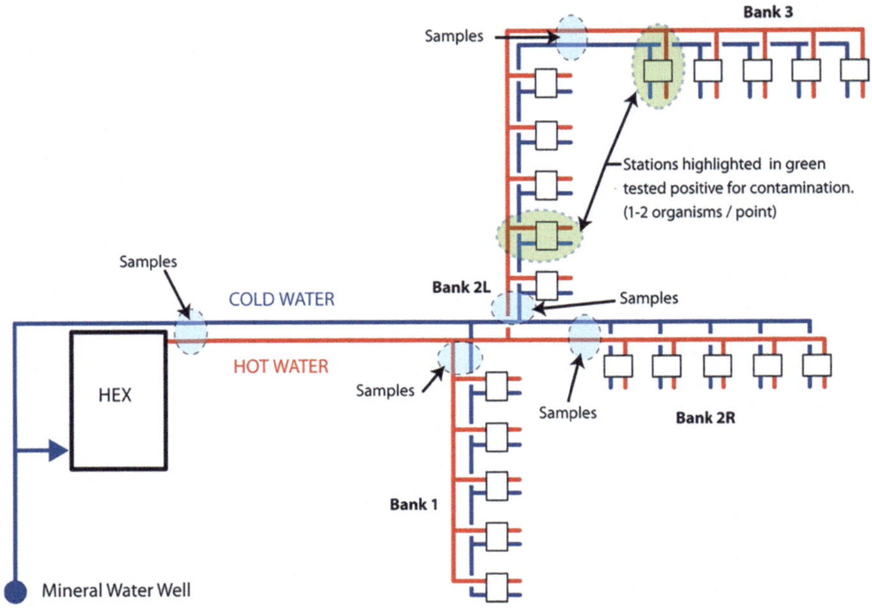

Figure 8.5 Location of sampling & contamination in water system (LuminUltra, 2017d).

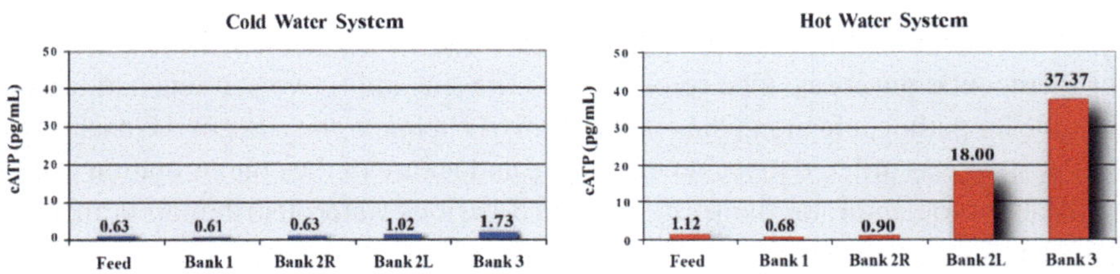

Figure 8.6 Audit results for water system (LuminUltra, 2017d).

8.6 CONCLUSIONS

Because of its speed, ease-of-use, and specificity to total living organisms, ATP monitoring serves as a valuable compliment to conventional water quality tests. The advancements made as part of Second Generation ATP technology enable rapid and accurate measurements in a significantly wider range of applications than what ATP tests have traditionally been applied to, both in terms of challenging sample matrices and sensitivity requirements. When dealing with potable water systems, it not only facilitates routine maintenance and troubleshooting but also helps maintain water quality by detecting microbiological contamination at the earliest signs so that they can be dealt with as quickly as possible. Similar concepts can be extended to other microbiological growth control applications such as product quality control, oil and gas, fuel systems, and manufacturing to immediately assess the level of microbiological contamination, determine if action is necessary, and assess the efficacy of mitigation activities in near-real-time.

8.7 REFERENCES

Ashbolt N. J., Grabow W. O. K. and Snozzi M. (2001). Indicators of microbial water quality. In: Water Quality: Standards and Health, L. Fewtrell & J. Bartram (eds), IWA Publishing, London, UK, pp. 289–316.

Bartram J., Cotruvo J., Exner M., Fricker C. and Glasmacher A. (2003). Heterotrophic Plate Counts and Drinking-Water Safety: The Significance of HPCS for Water Quality and Human Health. IWA Publishing, London, UK.

Berthold F., Herick K. and Siewe R. (2000). Luminometer design and low light detection. *Methods in Enzymology*, **305**, 62–87.

Birmele M. N., Roberson L. B. and Roberts M. S. (2010). Evaluation of an ATP Assay to Quantify Bacterial Attachment to Surfaces in Reduced Gravity. NASA, Florida, USA.

Cairns J. E., Nutt S. G., Afghan B. K. and Carlisle D. B. (1979). Application of the ATP Assay for Monitoring the Disinfection of Wastewater Streams. Environment Canada, Ottawa, Canada.

Cairns J. E., Whalen P. A., Whalen P. J., Tracey D. R. and Palo R. E. (2005). Dissolved ATP – a new process control parameter for biological wastewater treatment. WEFTEC Annual Conference 2005.

Cangleosi G. A. and Meschke J. S. (2014). Dead or alive: molecular assessment of microbial viability. *Applied and Environmental Microbiology*, **80**, 5884–5891.

Corrin E., Leong S. and Whalen P. A. (2009). Comparison of Biocides Using a New Microbial Detection Tool. Presented at 2nd International Symposium on Applied Microbiology and Molecular Biology in Oil Systems (ISMOS-2), Aarhus, Denmark, 17–19 June 2009. Danish Technological Institute, Aarhus.

Evans P., Smith J., LeChevallier M. L., Schneider O. D., Weinrich L. A. and Jjemba P. K. (2013). A Monitoring and Control Toolbox for Biological Filtration. Water Research Foundation, Denver, CO, USA.

Goodsell D. S. (1996). Our Molecular Nature. Springer, New York, USA.

Holm-Hansen O. and Booth C. (1966). The measurement of adenosine triphosphate in the ocean and its ecological significance. *Limnology and Oceanography*, **11**, 510–519.

Johnston I., Gaal B., Pires das Neves R., Enver T., Iborra F. and Jones N. (2012). Mitochondrial variability as a source of extrinsic cellular noise. *PLoS Computational Biology*, **8**(3), e1002416. https://doi.org/10.1371/journal.pcbi.1002416.

Knowles J. R. (1980). Enzyme-catalyzed phosphoryl transfer reactions. *Annual Review of Biochemistry*, **49**, 877–919.

L'Abbee J. and Ritchie R. (2008). Comparative study of the QGA test kit and Heterotrophic Plate Count (HPC) to detect microorganisms in water.

Laupland K. B. and Valiquette L. (2013). The changing culture of the microbiology laboratory. *Canadian Journal of Infectious Diseases and Medical Microbiology*, **24**, 125–128.

LuminUltra (2017a). Application Note: ATP Monitoring in Biologically Active Filters. LuminUltra Technologies, Fredericton, New Brunswick, Canada.

LuminUltra (2017b). Application Note: Mapping and Identifying the Source of Water Quality Issues. LuminUltra Technologies, Fredericton, New Brunswick, Canada.

LuminUltra (2017c). Application Note: Monitoring Biological Growth in Membrane Filters. LuminUltra Technologies, Fredericton, New Brunswick, Canada.

LuminUltra (2017d). Case Study: Water Distribution System Audit. LuminUltra Technologies, Fredericton, New Brunswick, Canada.

Maeir R., Pepper I., Gerba C. and Gentry T. (2009). Environmental Microbiology, 2nd Edn. Academic Press, Cambridge, MA, USA.

Sloan W. T., Quince C. and Custis T. P. (2008). The Uncountables. In: Accessing Uncultivated Microorganisms, K. Zengler (ed.), ASM Press, Washington, USA, Chapter 3.

Stoddart A. (2017). Field-Scale Evaluation of Drinking Water Biofiltration. PhD Thesis, University of Dalhousie, Halifax, Canada.

Stopa P. J. and Orahovec Z. (2002). *Technology for Combating WMD Terrorism* (Proceedings of the NATO Advanced Research Workshop on Technology for Combating WMD Terrorism, Hunt Valley, MD, USA, 19–22 November 2002) Kluwer, Netherlands.

Todar K. (2008). Todar's Online Textbook of Bacteriology. http://www.textbookofbacteriology.net/ (accessed 10 March 2018).

Travis B. and Tracey D. R. (2015). Long-term benefits of enhanced biological monitoring strategies on membrane filtration operations. *Membrane Technology*, **2015**(8), 7–10.

Vanderzwaag J. C, Bartlett K. H., Atwater J. W. and Baker D. (2009). Evaluation of field testing techniques used in a household water treatment study. *Water Quality Research Journal of Canada*, 22, 122–131.

Chapter 9

Counting totally matters – using GRUNDFOS BACMON for network monitoring

Lars S. Olesen, Bo Højris and Nestor B. Folia

9.1 INTRODUCTION

Water scarcity, ageing infrastructures, and emergence of new pollutants are all different factors resulting in the maintenance of water quality being one of the main challenges societies face in the 21st century. Degradation of water quality translates directly into threats to human health, limited food production, reduced ecosystem functions and into environmental and socio-economic challenges in general (UNESCO, 2015). Besides maintaining a good drinking water quality, a wish for better resource utilization has led to concepts like "Water Fit for Purpose" (Carpentier & Cole, 2018) and an increased focus on water reuse possibilities (Sgroi *et al.*, 2018; Olivier, 2018). Altogether, these challenges put larger and larger demands to how we collect, treat, distribute, use, dispose, clean, and discharge water. Grundfos has over the years moved a lot of water with pumps. However, the emerging challenges has led us – and our colleagues in the business – to supplement the pump itself with surveillance, demand driven distribution to reduce leakages, and a range of dosing and disinfection products to not only move the water but also treat it. Lastly, digitalization of equipment has given opportunities to optimize and operate entire systems in a much more structured and coherent way.

The chapter is from the book *Microbiological Sensors for the Drinking Water Industry,* Torben Lund Skovhus and Bo Højris (Eds.).
doi: 10.2166/9781780408699_0155

Yet, a desire to improve quality in all steps, reduce waste, and increase efficiency has increased the complexity in managing and operating water production and -treatment. A complexity that has put stronger requirements to how water quality is monitored, controlled, and maintained. Generally, the trend has gone towards seeing the entire network more holistically, exemplified by WHO introducing Water Safety Plans as such an approach: "*The most effective means of consistently ensuring the safety of a drinking-water supply is through the use of a comprehensive risk assessment and risk management approach that encompasses all steps in water supply from catchment to consumer*" (Bartram *et al.*, 2009).

Part of such a safety plan is of course to watch and control the water quality. Traditionally within water quality measurements, the results of water from yesterday are ready tomorrow. This is mainly due to the fact that culture-dependent methods are extensively used for water quality testing and these take time to obtain reliable results. It complicates the wish for water utilities to assert proactive actions in daily operations and hampers efficient tracing in case of unforeseen contamination events. Besides this, there is a broad distribution of pathogens that occur in water in a viable but non-culturable (VBNC) state such as *Campylobacter, Legionella, E. coli* and others (Li *et al.*, 2014), which thereby can lead to false negative results from such culture dependent methods (Zhao *et al.*, 2014).

In the other end of the temporal spectrum, faster measures do exist like conductivity and turbidity, which both operate in the order of seconds. Related to the concentration of ions in the water, conductivity gives information about the dissolved dissociated salts, while turbidity is a measure for how cloudy the water is and its ability to scatter light. However, neither of these faster methods tell anything about whether a change in water properties stems from microbiology. In Grundfos, we saw the need for methods that could help watch the water to inform operation of pumps, dosing, disinfection, and networks about the bacteria content of the water faster than by growing bacteria colonies in the laboratory. We thus saw a potential benefit in having a broad method that albeit maybe less specific than culturing methods could give rapid information about the general bacteria levels; i.e. if not in seconds, then at least in the order of minutes. After having searched for potential candidates without luck we decided to try and develop a solution ourselves. Being a mass producer and coming from the pumps and operations side, we were not as such looking for a complex laboratory instrument but rather an automatic solution applicable in the field next to pumps and other equipment and being relatively easy to operate.

A decade later came the first initial releases of GRUNDFOS BACMON sensor units into the market. By supplementing, and to some extent replacing, manual sampling, BACMON offers a way to monitor a network by means of total bacteria- and non-bacteria (abiotic particles) counts. Automated batch sampling technology delivering total bacteria counts in minutes without addition of chemicals can be installed at selected key points, connected, and compared directly online in a common interface. Before going into the entire functionality of the connected system, first a closer look at the sensor unit being installed at the sampling points.

9.2 TECHNOLOGY AND SOLUTION

So, how can bacteria be counted in the order of minutes? In essence, a BACMON sensor unit can be described as an intelligent microscope utilizing 3D image recognition techniques to count and classify objects. A water sample is held still between two windows in a flow cell and then more than a thousand pictures are taken of that water sample. With different imaging techniques, it is then possible to track and identify all objects in the images of the water sample. On each individual object, the sensor unit extracts a set of features, compares these to a learned set of characteristics, and from there classifies each object as belonging to either of two groups: bacteria or non-bacteria. After adding it all up, the count is reported, the flow cell flushed, and the system is ready for the next sample.

9.2.1 Mimicking the human brain a bit

One can compare the inner workings of the BACMON sensor unit to the human brain. As kids, we are presented with images of humans, animals, toys, cars, houses, and other objects again, and again, and again. Slowly our neural network learns the features of e.g. a human being: face, arms, legs. We learn to distinguish 'this is a person' and 'that thing over there is a dog'. Notably, without having been presented to a picture of every individual human being in the world we – over time – learn to generalize. By finding the right features, we can even classify persons we have never seen before as being just that: persons. This analogy can be used on BACMON. Here an artificial neural network has been trained by presenting it with a very large number of very carefully prepared samples of different bacteria, other objects, and various mixtures, in order to learn which objects are bacteria by looking at their features. More than 50 different features are extracted from each object in the image. Some are linked to spatial parameters of the objects like e.g. area, perimeter, how round, how long. Other parameters are related to image properties like contrast, how is light scattered around objects, how is light shining through etc. Details about bacteria types, suspensions, and mixtures used for training and validating the instrument can be found in (Højris *et al.*, 2016).

With persons we can, just from looking, not tell their name, and whether they live in the Netherlands, UK, or Lithuania but with relatively high probability we can still count them as persons. The same goes for BACMON; it cannot tell specifically which bacteria are present or whether they are alive or dead. This is comparable to the skills of a human being. By just quickly looking at a person lying on the floor, we can have a hard time determining whether we are dealing with a dead person or merely one sleeping. Of course, if the head is in one corner and the body in another, we are not in doubt. Same for BACMON, if the bacteria have disintegrated by e.g. chemical disinfection, and thereby undergone a change in morphology, the objects or fragments would count as non-bacteria. Likewise, as the sensor unit cannot deliver any specific pathogenicity information, sequencing or other advanced methods can be needed as sometimes even the same species can express different pathogenic properties (Alberts *et al.*, 2002). Instead, the sensor unit can give a total bacteria count in minutes.

The fact that BACMON operates by just 'looking' and thereby avoiding handling e.g. reagents with a potential shelf life, is seen as one of the clear benefits of the approach. Even though full specificity, viability, and pathogenicity assessed in seconds would be a long-term ideal to strive for, simplicity, practical use, and speed has deliberately been prioritized over specificity.

9.2.2 How does it compare?

The sensor unit simply counts all the particles in the water sample and classifies which ones are bacteria and which are non-bacteria. The result is an absolute number, telling how many there physically can be seen and not how many there are, compared to a baseline. This fact is important to remember when wanting to assess results from the individual sensor unit or when comparing across sensor units: one is dealing with actual numbers.

Validation of the method has been done up against e.g. total direct cell count (TDCC) through DAPI staining and fluorescent microscopy, which is a well known manual, technique to perform a total count of bacteria (Porter & Feig, 1980). By adding increasing amounts of yeast extract to act as nutrition for bacteria, and allowing an incubation time, one can get different growth dynamics and see how well BACMON compares to manually counted references. Results from such a test are depicted in Figure 9.1.

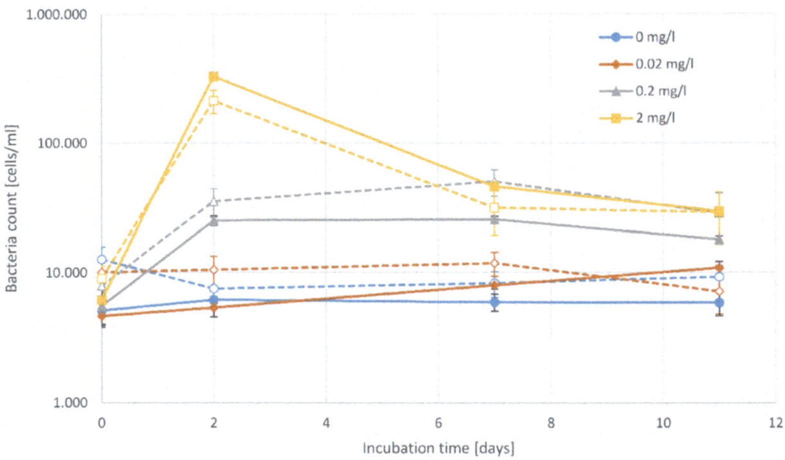

Figure 9.1 Different amounts of nutrient were added to tap water and growth patterns measured by BACMON (solid lines) and TDCC (dotted lines). The two methods correlate well, expressing the same dynamics. Importantly, also after a high count, BACMON follows the TDCC downwards, when bacteria start to decay (Illustration adapted from Højris *et al.*, 2016).

BACMON delivers results in bacteria per mL and can count as low as zero. It 'looks' at the water and if none of the objects in the analysed focus field in the flow cell are classified as bacteria, BACMON will report zero (0 bacteria per mL). However, the challenge is to ensure that the small water sample in focus is representative of the entire stream. This is difficult. The BACMON sensor unit analyses a volume of 6 μL in the flow cell. If one object is classified as bacteria, this will then lead to a count of 167 bacteria per mL (as 1 in 6 μL equals 167 in 1 mL). However, the sensor unit is designed for drinking water and here the numbers are normally much higher. In e.g. a total bacterial count analysis of more than 50 samples of drinking water from various distribution networks and rural communities in Finland led to a median of 55.784 cells per mL (Ikonen *et al.*, 2013). This is in line with the experience from BACMON readings across Europe, where, under normal conditions, the total bacterial counts can easily range from 1000 to 100,000 cells per mL depending on the location and source. What is worth noting, though, is that *within* the same source, the variation under normal conditions is much lower. How low depends on a lot of operational factors though.

With a BACMON sensor unit, there are more than 100 measurements the first day of installation and more than 1000 samples of total bacterial count in the first week. So, even though BACMON delivers an absolute number – and not a reference to a baseline – the effect is that after installation one relatively quickly gets an understanding of what *is* actually the baseline for the water in question at the specific sampling point. An example is a customer in EU, where sensor units were installed at different Slow Sand Filters (SSF) to monitor microbiological dynamics when skimming the filters. Despite being more or less 'the same' unit operation, one SSF consistently had more than double the number of bacteria per mL in absolute numbers when compared to the other; a comparison that then served as the entry point for further investigations into the cause and thereby a way for the customer to optimize the operation and implementing best practices across.

Many ask how to compare a BACMON reading to plate counts or CFU measurements done in the lab. This is not directly possible. A plate count is done on specific growth media and at temperatures favouring specific bacteria (e.g. 22°C or 37°C) whereas the BACMON sensor units count everything; also, those bacteria growing at 8°C or those that are viable but non-culturable. The result is a broader and more complete, *total* count.

Being an absolute measure, one can also compare individual BACMON sensors units. Yet, as shown in Figure 9.2, setting up five sensors to measure the 'same' water from the same water line will lead to small variations. This is important to have in mind when ensuring that the sampling point draws a sample as representative as possible from the water line. One cannot measure all water passing by, but always a sample. However, the fact that one can now measure e.g. total count much faster than traditional methods allows the method itself to even out these variations over time. Had the sensor units in Figure 9.2 been installed for example 10 km apart, it would be an interesting indication of water system integrity.

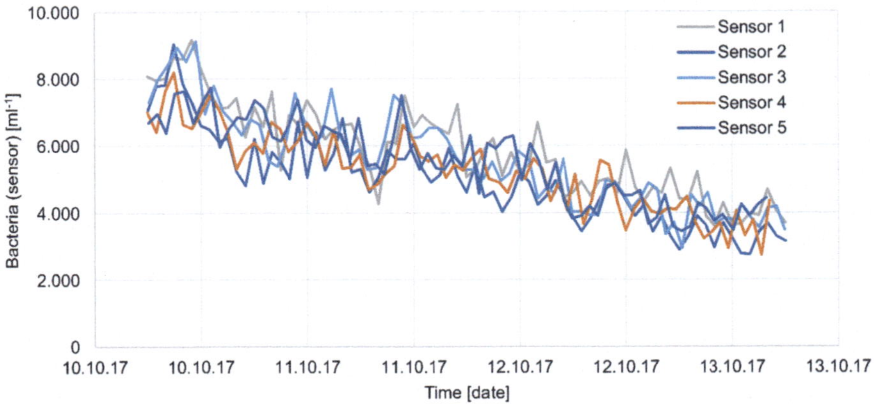

Figure 9.2 Five BACMON sensor units installed to sample from the same water source by means of a manifold distributing the water (Danish drinking water). The variations arise because one is not measuring on exactly the same water sample. However, the five units expressed the same trend, level and variation during the test.

Speed and comparisons of different sensor input from different sampling points can also be obtained with other methods of course. Turbidity is once such instrument which rapidly measures the cloudiness of water and has been used as an indicator for water quality for decades. BACMON has been compared to turbidity in various ways, where one is illustrated in figure 9.3. Experimentally trying to simulate a water contamination, clean water samples were spiked with different amounts of various contaminants. Depending on the source, the reading from the BACMON sensor unit reacts 10 to 100 times before the turbidity signal.

Besides a relatively higher sensitivity, one thereby gets the additional benefit of knowing, not only if the water is cloudy but also, whether that 'something in the water' is of microbiological origin or stems from suspended abiotic particles.

9.2.3 Air bubbles, biofilm, and fouling taken into account

With thousands of bacteria per millilitre being present in regular drinking water, fouling will occur on the inner surfaces of the flow cell used in the BACMON sensor units. When this happens, the (disposable) flow cell is exchanged and the sensor unit continues its operation. A status indicator is continuously reporting the present state of the flow cell giving the operator a chance to follow it and schedule an exchange, at a predefined level. Local conditions are determining how 'potent' the water is in terms of fouling. In non-chlorinated Danish drinking water, there are about 1 to 2 months between the need to visit the sensor unit and exchange the flow cell. Status of the flow cell can be given because there is a camera in BACMON that actually 'looks' at the water and the flow cell. If something is stuck on the window, it can be seen in the images. This is used to generate the flow cell status but also to take into account if other disturbances are present. Deliberately, the focus field for the imaging is set to be in the middle of the flow cell between the two windows. This means that if sediment has settled on the bottom, it is not in focus and can be 'looked through' and focus kept at the particles and objects held fixed under pressure in the middle of the flow cell. However, the fact that these disturbances can be seen also means they can be taken into account when interpreting the data. If an air bubble is rendering the image acquisition difficult in a part of the flow cell analysis, this area is consequently discarded and not included in the analyses; if needed additional pictures are be taken of areas without the air bubble, keeping good statistics behind the count reported.

		Bacteria	Abiotic particles	Turbidity	TDCC
Soil [mg_{dry}/l]	0,067	-	-	-	-
	0,2	-	-	-	-
	0,67	-	-	-	-
	2	-	-	-	-
	6,7	-	+	-	-
	20	+	+	-	+
	67	+	+	+	+
	200	+	+	+	+
Waste water [ml/l]	0,003	-	-	-	na
	0,01	-	-	(+)	na
	0,03	-	-	-	na
	0,1	-	+	-	(+)
	0,3	+	+	-	+
	1	+	+	-	+
	3	+	+	-	+
	10	+	+	+	+
Pidgin faeces [mg_{dry}/l]	0,017	-	-	-	-
	0,05	-	-	-	-
	0,17	-	-	-	-
	0,5	+	+	-	+
	1,7	+	+	-	+
	5	+	+	+	+
	17	+	+	+	+
	50	+	+	+	+

Figure 9.3 Experiments from various amounts of contaminants added to non-chlorinated Danish tap water. The boxes with a ' + ' sign indicate when the signals from the different methods are significantly higher than the background readings (Student's t-test with 98.5% confidence). Boxes marked '(+)' indicates responses significantly lower than the background. 'na' indicate samples unable to be enumerated due to too low concentrations. The Bacteria and Abiotic particles columns are from BACMON and it is seen that they respond at 10 to 100 times lower concentrations compared to the turbidity meter and at the same contamination levels as total direct cell counts (TDCC). (adapted from Højris *et al.*, 2018).

9.3 OVERALL CONCEPT

The BACMON solution is basically comprised of three elements (see Figure 9.4):

(1) A number of **Sensor Units** strategically placed e.g. at the water source, in the treatment train, and in the distribution network. The sensor units are sampling the water automatically and deliver their data to a server.

(2) A cloud-based **Water Quality Management** server, which manages the individual sensor units, provides integrity measures, and collects the data.

(3) The server allows for auxiliary data to be entered into the server; thereby allowing for information coming from outside the BACMON system to be stored, compared, analyzed, and accessed in combination with the existing data.

(4) Various **Interfaces**, which can then be used to view results, compare data, and work with the analyses in real time.

Figure 9.4 An overall illustration of a BACMON solution. A number of Sensor Units (1) send their results to a Water Quality Management server (2) into which other inputs in principle could be given (3) and from there extracted to different Interfaces (4).

Installation of the sensor unit of course requires access to the water and a suitable drain, but besides that only power and a data connection is needed to get started measuring. Typically, the water is sampled from an existing sampling valve or a new access point is created equivalently to the principles behind establishing representative sampling points.

The Water Quality Management server holds the data, provides easy access to historical views on each sampling point and allows for advanced, computationally demanding analyses to be performed; both on data from the individual sensor units and jointly across the entire installation for the network in question. Monitoring the results is done by means of extracting the data to an interface of preference (iPad app, web browser based tool, SCADA system, setting up SMS/ email alarms, etc).

Comparisons within a network will aid an operator to compare unit operations and optimize daily schedules and handling. Integration and use of other measures both online and offline will supplement and enrich the bacteria analyses and gradually it might help projecting and predicting from manual samples while at the same time perhaps guide where to best take the next manual sample for more detailed analyses to learn more and most. Eventually, the perspective will be to compare unit operations to unit operations, production to production, network to network, and broaden that view across borders and applications.

The benefit of having multiple sensor units combined in a single interface is, amongst others, to be able to get an overview of the bacterial levels across points in the entire network (see figure 9.5). With e.g. a simple 'traffic light' colour-coding system one can quickly see whether everything is ok (green), if something requires attention (yellow), or if a situation gives rise to an alert (red). The threshold levels associated with each colour code may be set individually in each case to reflect the 'normal' baseline situation and elevated alarm levels. In case of any unforeseen events, a distribution of online sensor units thereby also facilitates an assessment of the extent of the incident and aides a faster tracing and potentially confinement of the source or at least provides the possibility to focus in the search for a cause (source tracking).

To better compare against other existing measures (like pump and valve actions, pressure, temperature etc.), water utilities prefer often to incorporate the data in existing monitoring and operating systems. There is a multitude of different systems and the way one integrates BACMON data is via an API (Application Programming Interface) which, in a standard way, defines how one to read the data. An example of such integration is shown in Figure 9.6. The same interface is used when e.g. getting weather data from meteorological service servers and integrating those to anticipate flooding.

9.4 FROM DUSK TILL DAWN – APPLICATION CASES

Keeping the above comments about specificity in mind, what the BACMON solution *is* capable of delivering, is a total count of bacteria in minutes. This means in practice that one suddenly gets microbiological insight about the sampling points selected more than hundred times per day instead of perhaps once a day, week, or month. And why does this matter? It provides a broader, fuller picture of what is going on in the water.

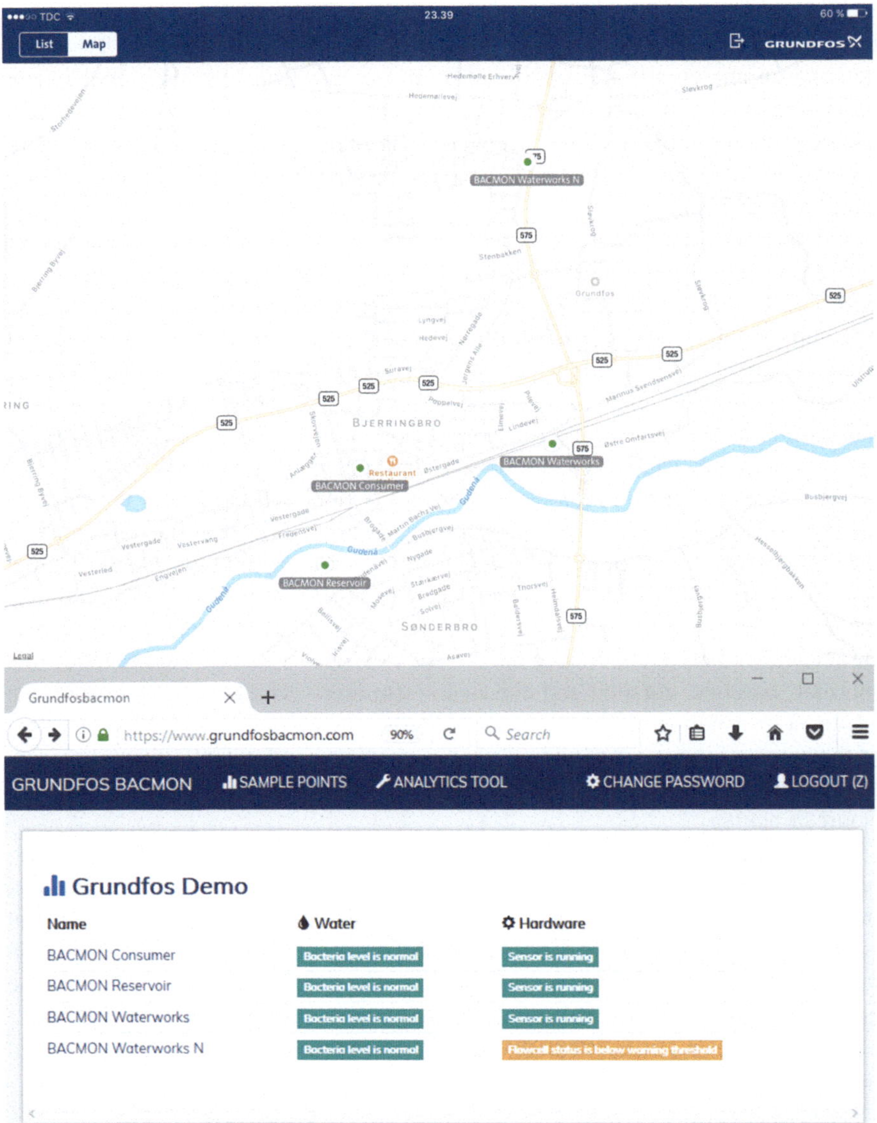

Figure 9.5 Top: Screenshot from the BACMON iPad App showing the location of distributed sensor units on a map along with an indicator of state (green, yellow, red) providing a quick overview of the situation and guides where to focus if something deviates (placement of sensor units for illustration purpose only). Bottom: Screenshot from the web interface showing a status screen of the individual sensors. In the example here, flow cell status was below 50% and the system has gone to a warning state, so the operator can plan for an exchange.

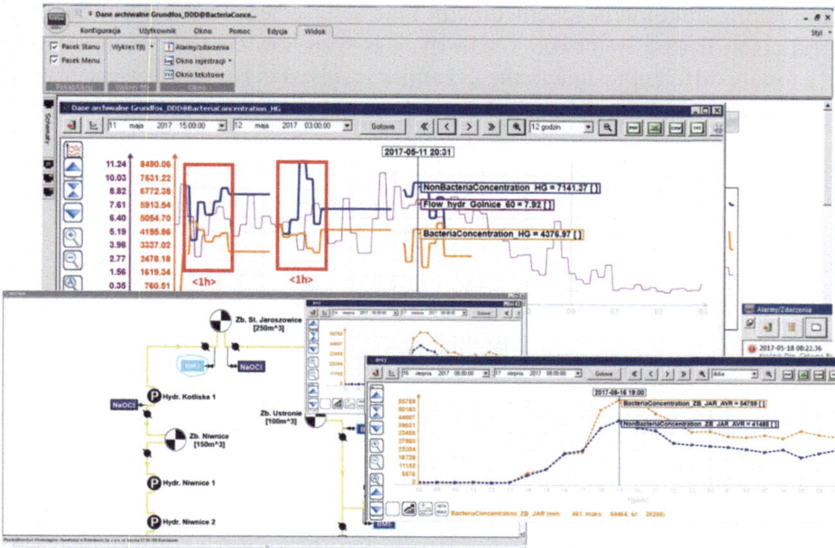

Figure 9.6 An example of integration of BACMON data directly into the SCADA system of a water utility. They use a number of different views/windows to operate their entire network and some are shown here, to illustrate how BACMON is integrated deeply into the individual views of the entire system and not just as a stand-alone window. Bottom left is a stylized map of their network and the sensor unit locations are visible as icons in this. The three windows upper right are different views used to manage the network where BACMON readings are integrated. In this way comparisons can be made directly to e.g. flow and other existing measures and thresholds and logic can be set up to alert the personnel in operation through the existing alarm schemes and protocols.

In 2017, the EU Commission had the 1998 Drinking Water Directive (DWD) under review. In their so-called 'REFIT Evaluation' they write: *"Special attention should be paid to the relevance of microbiological parameters, where 'new' pathogens not considered in the current DWD present real challenges."* (European Commission, 2016) It is not a claim that BACMON is the 'silver bullet' solely addressing these challenges but definitely it adds a perspective that previously has been difficult to get with current methods. It is fair to say that with its total count of bacteria in minutes, BACMON has not yet been installed without learning something new about the system in question. A few cases follow to exemplify such installations.

9.4.1 Flushing filters

In production of drinking water, sand filters of various types are widely used to reduce the number of bacteria and remove unwanted solids. Being an integrated part of the production, the water quality is most often assessed in totality and not around the sand filters specifically. In Figure 9.7 is an illustration of the total bacteria count from a waterworks. Clearly, it is seen how continuous monitoring gives a totally different insight to the dynamics involved in running the water work than infrequent grab sampling can deliver alone.

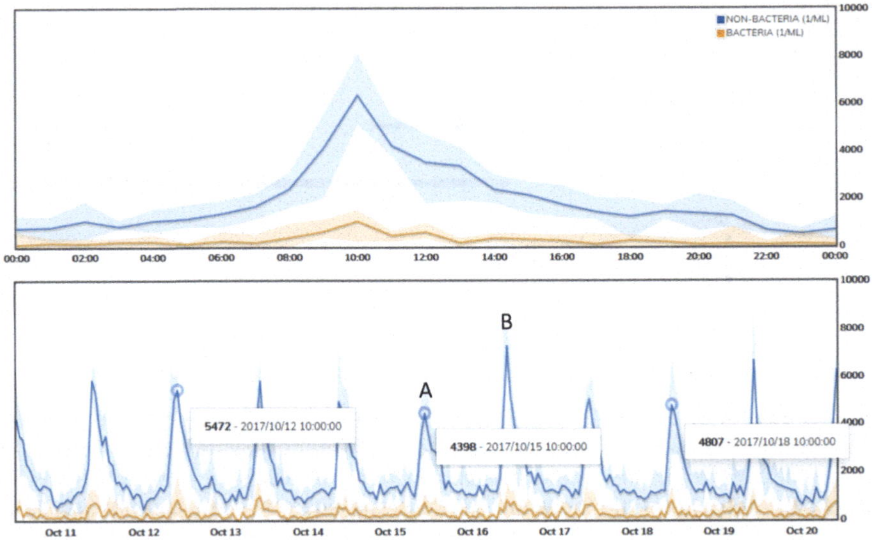

Figure 9.7 Top view: Seen as an isolated event, it might seem alarming at first glance, however, a (trained) water utility operator will recognize the time of day to match a filter flush. Bottom view: Even an untrained eye will see a clear pattern when looking further back in time; an effect one does not get by grab sampling once a day. If mapping against other information, like e.g. knowing the time for filter A and B being flushed respectively, one might learn something about the state of operation from the fact that the numbers are almost doubled when flushing the latter.

Individual filter flushed can be investigated and peak conditions, settling times, residual effects, and distribution of such events can be monitored in the system. Even more, when one starts to map filter flushes (as seen in Figure 9.7) to, for example, flow, filter number being flushed etc., one can get more information. Users of BACMON solutions have reported how they discovered that the sand media of certain filters needed replacement, as when these filters were flushed, the peaks were always higher than when other filters were flushed.

9.4.2 Network overview and optimization

In Figure 9.8 is a map of part of the network from a Polish water works where a number of sensor units are installed in a BACMON solution to monitor the daily operation. At some point, the water works had a planned renovation of a clean water tank in one of the water works and BACMON was used to follow the bacteria levels in the clean water tank while flushing before taking it back into operation. A view of the bacteria levels in the renovated tank can be seen in Figure 9.9.

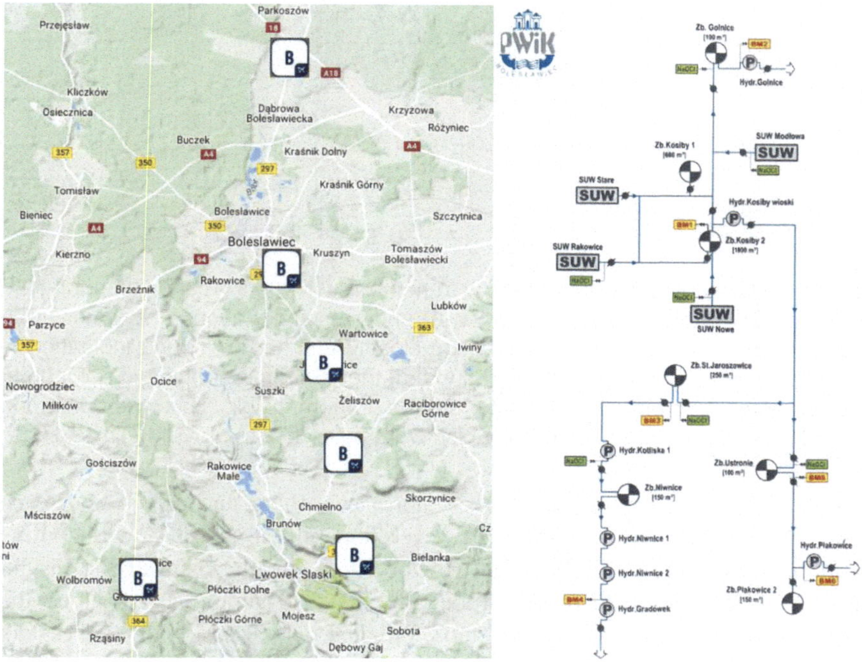

Figure 9.8 To the left an overview of how BACMON sensor are spread geographically to cover different areas of a water utility's production and distribution network. When coupling the geography with a Geographic Information System (GIS) and with sensor units installed in strategic important points in the distribution network, the water works has a better overview of how potential challenges might spread and it provides a way to isolate areas and better trace the origin in case of any unforeseen events. To the right, a detailed mapping of where the sensor units are installed in relation to the various process steps and a way for the water utility to get an overview of the entire operational parameters available.

Together with the flushing of the tank, the network was monitored while also adjusting chlorine dosing, to be in control of the distribution when recommissioning after the renovation.

The water works has the BACMON readings integrated into their SCADA system. By coupling the individual readings with the geographical and/or procedural placement in the network, they could follow the various bacteria levels, flow, conductivity, and other measures by direct comparison and use this to guide the operation and when manual samples were required to be taken, to release the renovated clean water tank to be taken back into production (see Figure 9.9).

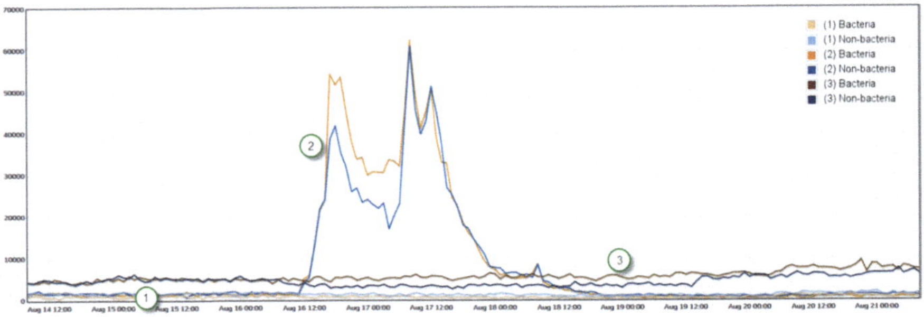

Figure 9.9 Watching the water coming from the water works (1) and comparing the reservoir (2) to downstream reservoirs (3) and pumping stations, the water utility could follow the effect of their renovation closely and have a strong indication of when the levels were back to normal and that the recommissioning did not lead to any unwanted side effects in the later stages of the network. This was confirmed by traditional laboratory samplings, which were guided by the BACMON readings.

Renovation of a clean water tank is not a new thing and can of course be done without a BACMON solution. However, hopefully it is clear how the online monitoring of bacteria levels in the network dramatically changed the insight into the effects and state of the production and distribution system along the way; thereby leading to a safer and more efficient recommissioning of the plant including documentation of the intermediate levels and system states. Also, the case hopefully serves as a general example of the strength in having a number of sensor units combined in a solution, to be able to follow, compare, trace, and watch how effects distribute and spread – or does not spread as in the renovation case – and thereby gives a holistic view of the entire network.

9.4.3 Red alert in food production

Besides being a general operational parameter to look at, an installation monitoring the total bacteria count several times an hour can also be used as an early warning or event detection assistant. In a Danish dairy processing plant, they have BACMON sensor units installed to monitor the intake water being used at several points in the production processes. At some point, a renovation of the distribution system outside the dairy plant went disturbingly wrong. The utility delivering the drinking water for the facility was renovating a pipe and by mistake during the renovation, dirty water was led 'the wrong way' through the installation and thereby into the inlet of the dairy processing plant. In Figure 9.10 it can easily be seen how suddenly rather low baseline levels was replaced by surprisingly high numbers.

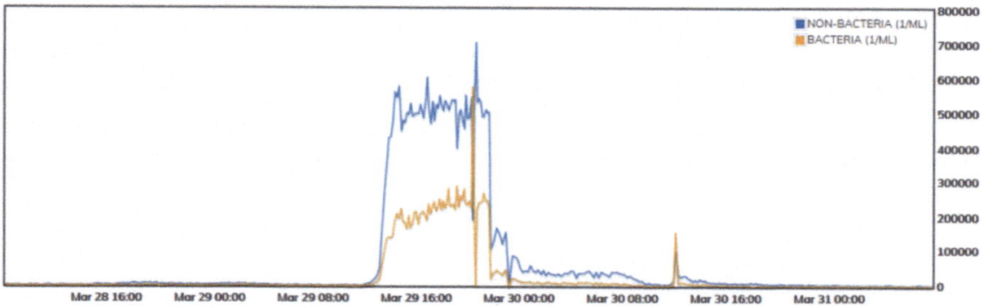

Figure 9.10 The concentrations from the BACMON sensor unit monitoring the inlet water quality at a dairy processing plant. A contamination stemming from pipe renovation external to the production site is evident and in the particular case so massive that the entire production had to be closed down and cleaned. Generally, the case is an example of the early warning and event detection potential in having a sensitive sensor unit installed monitoring the water quality in the order of minutes.

The dairy plant could see even with visual inspection that the water had gone bad. The site has three BACMON sensor units installed, one at the inlet and two in different sections of the production. Remains of the contamination were seen in the signals from the more remote units as well and it was decided that the entire factory needed to be closed down. A very unfortunate situation where water was then transported in tanker trucks to flush the lines and get everything cleaned before production was resumed. Even though the case in question was so massive that visual inspection of the water could tell the trouble, the example serves as a general illustration of how the temporal resolution of BACMON in many cases will enable the solution to function as an event detection mechanism and provider of an early warning.

9.5 FUTURE PERSPECTIVES

The pressures on our water resources are ever increasing and the demand for good water quality management practices follows. Broader and more frequent analyses are required to supplement grab sampling to ensure that we can maintain safe drinking water for many years to come.

The ability to analyse water quality and deliver total bacteria counts in minutes is a reality with BACMON: a relatively fast, simple, and intuitive way to get an absolute measure of the microbiological levels of the water. No doubt, the bacteria count in itself is a big help and improvement in operating the installations already being monitored out there. However, one must also recognize that BACMON is a novel solution and total bacteria count in minutes is a relatively unfamiliar measure to have available for water professionals. One can therefore only speculate how things can evolve from here. No doubt, changes on the individual sensor unit level can and will be analysed in greater detail as more and more units get into operation.

Levels, trends, anomalies, patterns, and behaviour in general will more and more be recognized and acted upon in an increasingly efficient way, the more water runs through these sensor units and the more historical data gets collected. A natural development would be to start integrating some of that collective knowledge 'back into the systems' so that installations can learn from each other; most likely anonymously, yet effectively.

9.6 REFERENCES

Alberts B., Johnson A., Lewis J., Raff M., Roberts K. and Walter, P. (2002). Introduction to Pathogens. In: Molecular Biology of the Cell, B. Alberts, A. Johnson, J. Lewis, M. Raff, K. Roberts and P. Walter (eds), 4th edn, Garland Science, New York, USA.

Bartram J., Corrales L., Davison A., Deere D., Drury D., Gordon B., Howard G., Rinehold A. and Stevens M. (2009). Water Safety Plan Manual: Step-By-Step Risk Management For Drinking Water Suppliers. World Health Organization, Geneva, Switzerland.

Carpentier C. and Cole J. H. (2018) Industry insights — imagine a fit-for-purpose water future. *Opflow*, **44**(2), 6.

European Commission (2016). Commission staff working document, REFIT EVALUATION of the Drinking Water Directive 98/83/EC, SWD(2016) 428 final. http://ec.europa.eu/environment/water/water-drink/pdf/SWD_2016_428_F1.pdf (accessed 5 March 2018).

Højris B., Christensen S. C. B., Albrechtsen H.-J., Smith C. and Dahlqvist M. (2016) A novel, optical, on-line bacteria sensor for monitoring drinking water quality. *Science Reports*, **6**, 23935. doi:10.1038/srep23935.

Højris B., Kornholt S. N., Christensen S. C. B., Albrechtsen H.-J. and Olesen L. S. (2018). Detection of drinking water contamination by an optical real-time bacteria sensor. Accepted for oral presentation at Singapore International Water Week 2018.

Ikonen J., Pitkänen T. and Miettinen I. T. (2013). Suitability of optical, physical and chemical measurements for detection of changes in bacterial drinking water quality. *International Journal of Environmental Research and Public Health*, **10**, 5349–5363. doi:10.3390/ijerph10115349.

Li L., Mendis N., Trigui H., Oliver J. D. and Faucher S. P. (2014). The importance of the viable but non-culturable state in human bacterial pathogens. *Frontiers in Microbiology*, **5**, 258. https://www.frontiersin.org/article/10.3389/fmicb.2014.00258 (accessed 10 April 2018).

Olivier L. (2018) Beyond NEWater: An insight into Singapore's water reuse prospects. *Current Opinion in Environmental Science & Health*, **2**, 26.

Porter K. G. and Feig Y. S. (1980). The use of DAPI for identifying and counting aquatic microflora. *Limnology and Oceanography*, **25**(5), 943–948.

Sgroi M., Vagliasindi F. G. A. and Roccaro P. (2018) Feasibility, sustainability and circular economy concepts in water reuse. *Current Opinion in Environmental Science & Health*, **2**, 20.

UNESCO (2015). International Initiative on Water Quality. United Nations Educational, Scientific and Cultural Organization Report, http://unesdoc.unesco.org/images/0024/002436/243651e.pdf (accessed 10 April 2018).

Zhao X., Lin C. W., Wang J. and Oh D. H. (2014). Advances in rapid detection methods for foodborne pathogens. *Journal of Microbiology and Biotechnology*, **24**(3), 297–312. doi: 10.4014/jmb.1310.10013

Chapter 10

Safety and quality control in drinking water systems by online monitoring of enzymatic activity of faecal indicators and total bacteria

Joep Appels, David Baquero, Belén Galofré, Marta Ganzer, Jaap van den Dries, Rubén Juárez, Clàudia Puigdomènech and J. Hein M. van Lieverloo

10.1 INTRODUCTION

10.1.1 The role of monitoring in water safety and quality

Water safety and quality should primarily depend on the protection of sources, the adequate design and operation of treatment, the structural integrity of production and distribution systems, and hygiene during manual operations.

The design of all parts of the system (infrastructure, automation and manual operations) should be validated by applying scientific methods. The World Health Organization has described the outlines of a safety system – Water Safety Plans – which drinking water companies can apply (Bartram *et al.*, 2009). Testing the provided quality is considered a verification step which does not protect consumers but is considered a last resort for detecting incidents when any or all of the preventive measures have failed. By the time tests are completed and results indicate the water is not safe to drink; people may have already consumed the water, get infected, develop illness or worse.

A variety of physical and chemical variables can be monitored at high frequency to detect deviations in water quality. The most commonly used variables are turbidity, pH and conductivity. Microbiological methods for detecting breakthrough incidents have long depended on collecting samples and quantifying

Editors: Torben Lund Skovhus and Bo Højris
doi: 10.2166/9781780408699_0171

the microorganisms after culturing, either by plate counting colony forming units or by presence / absence tests – including Most Probable Number (MPN) tests – assessing metabolic activity. The drawbacks of this strategy are:

(1) Acceptable concentrations of pathogens are much lower than the detection level of any current monitoring method (at best 1 per litre). In The Netherlands, for instance, pathogens should be absent in approximately 810000l ($<1.23 \times 10^{-6}$ l^{-1}) according to legislation (Anonymous, 2011), based on acceptable infection risks (exposure of <1 pathogen per 10000 persons per year), which is in line with intentions of EU and other countries regulations.
(2) The correlation between the presence and concentrations of pathogens and surrogate microbiological variables ('indicator bacteria') is insufficient, which means that ratios are highly variable (Van Lieverloo *et al.*, 2007b; Medema *et al.*, 2014).
(3) Sample size is limited by the water volume that can be filtered or concentrated due to other particles, although methods are available that filter large volumes.
(4) The time between sampling and the result of a standard analysis is too long (>24 h to days) for 'off-line' analysis to be an effective safety system. Problems also occur due to sample transport, storage and pre-treatment.
(5) The costs of sampling and analysis limit a) the number of sampling sites and b) the sampling frequency, resulting in very low probabilities (<< 5%) of detecting contamination incidents (Van Lieverloo *et al.*, 2007a).
(6) Bacteria can clump or attach to particles, leading to 'colony forming units' (cfu), an underestimation of actual numbers present in a sample (Parker & Darby, 1995).
(7) Most water microorganisms cannot be cultured, leading to an underestimation of the total population.

These points show, there is a serious drawback in any form of verification of drinking water quality and safety monitoring that depends on showing the absence of indicator bacteria, as absence creates a false sense of security for water companies, regulators and consumers alike. The majority of resources of a water company, therefore, have to be focused on preventive measures.

In the past few decades, a variety of online methods have evolved (Lopez-Roldan *et al.*, 2013) that may improve and speed-up the detectability of incidents which impair microbiological water quality and safety. The majority of these systems operate online, mitigating the two major drawbacks of existing microbiological methods (4: analysis time and 5b: analysis frequency). When these online monitors are applied in drinking water, the consumer is still not fully protected (the product has already been consumed), but the time between contamination and response can be limited considerably. Furthermore, these online systems provide a powerful tool for evaluating the effectiveness of processes in water treatment and distribution.

10.1.2 Rationale for monitoring enzymatic activity

Assessing the metabolic activity of microorganisms in water also mitigates other major drawbacks of traditional monitoring techniques listed in paragraph 10.1.1 (6: clumping/particle attachment and 7: culturability). Enzyme activity is also shown in viable but non-culturable bacteria (Zhao *et al.*, 2013). One of these systems assesses a variety of enzymatic activities of bacteria in any type of water (or other fluids). The chapter at hand presents the current capacities of this online monitor, the BACTcontrol supplied by microLAN (the Netherlands). This system can be applied to assess one or two different enzymatic activities intermittently, both at high frequency. In the chapter at hand the rationale for monitoring is described in different paragraphs for each of the four enzymatic reactions available.

10.1.2.1 Total bacteria ('total activity')

At the end of the 19th century, the Heterotrophic Plate Count (HPC) was considered an index for the effectiveness of sand filtration for water treatment (Koch, 1883; 1893). HPCs, also known as psychrotrophic total aerobic plate counts (Holvoet *et al.*, 2014), are found abundantly in many environments, including soil, air, water and food. Reasoner & Geldreich (1985) showed that allowing heterotrophic bacteria from drinking water samples to grow for a longer time, at low substrate concentrations, results in even higher counts. Generally, HPCs in (drinking) water are considered as originating from the source water or from multiplication in the treatment facility or distribution system (Bartram *et al.*, 2003).

Recent evaluations of more than 290000 samples, collected between 2005 to 2014 in The Netherlands, France and Australia (Van Lieverloo *et al.*, 2014; 2016) show a statistically significant relationship between the frequency of finding (faecal) indicators and the mean HPC in drinking water samples, not just after repairs and other mains operations in the distribution system (Figure 10.1), but also in periodic samples (not shown). With most drinking water suppliers, there was an exponential increase in the frequency of samples containing (faecal) indicators with increasing mean HPCs, in some companies the frequencies declined at extreme HPCs (not shown).

This means that a high HPC is an index of the probability of finding (faecal) indicators and therefore the probability of incidents that may impair water quality and safety. As bacteria common to the drinking-water ecosystem are part of HPCs, the presence of HPCs cannot be linked exclusively to incidents that impair water quality and safety. However, sudden increases may indicate such incidents.

Quantifying HPCs takes three days after sampling and is far too slow to be applicable as a warning system for possible incidents. Detection of a sudden increase is considered non-compliant with drinking water standards (Anonymous, 2011). Online detection of such incidents by high frequency monitoring, e.g. the (enzymatic) activity of bacteria, is cause for further action, including fast (inline) sampling to assess the exact nature of the incident.

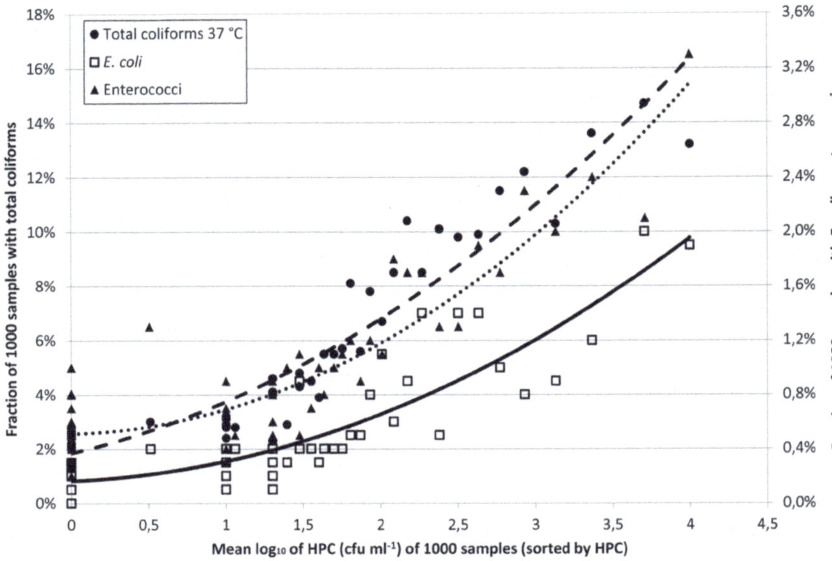

Figure 10.1 Example of correlations found world-wide between the frequency of indicator organisms and mean HPCs (data from one company). Regression of frequencies of detecting (faecal) indicators in groups of 1000 drinking samples collected after mains operations vs. mean HPCs in these samples (54440 in total). Lines (*E. coli*: uninterrupted line; enterococci: dotted line; total coliforms: dashed line) show statistically significant relations. Reprinted with permission from Van Lieverloo *et al.* (2016).

10.1.2.2 Total coliforms and (faecal) indicators

Online methods to monitor microbiological water quality target either specific indicator microorganisms, such as *Escherichia coli* (*E. coli*) and total coliforms, or measure total bacterial activity or concentration e.g. by ATP or direct cell counts. A large share of the available technologies are automated versions of the widely used 'Colilert' (Idexx) test kit, and measure *E. coli* and total coliforms by fluorescence/colour detection of enzymatic activity (Tatari *et al.*, 2016).

Online bacterial enzymatic monitoring can be used as a meaningful process variable that may improve and speed-up the detectability of incidents which impair microbiological water quality and safety. There is a lack of correlation between new methods like ATP, FCM and HPCs (Van Nevel *et al.*, 2016). The same can be seen for enzymatic monitoring, which was also shown in a short study by Al-Saffar *et al.* (2016). Research work from Cetaqua and microLAN showed that the BACTcontrol TA enzymatic method only detects living bacteria and shows a similar behaviour as intracellular ATP during a disinfection process (Appels *et al.*, 2017).

The challenge of implementing online enzymatic monitoring is to detect indicator organisms at the low concentration range relevant for drinking water. The detection level for total coliforms and *E. coli* with BACTcontrol has been shown to be 0.3 cfu/100 ml (former name 'Coliguard' in Lopez-Roldan *et al.*, 2013). But it is important to overcome the drawbacks mentioned in the introduction to make it possible that online methods can be implemented for continuous monitoring.

In a previous study Farnleiter *et al.* (2001) concluded that enzymatic monitoring (using β-glucuronidase) bears great potential for online monitoring of *E. coli* contamination. Servais *et al.* (2005) concluded that glucoronidase monitoring can serve as an alternative for plate counts and MPN methods for estimation faecal contamination in rivers. In two cases (raw water of a drinking water treatment plant and microbiological monitoring of a river) the enzymatic monitoring was able to detect pollution peaks. This demonstrates the potential importance of enzymatic monitoring as a tool to rapidly detect a potential deterioration of microbiological water quality allowing preventive actions in the treatment process to maintain the microbiological quality of the finished water. The cost in terms of consumables of the enzymatic monitoring is approximately the same than the cost of a culturable coliform enumeration by membrane filtration and is far less than a microplate MPN assay. The short reaction time (less than 1.5 hour compared to more than 18 hours) seems to be the main advantage of the enzymatic monitoring as it allows a fast and more or less "real time" warning in case of a pollution. Therefore, enzymatic monitoring seems to be a very useful and efficient tool for the monitoring and management of microbiological contamination.

10.2 EQUIPMENT DESIGN AND OPERATION

This section describes the design and operation of the BACTcontrol high-frequency, online monitoring system for bacterial enzymatic activity. The user can determine the filtrated sample volume for each measurement (from 10 to 1000 ml). During the validations of the method, a default volume of 100 ml was sampled (Farnleitner *et al.*, 2001; Ryzinska-Paier *et al.*, 2014). The complete BACTcontrol analyses (100 ml sample filtrations) are carried out within 1–1.5 hours, compared to plate counts which take about 24–48 hours. The system was available under another commercial name (Coliguard) until 2014 and has been improved considerably by microLAN through modifications as described in section 10.2.8.

10.2.1 Bacterial enzymes

BACTcontrol monitors the enzyme activities of β-glucuronidase (GLUC), β-galactosidase (GAL), β-glucosidase (GLUCAN) and alkaline phosphatase (ALP), enzymes of *E. coli*, coliforms, enterococci and total bacterial activity respectively. Alkaline phosphatase is a cell-surface bounded enzyme, abundantly present in nature and found in different human body tissues as well as animal and in many aquatic microorganisms.

Phosphatase enzyme hydrolyses phosphoric acid monoesters to produce phosphate and the corresponding alcohol (Millan, 2016). The importance of this enzyme in biological systems and its abundance in nature has made alkaline phosphatase activity assessments one of the most commonly performed enzymatic tests (Nalini *et al.*, 2015). More recently, alkaline phosphatase enzyme activity has been exploited as a rapid assay for the validation of milk product pasteurization without any cultivation step (Rankin *et al.*, 2010). In addition, alkaline phosphatase activity measurement has been applied for water quality assessment and as a sensitive indicator of organic pollutions in aquatic ecosystem in various studies (Bentzen *et al.*, 1992; Cotner & Wetzel, 1992; Awasthi, 2012; Tekaya *et al.*, 2013). One of the extensively employed assays to detect alkaline phosphatase activity is the fluorescence method. This method is based on measuring the fluorescence associated with the formation of 4-methylumbelliferone (MUF) from the fluorescent substrate 4-methylumbelliferyl-β-D-phosphate (MUP) upon binding with alkaline phosphatase enzyme (Yoshitomi, 2004).

10.2.2 Sampling the bacteria and their enzymes

A schematic overview of the BACTcontrol device is shown in Figure 10.2. The BACTcontrol device is patented under no. EP2165193 (Apparatus for monitoring water for microbial germs, microLAN BV). The main element of the device is the reactor, which consists of two chambers separated by a ceramic filter with a pore size of 0.45 μm. Here, the water sample is concentrated by the filter, the temperature and reaction are stabilized while the sample is constantly stirred by a magnetic stirrer and the enzymatic activities are measured (during the incubation period) by the fluorescence detector. Prior to each measurement, the water sample is pumped from the water source through the reactor chamber at flow rates from 1 to 24 ml per minute, the time needed for the filtering depends on the volume that has to be filtered and the condition of the filter. The sampled water volume is also measured by the pump during this process.

10.2.3 Enzymatic reaction

After the sample is pumped through the reactor and concentrated, the temperature is adjusted, and the reaction buffer is injected into the reactor. Different buffers are used for the detection of *E. coli*, coliforms, enterococci and total bacteria activity. They contain the different substrates that are hydrolysed to 4-methylumbelliferone (MUF)

- 4-methylumbelliferyl-β-D-glucuronide (MUG) by GLUC
- 4-methylumbelliferyl-β-D-galactopyranoside (MUGal) by GAL
- 4-methylumbelliferyl-β-D-glucopyranoside (MUGlu) by GLUCAN
- 4-methylumbelliferyl-β-D-phosphate (MUP) by ALP

Serial detection of *E. coli*, coliforms, enterococci and TA is possible.
MUF fluoresces after excitation via UV irradiation (λ_{ex} 360 nm; λ_{em} 450 nm).
A schematic view of the enzymatic reaction is shown in Figure 10.3.

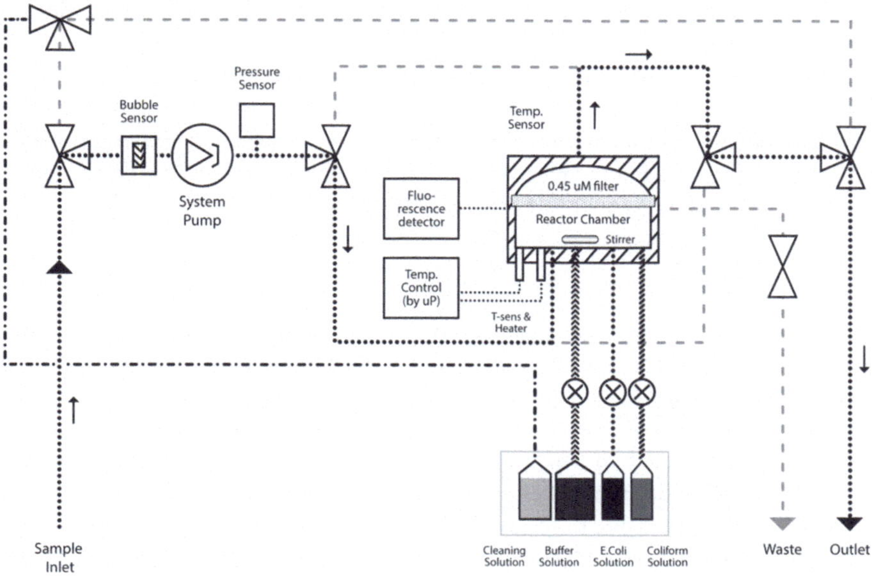

Figure 10.2 Schematic overview of the BACTcontrol system.

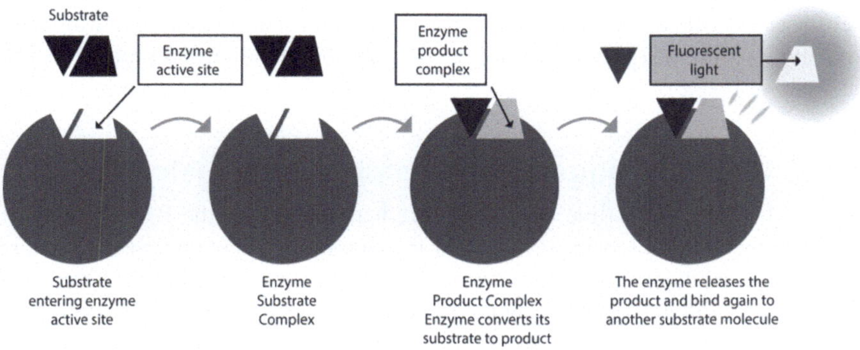

Figure 10.3 Schematic overview of the enzymatic reaction.

10.2.4 Quantifying enzymatic activity

After setting the temperature inside the reaction chamber to the optimum temperature (44 ± 0.1°C for GLUC, 36 ± 0.1°C for GAL, 37 ± 0.1°C for GLUCAN, 45 ± 0.1°C for ALP), the stabilization (with stirrer activated, 20 minutes period to stabilize the reaction mixture), followed by the actual measurement of the fluorescence intensity can take place during a 20 minute incubation period (during this period the stirrer is turned off).

The fluorescence intensity of the fluorometer has been calibrated using a standard with a concentration of 1000 nM MUF. This calibration allows the fluorometer to measure the production rate of MUF, which directly corresponds to the hydrolysis rate of the substrate, the fluorescence intensity is converted into MUF production per time and volume (pmol MUF × min × 100 ml^{-1}).

10.2.5 Data storage, processing and flow

The increase in fluorescence is automatically saved to the BACTcontrol computer and the slope of the signal in the steady state phase is used to calculate the enzymatic activity by ordinary least square linear regression analysis. Furthermore, the software calculates a limit of detection (DL) for each measurement performed. For this statistical approach, the measurement is regarded as significant if the average signal during the measurement exceeds threefold the standard deviation in relation to the theoretical zero line of the reaction. The DL calculation is determined after the stabilization period (only during 20 minutes of incubation), from where the slope of the regression curve is determined until the end of this phase.

10.2.6 Cleaning

After each measurement, a cleaning/disinfection procedure is performed by the device, which comprises the injection of a chlorine solution and a heating procedure within the reactor to eliminate residues of the measuring process within the system.

10.2.7 Speed: Frequency of sampling and analysis

Table 10.1 shows the outline of the measurement process of the BACTcontrol. It describes the different steps involved in the measurement including parameters such as time, volume, temperature and dosing of reagents. A total process depends on the sample volume and status of the filter and can vary between 1.5 to 15 hours.

10.2.8 Improvements since the last validation study

The BACTcontrol has been drastically improved in 2017 by microLAN BV. But the core patented technology regarding the concentration method using a reusable, ceramic filter and the enzyme detection has remained the same. Hardware improvements were mainly the following:

- The fluorescence detector is the heart of the system and the core technology of microLAN. This is also the backbone of the company and is used in other sensors for algae and toxicity detection.

- The electronics is now made in-house and controls the process by using a touchscreen PC and easy-to-use communication software.
- The addition of TA and enterococci detection offering maximum flexibility. This makes the instrument a platform for enzymatic detection of total and specific groups of bacteria. More developments for additional specific groups of bacteria can be expected in the near future.
- The integration into the microLAN product group using similar components for the instruments allowing easier production possibilities.

Table 10.1 Outline of the measurement steps.

Initiation	Filtration	Measurement	Cleaning
System filling	0.2–1 ml/sec.	Temp. 36–45°C 96.8–113°F	Temp. 65–70°C 149–158°F
Rinsing & cleaning	Volume depending	Heating of reactor	Heating of reactor
All pumping operations through filter in reverse direction	Filter condition depending	Buffer & reagent dosing & stabilisation. Stepwise measurement of fluorescent signal	Disinfection of system with chlorine solution

10.3 RESULTS OF VALIDATION STUDIES

Many validation studies of the applicability of (online) monitoring methods for water safety and quality focus on the comparison with the plate count methods of total coliforms and *E. coli*, as these are currently used as indicators of (faecal) contamination and the possible presence of pathogens. As argued earlier in this chapter, this standard indicator function suffers from a serious lack of representability and detectability. The applicability of online monitoring of bacterial enzymatic activity is mainly shown by case studies such as those published by Servais *et al.* (2005).

10.4 BARCELONA WATER WORKS CASE STUDY: TOTAL BACTERIAL ACTIVITY

To show the practical applicability of the BACTcontrol system, it was tested by Cetaqua in Barcelona (Spain) drinking water treatment plant (DWTP) Sant Joan Despí (SJD). The BACTcontrol system was operational in this treatment plant from March to August 2017.

10.4.1 Description of the treatment plant

Sant Joan Despí drinking water treatment plant (SJD DWTP) is one of the major suppliers of drinking water to Barcelona and its metropolitan area (roughly 3 million habitants), feeding the distribution system with 69200 m^3/year (data of 2015). The SJD DWTP water intake comes from two different types of water sources: surface water from the Llobregat River (Mediterranean-type river) and groundwater from extraction wells located across the low Llobregat aquifer system.

10.4.1.1 Surface water, the Llobregat River

The Llobregat River water has high mineral content, and a relatively high organic matter content as seen in Table 10.2. The surface raw water quality is highly affected by the sewage of wastewater treatment plants (WWTP) located along the river and also by drastic changes in water flow due to the Mediterranean climate (frequent, high-intensity droughts and floods). Accordingly, microbiological pollution, *Clostridium perfringens*, *E. coli* and *Enterococci* reach average concentrations of 3 log units. Therefore, an extensive disinfection treatment is required for the surface water.

Table 10.2 Raw quality data of the Llobregat River water, year 2015.

Parameter	*n*	Minimum	Average	Maximum
pH	7910	7.1	8.1	9.1
Conductivity (μS/cm)	8103	539	1385	2182
Temperature (°C)	8014	4.7	17.7	31.4
Turbidity (NTU)	8259	1.22	77	20000
TOC (mg C/l)	7583	<0.5	3.8	13
Ammonium (mg NH_4^+/l)	8028	<0.1	0.48	5.46
Sulphate (mg SO_4^{2-}/l)	52	115	160	226
HPC at 22°C (MPN/100 ml)	52	16000	60000	300000
Total coliforms (cfu/100 ml)	53	10000	100000	1700000
E. coli (cfu/100 ml)	52	700	4100	30000
C. perfringens (cfu/100 ml)	52	700	4100	30000
Enterococci (cfu/100 ml)	52	10	1600	15000

10.4.1.2 Groundwater

In terms of mineralization, groundwater characteristics are very similar to surface water, although it has a much more stable composition. There are two main contamination dangers to groundwater: chromium and halogenated solvents.

At the SJD DWTP, a combination of conventional and advanced treatment takes place. After a pre-treatment stage, including disinfection with ClO_2, coagulation-flocculation, sedimentation and sand filtration, water is divided into two streams that are treated with different technologies (conventional treatment and advanced membrane treatment). Conventional treatment consists of a first disinfection step with ozone followed by a Granular Activated Carbon (GAC) filtration and advanced membrane treatment includes a first step of ultrafiltration followed by a reverse osmosis. After these treatment schemes, a mix of both outlets is finally disinfected with chlorine (see Figure 10.4).

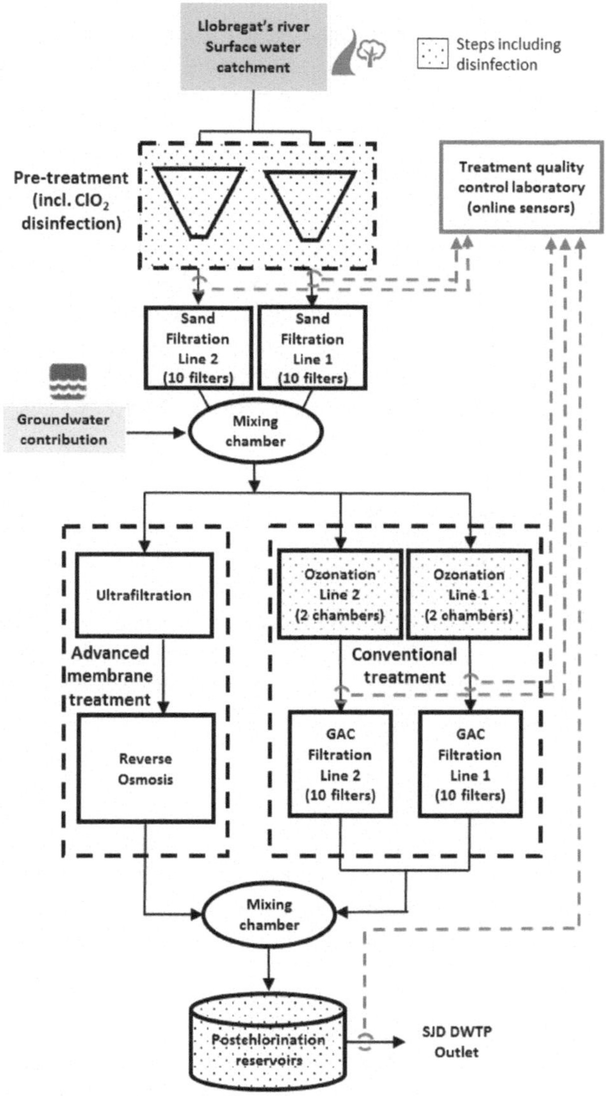

Figure 10.4 Sant Joan Despí drinking water treatment flow sheet. Treatment steps are indicated in boxes and those including disinfection agents have dotted background. Path of samples monitored by the BACTcontrol system is indicated in grey broken lines.

10.4.2 Monitoring points

BACTcontrol TA was tested in three water types: sand-filtered water, GAC-filtered water and finished water (after final chlorination). Water is pumped from the outlet of every process to the laboratory, used for routine analyses and for feeding online sensors installed in the laboratory building, thus reducing time for taking samples or maintenance. The BACTcontrol system was installed inside the laboratory and used this sample transportation system to test the three afore-mentioned matrices (indicated in grey broken lines in Figure 10.4).

10.4.2.1 Sand-filtered water

This water is pumped from the central channel that collects filtered water of each operating sand filter in Line 1. There are ten filters in total although one or two are usually out of service. Each filter is backwashed almost daily (after a filtrated volume set point), so there are between eight and ten backwash operations per day in Line 1. Sand filters receive water from pre-treatment steps, which include chlorine dioxide addition to carry out a first disinfection of surface water. Sand-filtered water TA was measured from June to August 2017.

10.4.2.2 GAC-filtered water

This water is pumped from the pipe that collects filtered water from the GAC-filters in Line 1. This line holds ten filters in total, although only seven were in service during the period of the study. Backwashes of each filter are performed after a filtrated volume set point. Each filter is backwashed with chlorinated water approximately every 3 days. GAC-filters inlet comes from ozonation chambers, where pre-treated surface water is oxidized and partially disinfected. GAC-filtered water TA was measured from April to June 2017.

10.4.2.3 Treated water

The DWTP outlet water, ready for distribution and consumption, consists on a mix of two streams, coming from the conventional and advanced treatment lines (see Figure 10.4), after mixing water is chlorinated in two tanks with approximately 1 hour of contact time. Treated water TA was measured during March 2017.

10.4.3 Quality events

Surface water is the primary source and groundwater is only used when river water quality parameters do not meet the quality standards. Non-compliance is typically caused by rainfall events that lead to increases in microbiological, organic and inorganic loads, mainly due to general runoff. The operators of the DWTP monitor a set of quality parameters to control the quality of surface water, turbidity and ammonium being the most important event indicators. When one of these parameters exceeds set limits, surface water intake, pre-treatment and sand filtration are suspended. To maintain the treated water supply, groundwater is pumped after the sand filtration process. When treatment is operated under this scheme, there is no injection of ozone at the conventional treatment (see Figure 10.4) to avoid the formation of disinfection by-products such as bromates.

10.4.4 Verification strategy and methods

On becoming available, online systems for microbiological monitoring are starting to be used in DWTPs, mainly those technologies that translate conventional laboratory analyses required by legislation into in-situ automatic detection. However, microbiological TA systems have not yet been widely explored for real environments and their applicability needs to be determined according to the monitored process.

BACTcontrol TA was verified in accordance with "verification" definition of the WSP Manual (Bartram *et al.*, 2009): "*The application of methods, procedures, tests and other evaluations, to determine compliance with the WSP, i.e. checking whether the system delivers water of the desired quality and whether the WSP is implemented in practice*"

The aim of the verification tasks was not only to check the robustness and autonomy of the online system but determine what information it can offer to DWTP operators to integrate with current process control and optimization procedures. To do so, the verification strategy included comparison of TA values with:

- *Microbiological parameters*: HPC at 22°C (Water ISO Plate Count Agar, 72h at 22°C, ISO 6222 (1999)), as it is a well known legislated parameter, Frequency: 2 laboratory analyses per week; ATP viability assay using reagent Bactiter-Glo™ (Promega, USA), as it is a molecular technique for bulk microbiological analysis, Frequency: 4 laboratory analyses per week.
- *Physicochemical parameters*: temperature, TOC, UV254, turbidity, particle count, conductivity and pH (continuous data from online analysers at SJD DWTP). These parameters are currently used for basic characterization of water sample by the SJD DWTP operators.
- *Operational parameters*: % of groundwater injected, ozone dose, sand and GAC filter backwash schedule. These parameters are known by the SJD DWTP operators to possibly have an impact in the quality of water.

10.4.5 Results

In this section, an overview of results is given for each studied process. As TA is an innovative parameter not commonly used in for treatment process monitoring, this study was focused on explaining baseline and peaks and assessing how known quality and operational events affect the response of an online microbiological monitor. Although the influence of several parameters was studied, TA variation could be mainly explained with operational actions such as the inclusion of groundwater or filters backwashes, the use of disinfectant agents, indicators of quality events such as rainfall and progressive temperature changes within weeks that led to changes in baseline values. ATP has been used as the main reference parameter in graphs and HPC was discarded as no correlation was found with TA and ATP, except for chlorinated matrices.

Figure 10.5 shows boxplots with TA measured in the effluents of subsequent processes in SJD DWTP and one point of the distribution network. It shows the median (solid line), the 25 and 75 percentiles (box) and outliers (crosses).

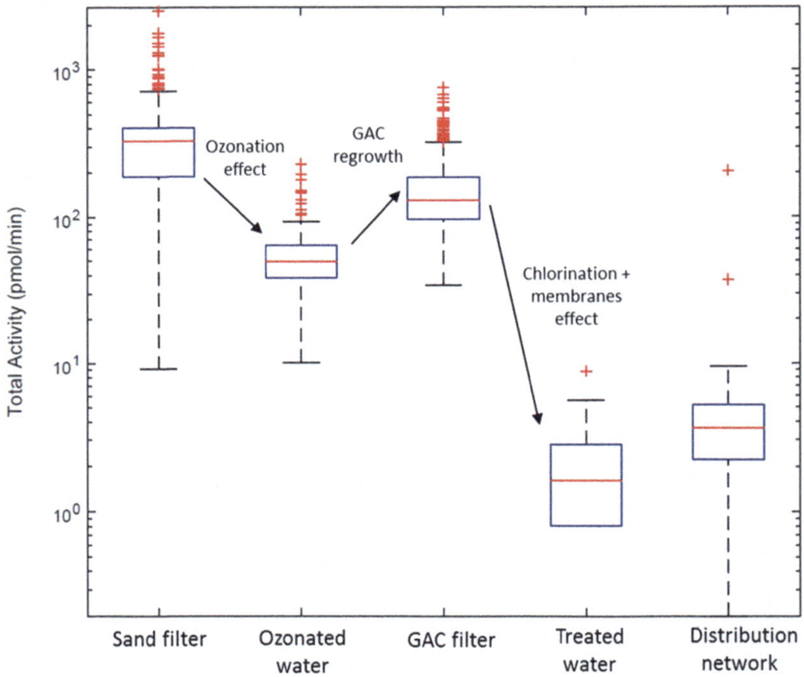

Figure 10.5 Total activity boxplot TA measured in the effluents of subsequent processes in SJD DWTP and one point of the distribution network.

This shows that TA method can be used that the system is very well applicable for assessing removal effects of bacteria. In this way the system is very well applicable for assessing removal effects throughout a treatment facility.

10.4.5.1 Sand filters

Water from ten open sand filters was analysed over more than two months (Figure 10.6). TA values ranged from 50 to 1200 pmol/min with a baseline varying between

100 and 400 pmol/min and several intense peaks increasing to a level of 200 pmol/min higher than the baseline.

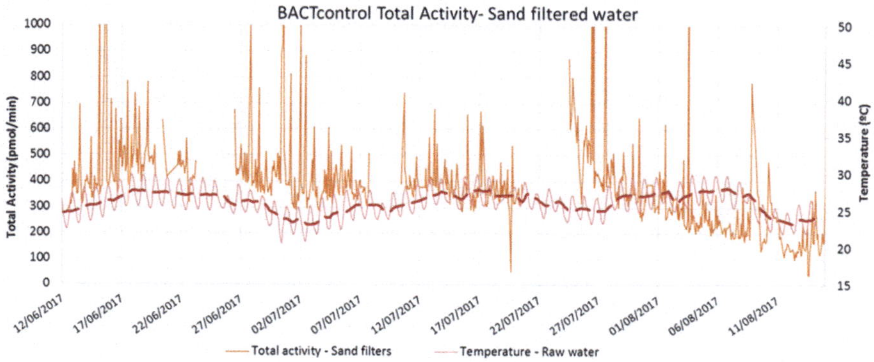

Figure 10.6 Sand filtration monitoring overview. Online total activity: solid line; Average temperature of raw water (surface water abstraction); hourly temperature: dotted line; average daily temperature: solid semi-discontinuous lines).

TA seemed to follow the trend in water temperature during the first half of the verification, thus suggesting variations in bacterial growth within sand beds. A relation was found between the baseline TA values (excluding the recurrent peaks) and daily average temperature. The 5 to 10°C fluctuation between day and night is, based on the SJD DWTP operators' experience, quite usual and significant in both raw surface water and sand-filtered water. At the end of July, some rainfall had an impact on surface water quality, leading microbiological parameters to increase and changing its behavioural pattern on subsequent days, being then independent of temperature and showing a constant decrease.

During the last days of the verification period in the sand filters, another rainfall event occurred, and the baseline again followed the temperature tendency. ATP concentration values ranged from 80 to 160 pg/ml in normal conditions and increased up to 550 pg/ml during a quality event, although its variability was not similar to the TA baseline.

Sand filters backwashes are thought to be the cause of short high peaks in activity, based on a report found in literature on/regarding an increase in parameters such as turbidity and particle count in sand filters during filter ripening after a filter backwash (Colton *et al.*, 1996; Logsdon *et al.*, 2002). Therefore, no clear correlation between TA values and backwashes was found, probably because the water is collected from ten sand filters that are backwashed at least once per day and BACTcontrol system only took in water during 10 minutes in every 90. Single-filter studies should reveal a better correlation. Two quality events had great impact (periods marked green in Figure 10.6). Details of the second event are shown in Figure 10.7. Intense rainfall caused increase of both organic and inorganic matter, turbidity and even ammonium in the river water due to wastewater system overflows. This event showed three stages: (1) river water polluted with wastewater overflows reaches the DWTP catchment, causing a short ammonium peak; (2) arrival of upstream rainwater carrying more particles and organic matter, causing long turbidity and ammonium peaks; (3) a second bout of rainfall occurred, causing a long episode of ammonium and a limited increase in turbidity. Both episodes resulted in an increase of BACTcontrol TA levels in the sand filter effluent.

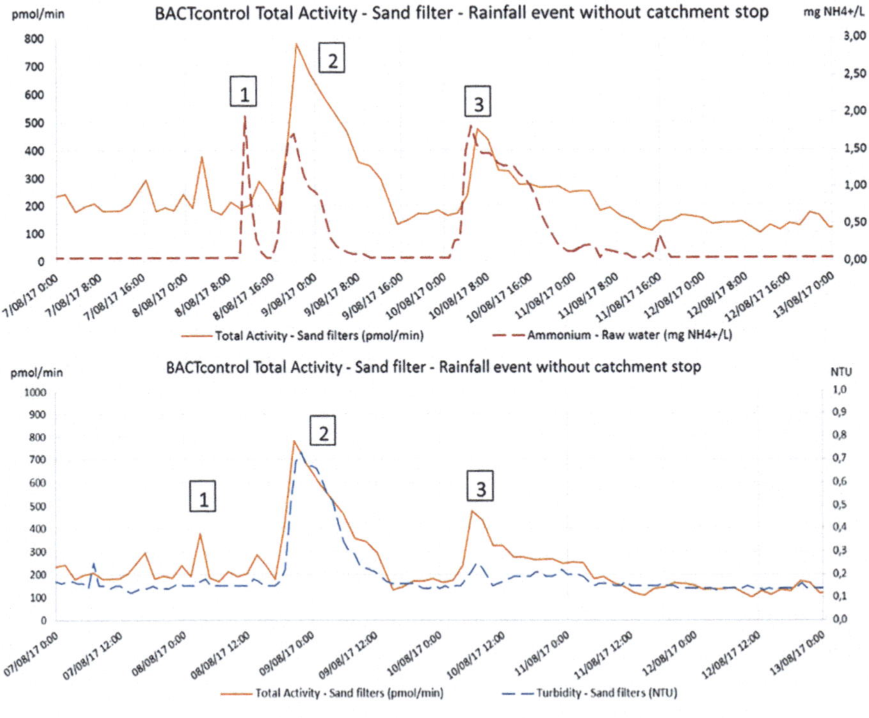

Figure 10.7 Total activity during diminished quality of river water. Upper graph: Total activity (solid line) in sand filters effluent and ammonium (NH_4^+) in raw water (broken line) (indication of wastewater overflow discharges to the river). Lower graph: Total activity (solid line) and turbidity in sand filters effluent (broken line).

BACTcontrol TA can detect episodes that compromise water quality in sand filter effluent, tracking the quality event even better than turbidity (the second turbidity peak is barely higher than the baseline). It is also well related to ammonium concentrations in river water but not simultaneous nor proportional as the water takes 1 to 2 h to reach the sand filters. However, relationship between TA and ammonium should not be expected in other monitoring sites, as ammonium is site-specific indicator for SJD DWTP surface intake pollution. BACTcontrol TA was shown to be as useful as current online analysers for quality event detection.

10.4.5.2 GAC filters

GAC filtration was also studied over two spring months (April and May 2017). In this case, water reaching the analyser came from a line of 10 GAC beds though only 7 were used at the same time. TA values ranged from 50 to 750 pmol/min, having a baseline varying between 100 and 200 pmol/min (see Figure 10.8). It can be seen that values during normal operation followed the same trend as temperature, which gradually increased over the testing period, as happened with sand filtration. Therefore, it strengthens the ability of BACTcontrol TA to monitor bacterial growth in bed-based filters.

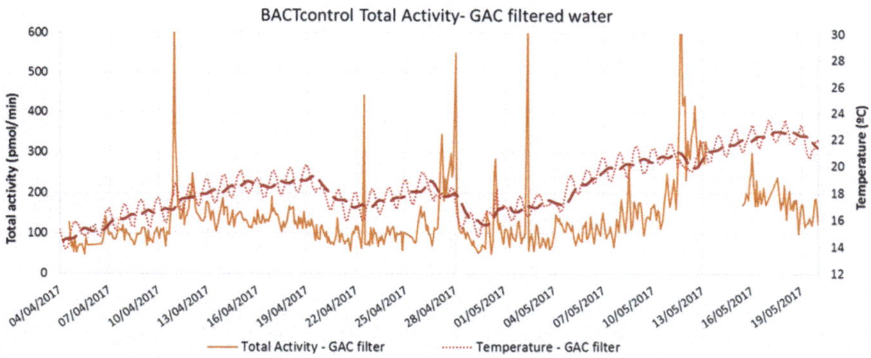

Figure 10.8 GAC filtration monitoring overview. Online total activity: solid line; Average temperature of raw water (surface water abstraction); hourly temperature: dotted line; average daily mean: solid semi-discontinuous lines).

Short-term variation of TA cannot be explained by quality parameters, but operational events such as filters backwashing and inlet of non-ozonated groundwater. Figure 10.9 shows the percentage of groundwater reaching the GAC filters, which is 0% in normal operation and 100% when surface catchment is stopped due to severe quality episodes. It is shown that operation with 100% of groundwater led TA to increase over the baseline, thus explaining most of high values obtained for this matrix. There is no ammonium presence in groundwater, thus a correlation between TA and ammonium concentration cannot be assumed.

Two other short peaks were obtained with 0% of groundwater and they are thought to be caused by single GAC filter backwashes, although, as it occurred in sand filtration, operational data did not allow a conclusive demonstration to be drawn.

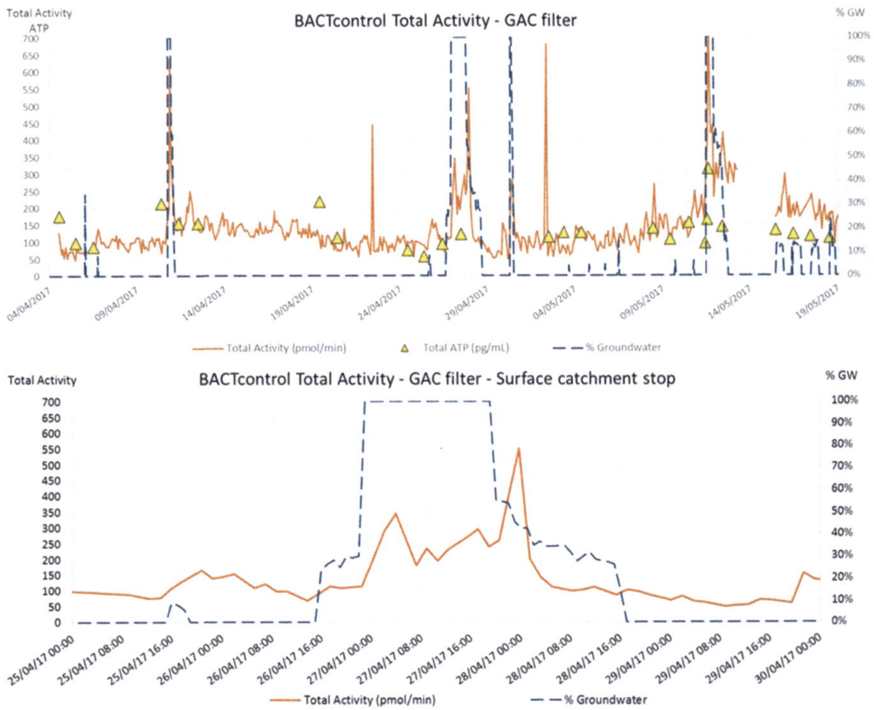

Figure 10.9 Effect of water source on total activity in the GAC effluent. The lower graph is an excerpt of the middle of the upper graph (period from 24 April to 30 April 2017). Total activity (solid line), % groundwater (broken line), ATP (triangles).

Finally, another observation can be extracted from this study. Some surface water catchment stops in SJD DWTP can last for more than a day, due to long rainfall episodes or maintenance tasks carried out before restarting normal operation. These stops involve water stagnation at pre-treatment leading to microbiological growth in sand beds.

This exceptional occurrence showed impact on TA in the case of the 26 April event (see the lower graph in Figure 10.9, an excerpt of the central part of the upper graph): values increased from 100 to 200–300 pmol/min during catchment stop and a higher peak was obtained when it was started up, i.e. when the groundwater proportion fell to 35%.

BACTcontrol TA can detect episodes of operational stops and switching procedures between water types with effect on the GAC filters. ATP again did not show similar effects but this might be the result of the low-frequency manual sampling procedures.

10.4.5.3 Treated water

BACTcontrol TA was also used for finished water bacterial monitoring, as shown in previous studies (Appels *et al.*, 2017). TA monitor only detects live bacteria and chlorine does not cause interference as it is drinking water chlorinated up to 1 ppm, historical values for HPC are between 0 and 5 CFU. Therefore, obtained TA values were as low as expected (Figure 10.10). TA and HPC values were practically zero due to chlorination, but TA was near detection limits. High values of total ATP were obtained for drinking water, probably due to extracellular ATP released from microbiological cells during disinfection.

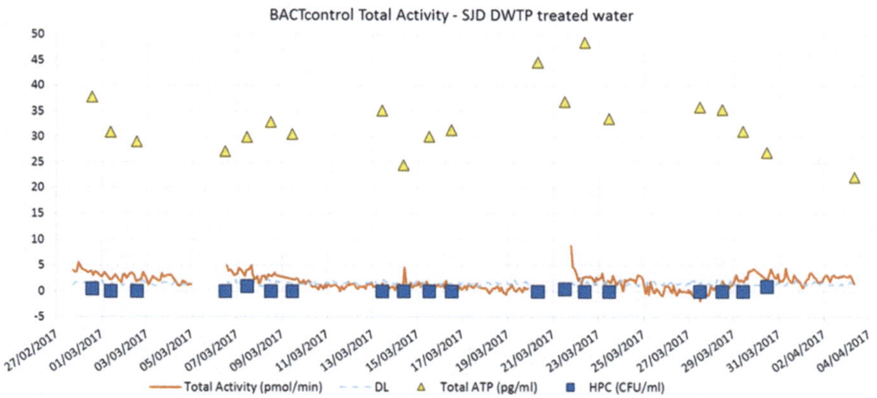

Figure 10.10 Monitoring of finished water. Total activity (solid line), DL (broken line), ATP (triangles) and HPC (squares).

BACTcontrol gathered data over a month and TA values were around a baseline of 0–5 pmol/min, proving the good performance of final disinfection in the plant and thus the low bacterial activity.

In Figure 10.11, a moderate quality event including an increase in organic matter and ammonium (see peak on 2nd of April for mixing chamber water before chlorination) was tracked from the river catchment to the chlorination step. There, ammonium and microorganisms avoiding upstream treatment barriers were oxidized with chlorine, leading to low values of TA at the outlet of the chlorination tanks.

Total ATP concentration was found to be between 20 and 50 pg/ml although other parameters were close to zero. However, to find out the cause, chlorinated samples were filtered before ATP analysis and it was seen all ATP content was extracellular, thus corresponding to recently inactivated bacteria. This made sense since treated water samples at the outlet of the plant tend to present chlorine contact times of 1–2 hours and, in contrast, distribution network samples, which have between one and several days of contact time showed ATP to be degraded.

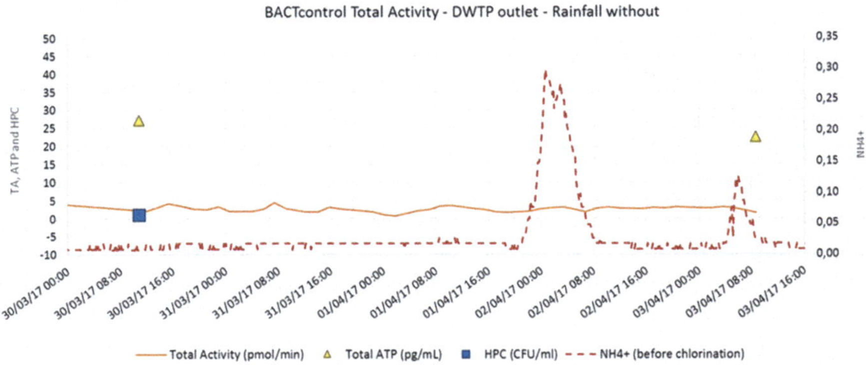

Figure 10.11 Impact of quality events on total activity in finished water. Total activity (solid line), ATP (triangles), HPC (squares) and NH4+ (broken line).

Total activity monitoring is not affected by chlorine and measures the enzyme activity only of living cells. Thus, this system can be used as an early warning monitor when upstream disinfection steps show failures and in this way, can contribute to the microbiological quality of the treated water and the safety of the users and customers of the water utility.

10.4.6 Total activity vs. other verification methods

Table 10.3 shows the matrix of Pearson's correlation coefficients (*r*) among all methods of total microbiological activity for GAC and sand filtration processes in SJD DWTP. Note that TA values correspond to online analyses and ATP and HPC were analysed in a laboratory close to the plant.

10.4.6.1 GAC filtration

Very low correlations were obtained. Different treatment scenarios affect GAC filters: 100% surface water treatment, 100% groundwater treatment and blends of both sources. This feature could explain the poor correlation between techniques when the whole data set is treated at once. Therefore, further statistical studies could be carried out to detect factors causing deviations between methods.

10.4.6.2 Sand filtration

This process is more stable than GAC filtration since its inlet flow is only surface water and it is stopped in case of a quality event. This seemed to lead to higher correlations, as can be seen for HPC in 3 and 7 days (more homogeneity in microbiota and thus in bacterial growth). However, TA is poorly correlated to any other method, which could be explained by two main differences: TA was obtained in situ, avoiding deviations caused by grab sampling, and it corresponds to an enzymatic procedure very different to culture methods and to ATP strategy. On the other hand, ATP values showed low dispersion and seemed to be somehow correlated to plate counts, taking into account that both techniques used the same grab samples in the laboratory.

Table 10.3 Correlation matrix for analysed parameters in GAC and SF treatment processes.

GAC Filter	Total Activity	Total ATP	HPC 3 Days	HPC 7 Days
n	36	64	30	30
Total Activity	1			
Total ATP	0.097	1		
HPC 3 days	−0.066	0.075	1	
HPC 7 days	−0.229	0.022	0.723	1
Sand Filter	**Total Activity**	**Total ATP**	**HPC 3 Days**	**HPC 7 Days**
n	41	46	22	22
Total Activity	1			
Total ATP	0.464	1		
HPC 3 days	0.676	0.951	1	
HPC 7 days	0.684	0.951	0.998	1

10.4.7 User feedback of BACTcontrol

During the period that Cetaqua and Aigües de Barcelona tested BACTcontrol, the online analyser showed a friendly interface to the user and was relatively easy to adapt at the different sampling lines. The system showed itself to be robust enough to correctly monitor the three tested matrices. The BACTcontrol detected some natural and operational events, such as rainfall episodes and source water switches from surface water to groundwater. The maintenance of the online monitor was acceptable with a required maintenance frequency of 3 times per month, 2 months between reagent replacement and 15 to 30 days between filter replacements. An improvement in data accessibility and extraction, as well as MODBUS communication, would improve the overall rating of the online device.

10.5 DISCUSSION

Most validation studies of new methods compare their performance with old methods. The focus of these studies commonly is the limit of the detection level for single samples. At low concentrations of microorganisms, there is a low correlation between enzymatic activity and plate count. However, the reproducibility of low plate counts in duplicate samples generally is low as well, which can be explained from the relatively high variability of low counts that are known to follow the Poisson distribution (where variance is equal to the mean).

Most importantly, however, the validity of using online sampling techniques in general, including enzymatic activity, should be compared to their value in guaranteeing water safety and quality. It is known that the collection of periodic samples is not good at detecting short-lived events of diminished safety and quality and usually are very late in detecting any event, including longer-lasting events (Van Lieverloo *et al.*, 2007a).

When comparing enzymatic activity monitoring to other online variables, such as particle counting and counting bacteria, the advantage of monitoring enzymatic activity is that it is able to distinguish active bacteria, allowing the effect of inactivation steps that do not remove the bacteria from the water to be evaluated (see Figure 10.5).

10.6 CONCLUSIONS

Online monitoring of enzymatic bacterial activity has been shown to be a valuable addition to the palette of microbiological variables available to water companies. Major advantages of the online TA are:

- High frequency of sampling (every 90 minutes).
- A high velocity of analysis (90 minutes).
- Ability to assess activity of all bacteria, including non-culturable fractions.
- A limit of detection that is close to that of heterotrophic plate counts.
- *Flexibility*: the system can be used for both important faecal indicators (e.g. *E. coli*, coliform and enterococci) as well as process parameters.
- *Added value*: this method can be used to create a fingerprint to show the log-removal of different disinfection steps throughout a treatment facility.
- *Costs*: depending on the number of samples taken per day, the cost of online TA can be lower than similar HPC samples.
- *Automation*: fully automated online TA monitoring offers unique perspectives with respect to increased security and protection of the public.

Conclusions of the use of the BACTcontrol at Aigües de Barcelona:

- TA module can detect episodes of diminished quality in sand filter effluent even better than turbidity or ammonium, the current indicators of river pollution.
- It can detect episodes of operational stops and switching procedures between water types with effect on the GAC filters.
- Its measurement is not affected by chlorine and measures the enzyme activity only of living and active cells.

- Overall it can be said that the BACTcontrol can be used as an early warning monitor when upstream disinfection steps show failures and in this way, can contribute to the microbiological quality of the treated water and the safety of the users and customers of the water utility.

As was the case after the introduction of verification methods in the late 19th and early 20th centuries, the value of new methods in increasing the detectability of events impairing water safety and quality can only be proven by widespread and long-term testing, alongside the methods, flawed as they may be, commonly used and accepted by regulators.

10.7 ACKNOWLEDGEMENTS

BACTcontrol TA was tested for the first time in a large drinking water treatment plant within the programme of the EU-granted project Aquavalens (FP7 grant number 311846, www.aquavalens.org).

10.8 REFERENCES

Al-Saffar Z., Appels J. and Nnezev A. (2016). Alkaline phosphatase enzyme activity measurement as a robust on-line monitoring assay for water quality assessment., Bachelor Biotechnology and Medical Research, Avans University of Applied Sciences, Breda, the Netherlands, 17 pp.

Anonymous (2011). Besluit van 23 mei 2011, houdende bepalingen inzake de productie en distributie van drinkwater en de organisatie van de openbare drinkwatervoorziening (Drinkwaterbesluit) (Decision of 23 May 2011, laying down rules on the production and distribution of drinking water and the organization of public drinking water (Drinking Water Act)). In Dutch. Staatsblad van het Koninkrijk der Nederlanden 293: 1–90.

Appels J., van den Dries J., van Strien C., Puigdomènech C., Baquero D., Pla A., Juárez R., Saucedo G., Galofré B. and Arnedo M. J. (2017). Estimation of dead or live bacteria using online bacteria activity monitoring. *Poster presented at "How dead is dead? Exploring the edge of bacterial life" Conference*, Vienna, Austria, 6–8 September 2017.

Awasthi M. (2012). Relevance of alkaline phosphatase activity of immobilized green algae and cyanobacteria for heavy metal toxicity monitoring. *Journal of Materials and Environmental Science*, **3**, 446–451.

Bartram J., Cotruvo J., Exner M., Fricker C. and Glasmacher A. (2003). Heterotrophic Plate Counts and Drinking-water Safety. The Significance of HPCs for Water Quality and Human Health. World Health Organization, IWA Publishing, London, United Kingdom, p. 256.

Bartram J., Corrales L., Davison A., Deere D., Drury D., Gordon B., Howard G., Rinehold A. and Stevens M. (2009). Water Safety Plan Manual: Step-By-Step Risk Management for Drinking-Water Suppliers. World Health Organization, Geneva, Switzerland.

Bentzen E., Taylor W. D. and Millard E. S. (1992). The importance of dissolved organic phosphorus to phosphorus uptake by limnetic plankton. *Limnology and Oceanography*, **37**, 217–231.

Colton J. F., Hillis P. and Fitzpatrick C. S. B. (1996). Filter backwash and start-up strategies for enhanced particulate removal. *Water Research*, **30**(10), 2502–2507.

Cotner J. B. and Wetzel R. G. (1992). Uptake of dissolved inorganic and organic phosphorus compounds by phytoplankton and bacterioplankton. *Limnology and Oceanography*, **37**, 232–243.

Farnleitner A. H., Hocke L., Beiwl C., Kavka G. G., Zechmeister T., Kirschner A. k. T. and Mach R. L. (2001). Rapid enzymatic detection of *Escherichia coli* contamination in polluted river water. *Letters in Applied Microbiology*, **33**, 246–250.

Holvoet K., Sampers I., Seynnaeve M. and Uyttendaele M. (2014). Relationships among hygiene indicators and enteric pathogens in irrigation water, soil and lettuce and the impact of climatic conditions on contamination in the lettuce primary production. *International Journal of Food Microbiology*, **171**, 21–31.

ISO 6222:1999. Water quality - enumeration of culturable micro-organisms. Colony count by inoculation in a nutrient agar culture medium.

Koch R. (1883). Über die neuen Untersuchungsmethoden zum Nachweis der Mikrokosmen in Boden, Luft und Wasser (About the new investigative methods to determine micro-organisms in soil, air and water). Presentation at the XI. German physician day in Berlin on 23 June 1883. From: *Ärztliches Vereinsblatt für Deutschland*, 1883, Nr. 137. Kommissions-Verlag von F. C. W. Vogel, Leipzig. In: J. Schwalbe (1912), *Collected Works of Robert Koch*. Vol. 1, (in German) Verlag von Georg Thieme, Leipzig, Germany, p. 274–284.

Koch R. (1893). Wasserfiltration und Cholera (Water filtration and cholera). Presentation at the XI. German physician day in Berlin on 23 June 1883. *Zeitschrift für Hygiene und Infektionskrankheiten*, 1893, Bd, XIV. In: J. Schwalbe (1912) *Collected Works of Robert Koch*. Vol, 2.1, (in German). Verlag von Georg Thieme, Leipzig, Germany, p. 183–206.

Logsdon G. S., Chipps M. J., Hess A. F. and Rachwal A. J. (2002). Filter Maintenance and Operations Guidance Manual. American Water Works Association. Lslf edition (October 1, 2002), ISBN 1–58321–234–5. http://www.waterrf.org/PublicReportLibrary/RFR90908_2002_2511.pdf (accessed 4 May 2018).

Lopez-Roldan R., Tusell P., Courtois S. and Cortina J. L. (2013). On-line bacteriological detection in water. *TrAC Trends in Analytical Chemistry*, **44**, 46–57.

Medema G., Smeets P., Blokker M. and Van Lieverloo H. (2014). Safe distribution without a disinfectant residual. Chapter 5 In: Microbial Growth in Drinking-Water Supplies. D. Van der Kooij and P. W. J. J. Van der Wielen (eds), IWA Publishing, London, UK, pp. 95–125.

Millan J. L. (2006). Alkaline phosphatases, structure, substrate specificity and functional relatedness to other members of a large super family of enzymes. *Purinergic Signalling*, **2**, 335–341.

Nalini P, Ellaiah P., Prabhakar T. and Girijasankar G. (2015). Microbial alkaline phosphatases in bioprocessing. *International Journal of Current Microbiology and Applied Sciences*, **4**, 384–396.

Parker J. and Darby J. (1995). Particle-associated coliform in secondary effluents: shielding from ultraviolet light disinfection. *Water Environment Research*, **67**, 1065–1075.

Rankin S. A., Christiansen A., Lee W., Banavara D. S. and Lopez-Hernandez A. (2010). Invited review: the application of alkaline phosphatase assays for the validation of milk product pasteurization. *Journal of Dairy Science*, **93**, 5538–5551.

Reasoner D. J. and Geldreich E. E. (1985). A new medium for the enumeration and subculture of bacteria from potable water. *Applied and Environmental Microbiology*, **49**, 1–7.

Ryzinska-Paier G., Lendenfeld T., Correa K., Stadler P., Blaschke A. P., Mach R. L., Stadler H., Kirschner A. K. and Farnleitner A. H. (2014). A sensitive and robust method for automated online monitoring of enzymatic activities in water and water resources. *Water Science & Technology*, **69**, 1349–1338.

Servais P., Garcia-Armisen T., LePeuple A. S. and LeBaron P. (2005). An early warning system to detect faecal contamination of river waters. *Annals of Microbiology*, **55**, 151–156.

Tatari K., Corfitzen C. B., Albrechtsen H.-J. and Christensen S. C. B. (2016). Sensors for microbial drinking water quality. Report. DTU Environment, Technical University of Denmark, pp. 43.

Tekaya N., Saiapina O., Ben Ouada H., Lagarde F., Ben Ouada H. and Jaffrezic-Renault N. (2013). Ultra-sensitive conductometric detection of heavy metals based on inhibition of alkaline phosphatase activity from *Arthrospira platensis*. *Bioelectrochemistry*, **90**, 24–29.

Van Lieverloo J. H. M., Mesman G. A. M., Bakker G. L., Baggelaar P. K., Hamed A. and Medema G. J. (2007a). Probability of detecting and quantifying faecal contamination of drinking water by periodically sampling for *E. coli*; a simulation model study. *Water Research*, **41**, 4299–4308.

Van Lieverloo J. H. M., Blokker E. J. M. and Medema G. J. (2007b). Quantitative microbial risk assessment of distributed drinking-water using faecal indicator incidence and concentrations. *Journal of Water and Health*, **5**(Supplement 1), 131–149.

Van Lieverloo J. H. M., Donocik A., Schaap P. and Hoogenboezem W. (2014). Heterotrophic Plate Counts as an index of the probability of contamination of drinking water in distribution systems. *Oral presentation on 25 September 2014 at the IWA World Water Congress*, Lisbon, Portugal, 21–26 September 2014.

Van Lieverloo J. H. M., Donocik A., Sheehan D., Prevos P., Schaap P., Bakker G. L., Enault J., Nijdam Y., Lut M. and Hoogenboezem W. (2016). Internationally observed correlation between frequencies of (faecal) indicators and mean heterotrophic plate counts in drinking water. Poster presented at the IWA World Water Congress, Brisbane, Australia, October 2016.

Van Nevel S., Buysschaert B., De Gusseme B. and Boon N. (2016). Flow cytometric examination of bacterial growth in a local drinking water network. *Water and Environment Journal*, **30**, 167–176.

Yoshitomi K. (2004). Alkaline phosphatase activity in cheeses measured by fluorometry. *International Journal of Food Science and Technology*, **39**, 349–353.

Zhao F., Bi X., Hao Y. and Liao X. (2013). Induction of viable but nonculturable *Escherichia coli* O157:H7 by high pressure CO_2 and its characteristics. *PLOS One*, **8**, 1–9.

Chapter 11

Mean oxidation state of organic carbon: A novel application to evaluate the extent of oxidation of natural organic matter in drinking water biological treatment

Bofu Li, Amina K. Stoddart and Graham A. Gagnon

11.1 INTRODUCTION

Understanding of the oxidation of organic matter by chemical and biological reactions is crucial in water treatment. Much of the literature published on drinking water utilizes total organic carbon (TOC), dissolved organic carbon (DOC) or related parameters to quantify natural organic matter (NOM). In practice, TOC and DOC are used as key indicators of bulk NOM concentration. However, measurement of NOM by TOC/DOC overlooks a key feature of biological stability: organic carbon is an electron donor (Rittmann & Huck, 1989). For example, the same TOC concentration of formic acid (HCOOH) and methanol (CH_3OH) would not represent the same amount of electrons because the oxidation state of carbon in methanol is –2 while that in formic acid is +2 (Rittmann & Huck, 1989). Theoretically, a highly reduced organic carbon would contribute more electrons than a relatively oxidized organic carbon. Therefore, TOC and DOC cannot always provide sufficient information on the transformation of NOM from various drinking water treatment processes, due to the incomplete oxidation of organic carbon. Incomplete oxidation widely exists in biochemical reactions. In incomplete oxidation, NOM cannot be completely converted to the final oxidation product (e.g., carbon dioxide) (Rieger

doi: 10.2166/9781780408699_0197

et al., 1983). The intermediates of incomplete oxidation would be quantified as TOC/DOC. This disadvantage limits further understanding about the transformation of NOM, especially during biological treatment of drinking water (e.g., biofiltration).

In both complete oxidation and incomplete oxidation, the formation of bonds between oxygen and carbon and the deformation of bonds between hydrogen and carbon increase the oxidation state of organic carbon (Kroll et al., 2011). Combining chemical oxygen demand (COD) with TOC/DOC measurement can provide valuable information on the oxidation state of organic carbon during drinking water treatment. However, the conventional dichromate COD method is not sensitive enough to measure COD in surface water (Rittman & Huck, 1989; Stoddart & Gagnon, 2014). Advancements in sensor development for the determination of COD in water (Zhang *et al.*, 2004) have allowed researchers to rapidly quantify COD during drinking water treatment using photoelectrochemical chemical oxygen demand (peCOD) (Stoddart & Gagnon, 2014). When applied in a full-scale biofiltration drinking water treatment plant, peCOD removal was greater than TOC/DOC removal (Stoddart & Gagnon, 2014), which indicated peCOD might be more sensitive than TOC/DOC analysis for understanding biological treatment performance. Even though peCOD indicated a promising application in drinking water treatment performance monitoring, the lack of understanding of the relationship between peCOD and TOC/DOC in drinking water limits its application. This is due in part to many years of TOC/DOC data which supports our understanding of NOM in drinking water. There is a need to construct a "bridge" to connect the conventional TOC/DOC evaluation system and novel approaches such as the peCOD evaluation system.

Therefore, a concept of mean oxidation state (MOS) of organic carbon ($\overline{\text{Cos}}$), combining TOC/DOC and peCOD, is introduced in this study. The major advantages of this method are: (i) $\overline{\text{Cos}}$ could provide more information about the transformation of NOM than TOC/DOC, especially when incomplete oxidation dominates the biochemical reactions of treatment; (ii) theoretically, $\overline{\text{Cos}}$ only responds to oxidation-reduction reactions, which means that physical removal (e.g., filtration, precipitation and adsorption) does not change the value of $\overline{\text{Cos}}$; (iii) $\overline{\text{Cos}}$ is a dimensionless number that is not affected by the concentration fluctuation of NOM and only represents the oxidation potentials of organic carbon. In addition, it may be possible to correlate $\overline{\text{Cos}}$ with biomass concentration, as determined by measurements such as adenosine triphosphate (ATP), to further understand NOM removal mechanisms during biological drinking water treatment processes, such as biofiltration.

11.2 BACKGROUND

11.2.1 Quantifying natural organic matter by Photoelectrochemical Chemical Oxygen Demand (peCOD)

Photoelectrochemical chemical oxygen demand (peCOD) is a rapid and environmentally friendly COD measurement method that holds promise for the drinking water industry (Stoddart & Gagnon, 2014). The determination of peCOD involves the mineralization of organic matter by titanium dioxide (TiO_2) under ultraviolet (UV) irradiation.

Photo-generated holes, formed by the illumination of TiO_2 with photons, have a powerful oxidation capacity for almost all organic matter in water (Qiu *et al.*, 2012). However, oxygen concentration (e.g., less than 10 ppm at 25°C, 1 atm) could inhibit the reaction rate of mineralization process (Equation (1)) in a conventional TiO_2 photocatalysis system.

$$C_xH_yO_zX_q + \left(x + \frac{y - 2z - q}{4}\right)O_2 \rightarrow xCO_2 + xH^+ + qX^- + \left(\frac{y - x}{2}\right)H_2O \quad (1)$$

The peCOD method utilizes immobilized nanostructured TiO_2 and an electric field to improve the oxidation efficiency. In a peCOD system, the nanostructured TiO_2 is immobilized on conducting materials which functions as the working electrode. A force generated by the electric field separates the photoelectron from TiO_2 particles (Zhang *et al.*, 2004). Therefore, the reduction reactions on the auxiliary electrode replace the function of adsorbed O_2 (Qiu *et al.*, 2012). The generated electron can be quantified and converted to COD concentration by Faraday's law.

11.2.2 Mean Oxidation State (MOS) of organic carbon (Cos)

During the biological oxidation of NOM in drinking water treatment, not all organic carbon is converted to carbon dioxide (CO_2) or other inorganic carbon (e.g., CO_3^{2-} and HCO_3^{1-}). Accordingly, TOC/DOC measurements cannot quantify intermediate compounds that are formed during oxidation. Previous studies indicated that $\overline{\text{Cos}}$ could be used to investigate the oxidation potential of carbon in advanced oxidation processes (AOPs) (Vogel *et al.*, 2000; Mantzavinos *et al.*, 1996). The following assumptions were used to derive the concept of $\overline{\text{Cos}}$ in drinking water:

(i) NOM is the most abundant organic compound in treated drinking water.
(ii) The value of peCOD can represent the amount of NOM.
(iii) In treated drinking water, the number of organically bound heteroatoms (i.e., S, P, and Cl) is negligible compared to C, H, O, and N. In other words, Cos is determined mostly by the ratios among C, H, O, and N.
(iv) H has an oxidation state of +1, O has an oxidation state of –2, and N has an oxidation state of –3.

Zhao *et al.* (2004) described the stoichiometric photoelectrochemical oxidation of organic compounds by Equation (2).

$$\begin{aligned} &C_yH_mO_jN_kX_q + (2y - j)H_2O \\ &\rightarrow yCO_2 + qX^- + kNH_3 + (4y - 2j + m - 3k)H^+ + (4y - 2j + m - 3k - q)e^- \end{aligned} \quad (2)$$

where *X* represents a halogen atom; the numbers of C, H, O, N and halogen atoms are represented by *y*, *m*, *j*, *k* and *q*.

Therefore, peCOD is expressed by Equation (3). $\overline{\text{Cos}}$ can be expressed by TOC (or DOC) and total peCOD (or dissolved peCOD) as shown in Equation (4) (Mantzavinos *et al.*, 1996).

$$\text{peCOD (mol O}_2\text{)} = \frac{(4y - 2j + m - 3k - q)}{4} \tag{3}$$

$$\overline{\text{Cos}} = \frac{\sum_1^i n_i \text{OSC}_i}{\sum_1^i n_i} = \frac{2j - m + 3k + q}{y} = \frac{4(\text{TOC} - \text{total peCOD})}{\text{TOC}} \tag{4}$$

where n_i is the molar concentration; OSC_i is the oxidation state of organic carbon for individual species *i*; TOC is in mol C/L; total peCOD is in mol O_2/L.

11.2.3 Biomass Adenosine Triphosphate (ATP)

Adenosine triphosphate (ATP) carries the energy in cellular metabolism and is the primary compound for most biochemical enzymatic reactions (Tifft & Spiegel, 1976). In general, ATP is regarded as a key indicator of biochemical reactions and can be used to indicate the presence and abundance of active/viable biomass (Berney *et al.*, 2008). ATP concentration is quantified by the light produced from luciferin/luciferase enzyme with Mn^{2+} and O_2. The reactions between ATP, adenosine 5′-monophosphate (AMP) and the luciferin/luciferase enzyme are shown in Equations (5) and (6) (Marques & Esteves da Silva, 2009).

$$\begin{aligned}&\text{Luciferin} + \text{Luciferase} + \text{ATP} - \text{Mg}^{2+}\\ &\quad \leftrightarrow \text{Luciferase} \cdot \text{Luciferin} - \text{AMP} + \text{Pyrophosphate} - \text{Mg}^{2+}\end{aligned} \tag{5}$$

$$\begin{aligned}&\text{Luciferase} \cdot \text{Luciferin} - \text{AMP} + \text{O}_2\\ &\quad \rightarrow \text{Luciferase} + \text{AMP} + \text{CO}_2 + \text{Oxyluciferin} + \text{Photon}\end{aligned} \tag{6}$$

Reaction with luciferase enzyme and quantification of the photon emitted in relative light units (RLU) using a luminometer enable ATP measurements using rapid test procedures (i.e., approximately 5 min per sample). Commercial ATP test kits with favourable precision, accuracy, span, representativeness and selectivity/specificity (Evans *et al.*, 2013b) are now available for the water industry. The ability to rapidly measure ATP has notable applications in the water industry where conventional biomass quantification techniques require hours – if not days – to complete (i.e., heterotrophic plate counts [HPC]).

Many studies have found correlations between ATP and conventional biomass analysis methods in drinking water treatment. Delahaye *et al.* (2003) found that ATP concentration was correlated with HPC by investigating a drinking water distribution system in Paris. Dowdell (2012) reported a correlation coefficient of

0.91 between biomass ATP and phospholipid content in drinking water biofilters. Furthermore, ATP measurements on granular activated carbon (GAC) have been shown to correlate with total direct bacterial cell counts (Magic-Knezev & van der Kooij, 2004).

Currently, there is a specific need for monitoring tools for drinking water biofiltration that can provide insight on biofilter biomass to improve filter operation and ensure process optimization (Evans *et al.*, 2013a). ATP measurements have been used to assess when GAC (Velten *et al.*, 2011) and anthracite-sand (Stoddart & Gagnon, 2015; Stoddart *et al.*, 2016) filters have become biologically active. An evaluation of several full-scale drinking water biofilters conducted by Evans *et al.* (2013a) found that order of magnitude changes in ATP concentration over time is considered significant. For example, Stoddart *et al.* (2016) converted a full-scale conventional filter to biofiltration and demonstrated that biomass ATP had an initial concentration of 60 ng ATP/cm^3 media and matured to 270 ng ATP/cm^3 media once steady-state was achieved (approximately 220 days after conversion to biofiltration).

11.3 MATERIALS AND METHODS

11.3.1 Full-scale drinking water biofilters

The investigation of full-scale drinking water biofilters was conducted in J. D. Kline Water Supply Plant, Halifax, Nova Scotia, Canada. J. D. Kline Water Supply Plant is a direct filtration water treatment plant with a designed capacity of 227 ML/day (Figure 11.1). The source water comes from Pockwock Lake, which is characterized as low-turbidity (~0.4 NTU), low-alkalinity (<1mg/L), low-pH (~5.5), and low-organic carbon (~2.5 mg DOC/L; ~2.5 mg TOC/L) (Stoddart *et al.*, 2016; Stoddart & Gagnon, 2015; Vadasarukkai *et al.*, 2011).

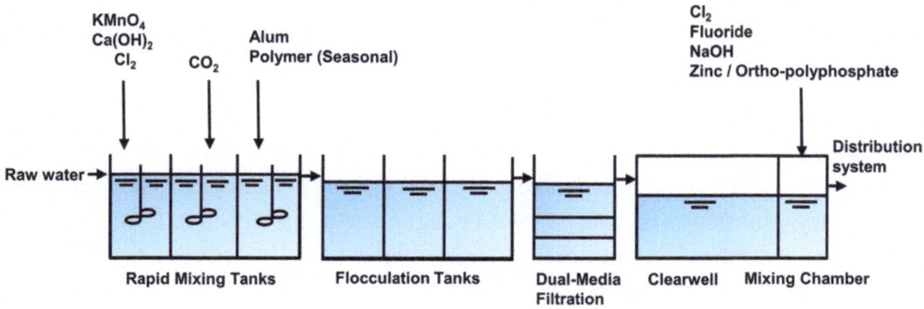

Figure 11.1 Schematic of J.D. Kline Water Supply Plant, Halifax, Nova Scotia, Canada.

In the J. D. Kline Water Supply Plant, the pre-mix process involves the addition of calcium hydroxide and potassium permanganate for the oxidation of metal, pH adjustment by carbon dioxide, and coagulation with aluminium sulfate. Flocculation is conducted using three-stage hydraulic mixing. Eight dual media anthracite-sand filters (600 cm of anthracite, 300 cm of sand) are used as biofilters. Detailed information concerning the biofiltration performance for this facility is described elsewhere (Stoddart *et al.*, 2016; Stoddart & Gagnon, 2015).

Two biofilters (biofilter A and biofilter B) were operated in parallel for the duration of the study. Biofilter influent water (common to both biofilters), biofilter A effluent, biofilter B effluent and filter media from the tops (0–10 cm) of both filters were sampled at various points during a 53 h filter cycle.

11.3.2 Analytical methods

11.3.2.1 Total organic carbon (TOC)/dissolved organic carbon (DOC)

TOC and DOC were monitored on influent and effluent of biofilters. DOC samples were filtered through preconditioned 0.45 μm polyethersulfone filter membranes and stored in headspace-free 40 mL vials, preserved at 4°C to pH < 2 by phosphoric acid addition. TOC and DOC samples were analyzed with a TOC-V CPH analyzer (Shimadzu Corp, Kyoto, Japan).

11.3.2.2 Photoelectronchemical chemical oxygen demand (peCOD)

Total peCOD and dissolved peCOD (0.45 μm polyethersulfone filter membrane) were monitored on influent and effluent of biofilters. All samples were preserved at 4°C to pH < 2 by addition of sulfuric acid. Samples were measured using a PeCOD® L100 AssayPlus™ analyser (Mantech Inc., Guelph, Canada) after neutralizing to pH 7.0 with sodium hydroxide. The water samples were analysed as described by Stoddart and Gagnon (2014).

11.3.2.3 Adenosine triphosphate (ATP) of biomass

Filter media samples from two paired biofilters were taken from the top of each filter bed using a sampling pole fitted with a wide mouth polypropylene bottle. Biomass ATP was measured immediately using 1 g (wet weight) subsamples. The biological activity of biomass was measured by ATP using surface analysis test kits (Deposit Surface Analysis test kit, LuminUltra Technologies Ltd., Fredericton, Canada) and a luminometer (PhotonMaster™ Luminometer, LuminUltra Technologies Ltd., Fredericton, Canada). In short, this method involved biomass extraction from filter media using chemical removal, followed by a dilution of the extracted biofilm suspension. Luciferin/luciferase enzyme was used to react with the ATP of diluted biofilm suspension to emit light. The biomass ATP concentration was quantified by the intensity of light using a luminometer.

11.3.2.4 Data analysis

The growth phase of biomass was analysed as described by Baranyi & Roberts (1994) and Stoddart *et al.* (2016). Statistical summary and local polynomial regression were performed using the R programming language (version 3.3.3).

11.4 RESULTS AND DISCUSSION

11.4.1 Mean oxidation state of organic carbon before/after biofiltration

The evolution of $\overline{\mathrm{Cos}}$ was investigated in two paired full-scale drinking water biofilters. After biofiltration, average TOC decreased from 3.2 mg/L (biofilter influent) to 2.2 mg/L (effluent of biofilter A) and 2.2 mg/L (effluent of biofilter B) (Table 11.1). Average TOC concentrations of both biofilter effluents were almost the same as the biofilter effluent DOC of 2.1 mg/L (biofilter A) and 2.1 mg/L (biofilter B). No significant decrease in DOC occurred from biofiltration. This indicated that primarily particulate TOC (e.g., the floc formed from coagulation and flocculation) had been removed by filtration and that DOC represented the vast majority of residual organic carbon in the biofilter effluent.

Table 11.1 Summary of water qualities in influent and effluents of biofilters.

	Avg. TOC (mg/L)	Avg. DOC (mg/L)	Avg. Total peCOD (mg/L)	Avg. Dissolved peCOD (mg/L)
Influent ($n = 7$)	3.2 ± 0.1	2.2 ± 0.1	8.3 ± 0.6	7.3 ± 0.5
Biofilter A ($n = 11$)	2.2 ± 0.2	2.1 ± 0.1	5.1 ± 1.0	4.4 ± 0.6
Biofilter B ($n = 11$)	2.2 ± 0.2	2.1 ± 0.1	5.1 ± 0.9	4.4 ± 0.8

Significant removal of peCOD was observed after biofiltration; total peCOD and dissolved peCOD decreased 39% and 40%, respectively. Total and dissolved $\overline{\mathrm{Cos}}$ increased after biofiltration (Figure 11.2). Total $\overline{\mathrm{Cos}}$ increased from 0.12 to 0.46 (biofilter A) and 0.49 (biofilter B). Dissolved $\overline{\mathrm{Cos}}$ increased from −1.2 to 0.79 (biofilter A) and 0.83 (biofilter B). The increased dissolved $\overline{\mathrm{Cos}}$ and relatively consistent DOC before/after biofiltration suggests that the majority of dissolved organic carbon underwent incomplete oxidation and still existed in water as DOC, while a small amount of DOC (0.08 to 0.09 mg/L) was completely oxidized to inorganic carbon. The difference of total $\overline{\mathrm{Cos}}$ before/after biofiltration was lower than that of dissolved $\overline{\mathrm{Cos}}$ which may be because the particle TOC cannot be completely oxidized by the peCOD system without digestion.

The increase in $\overline{\mathrm{Cos}}$ indicated the incomplete oxidation of NOM provided electrons for biomass growth in both biofilters. It proved suggested that, during biofiltration, biomass utilized NOM as substrate to support metabolism. Some NOM was completely oxidized as inorganic carbon, which represents the removal of DOC. On the other hand, the increased $\overline{Cos}$ means the source of electrons comes not only from complete oxidation of NOM but also from incomplete oxidation of NOM. The application of $\overline{Cos}$ provides another method to evaluate the performance of biofiltration, because TOC/DOC overlooks the electrons that come from the incomplete oxidation of organic carbon. The combination of TOC/DOC and $\overline{Cos}$ would provide a comprehensive understanding of the transformation of NOM during the biofiltration.

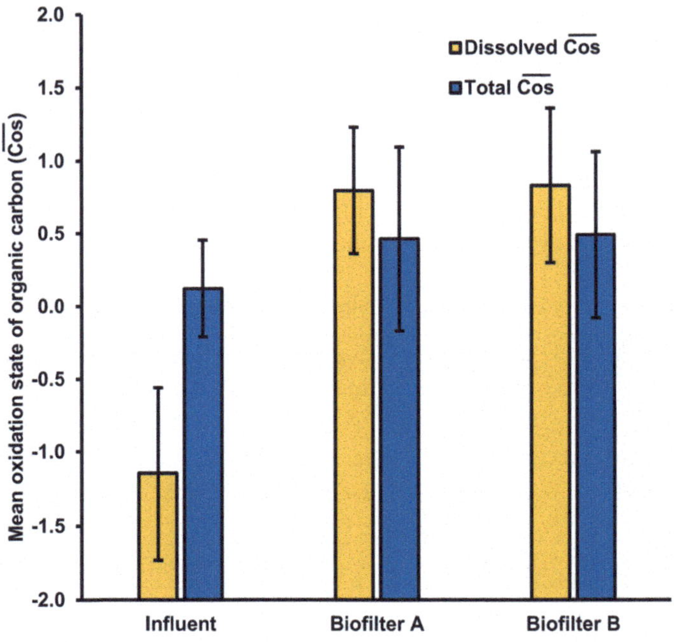

Figure 11.2 Mean oxidation state of dissolved organic carbon (dissolved $\overline{Cos}$) and total organic carbon (total $\overline{Cos}$) in influent and effluents of biofilter A and biofilter B.

11.4.2 Evolution of biomass ATP and mean oxidation state of organic carbon

In both biofilters, evolutions of $\overline{Cos}$ and biomass ATP concentration demonstrated a similar trend (Figure 11.3). In biofilter A, the increased biomass ATP concentration at 24 to 47 h corresponded to the increased $\overline{Cos}$. In biofilter B, the rapid increase of biomass ATP concentration at 3 to 22 h and 29 to 53 h corresponded to the increased $\overline{Cos}$ at 5 to 22 h and 29 to 53 h. In addition, the rapid decrease of $\overline{Cos}$ seems to have followed the decrease of biomass ATP. The decrease of $\overline{Cos}$ at 2 to 5 h and at 22 to 24 h in biofilter A, and that at 22 to 29 h in biofilter B also corresponded to the rapid decrease in biomass ATP. The local polynomial regression of the biomass ATP and $\overline{Cos}$ (Figure 11.4) indicated that the extent of oxidation increased with the maturation of biomass. Based on the biomass growth model described by Baranyi and Roberts (1994) and Stoddart *et al.* (2016), the exponential phase was regarded as the first few hours of biofilter operation. This corresponded to ATP ≤ 370 ng/g during the exponential phase and ATP > 370 ng/g during the steady phase. $\overline{Cos}$ increased gradually during the exponential phase. Then, total $\overline{Cos}$ of the fitting curve increased from 0 to 1.5 and dissolved $\overline{Cos}$ of the fitting curve increased from 0.5 to 1.5, when the biomass ATP was greater than approximately 380 ng/g during the steady growth phase.

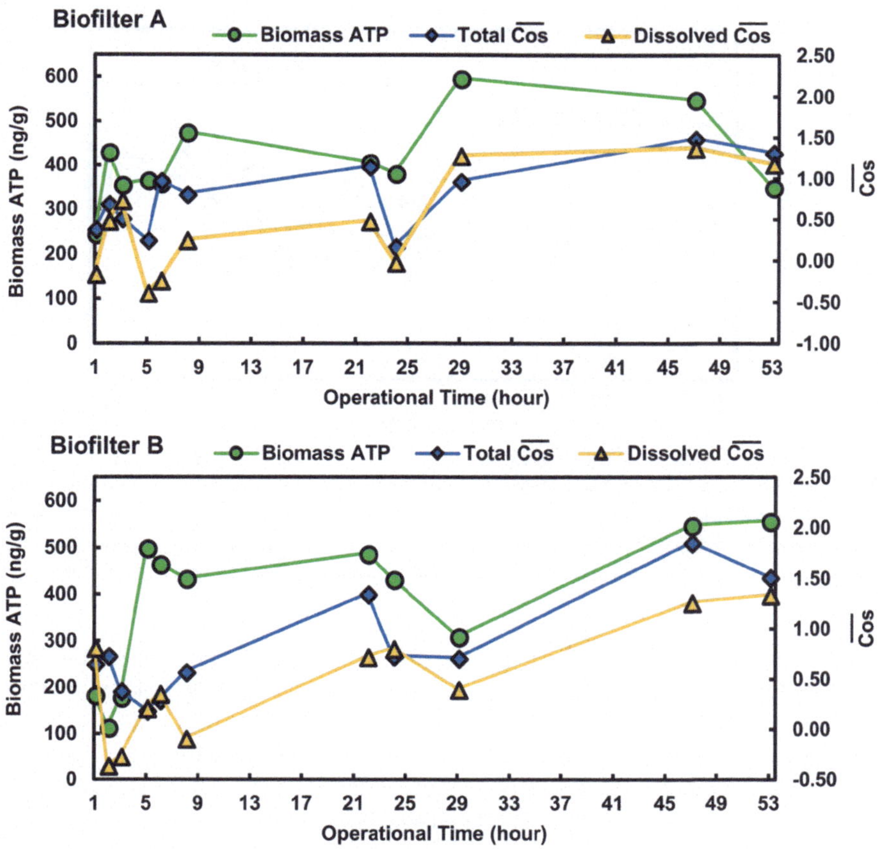

Figure 11.3 Biomass ATP and mean oxidation state of dissolved organic carbon (dissolved $\overline{Cos}$) and total organic carbon (total $\overline{Cos}$) in effluent of biofilter A (top) and biofilter B (bottom) within one filter cycle.

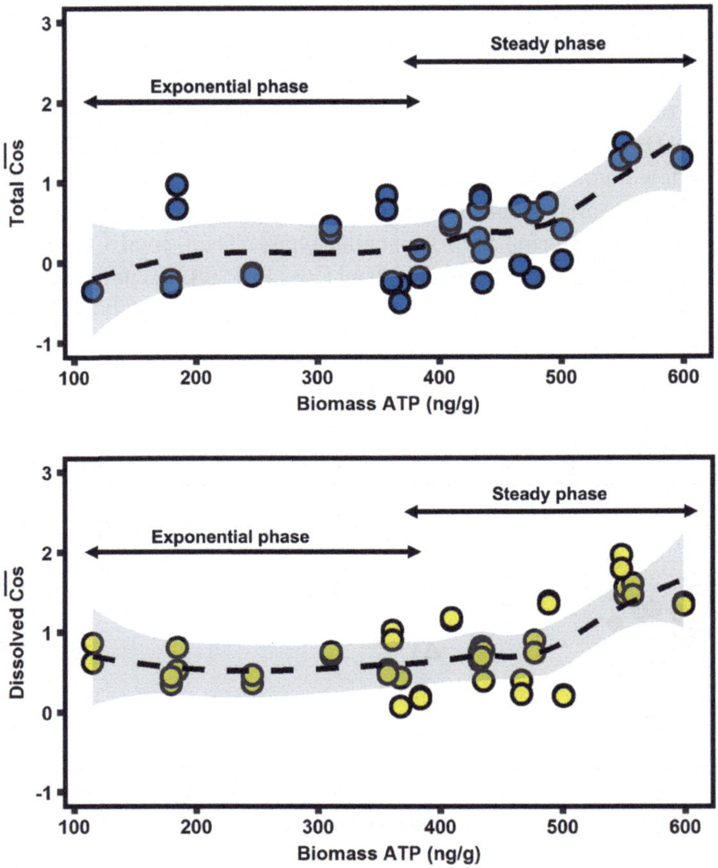

Figure 11.4 Local polynomial regression of biomass ATP and mean oxidation state of organic carbon (total $\overline{\text{Cos}}$ and dissolved $\overline{\text{Cos}}$). The grey background is the 99% confidential interval. The exponential phase and steady phase were determined by the growth model described by Baranyi and Roberts (1994) and Stoddart *et al.* (2016).

The same trend between $\overline{\text{Cos}}$ and biomass ATP indicated the possible relationship between food substrate and biomass growth. Both complete oxidation and incomplete oxidation of NOM contribute electrons to biomass growth, which results in the increase/decrease in biological activity (i.e., biomass ATP). Due to the limitation of TOC/DOC analysis, biomass ATP is not necessarily related to TOC or DOC removal (Pharand *et al.*, 2014). The possible relationship between $\overline{\text{Cos}}$ and biomass ATP was shown in this study. It is noted that not all data points were fitted within the 99% confidence interval, especially when the biomass ATP was in the range of 350 to 400 ng/g. The ATP growth model derived from Stoddart *et al.* (2016) in the same biofilters indicated that the biomass growth shifted from exponential phase to steady phase when biomass ATP was around 370 ng/g. The change in biomass concentration during this period provides a possible explanation for the fluctuation in effluent water quality. These data point to the need for a greater dataset and use of rapid test procedures such as ATP and peCOD. Although the data set presented in this Chapter was small in size, the relationships point to an opportunity for integrating rapid tests procedures to improve decision making in biological treatment processes.

11.4.3 Implications for drinking water utilities

The present work demonstrates that Cos can be used an indicator to evaluate the transformation of NOM in drinking water treatment processes with low oxidation efficiency (e.g., biofiltration). The results of full-scale biofilters provide insights that Cos could be useful for monitoring the performance of drinking water biofilters when coupled with other measurements (e.g., biomass ATP). In drinking water treatment, conventional TOC/DOC only provides information about physical removal and complete oxidation of organic carbon. Except for physical removal (e.g., filtration or adsorption), in a low-oxidation efficiency system (e.g., biofiltration), incomplete oxidation of organic carbon may play a major role which would not result in the significant removal of TOC/DOC. If NOM transformation is not detected, TOC/DOC measurement might not provide information about the performance of biological treatment in biofilters. For example, TOC/DOC removal results may not correlate to water quality objectives (e.g., reduction of disinfection byproduct formation). However, Cos, combining peCOD and TOC/DOC analyses, is an easily accomplished method to evaluate the incomplete oxidation of organic carbon in biological treatment. Through the monitoring of Cos, drinking water utilities would obtain more information about the reactivity of NOM and performance of biological treatment.

In addition, the mass balance between substrate and biomass is the foundation of biological treatment. However, in drinking water biofiltration, the relationship between substrate (i.e., TOC/DOC) and biomass (i.e., ATP) is still unclear. The possible reason is, besides the complete oxidation of organic carbon, the incomplete oxidation of organic carbon also contributes electrons to the metabolism of microorganisms. In this study, a possible relationship between Cos and biomass ATP has been observed in full-scale biofilters. Therefore, Cos could close the gap in the relationship between substrate and biomass in drinking water treatment. Finally, many years TOC/DOC built our understanding of NOM in drinking water treatment; however peCOD is gradually being applied in more and more research studies and industrial projects (e.g., pulp and paper, and breweries). TOC/DOC and peCOD are two different methods to quantify NOM in drinking water, which result in different measurement results (e.g., percentage removal). Cos is a connection between these two parameters, which would help us to have a better understanding of the conventional TOC/DOC and unconventional peCOD measurements.

11.5 CONCLUSION

Mean oxidation state (MOS) of organic carbon ($\overline{\text{Cos}}$) is a promising indicator to evaluate the extent of oxidation of NOM in drinking water treatment. The application of $\overline{\text{Cos}}$ provided additional valuable information about the transformation of NOM during drinking water biofiltration, which solved the limitation of conventional TOC/DOC analysis in relation to incomplete oxidation. In a low-energy/oxidant consuming treatment scenario (i.e., biofiltration), TOC/DOC could not quantify the incomplete oxidation of NOM. Incomplete oxidation of organic carbon increases the oxidation state of carbon and does not improve the removal of TOC/DOC. $\overline{\text{Cos}}$ could represent the incomplete oxidation NOM to give a more comprehensive evaluation of the performance of biofiltration. In this case study, the increased $\overline{\text{Cos}}$ and relatively consistent DOC concentration indicated that incomplete oxidation was the dominant biochemical function in the studied biofilters. Notably, $\overline{\text{Cos}}$ was correlated with biomass ATP; $\overline{\text{Cos}}$ and biomass ATP indicated the same trend within a filter cycle, which has not been observed before.

Since this study only investigated two full-scale biofilters within one filter cycle, further studies and long-term investigations are required to validate the relationship between biomass ATP concentration and $\overline{\text{Cos}}$ in full-scale drinking water biofilters identified in this case study.

11.6 REFERENCES

Baranyi J. and Roberts T. A. (1994). A dynamic approach to predicting bacterial growth in food. *International Journal of Food Microbiology*, **23**, 277–294.

Berney M., Vital M., Hülshoff I., Weilenmann H. U., Egli T. and Hammes F. (2008). Rapid, cultivation-independent assessment of microbial viability in drinking water. *Water Research*, **42**, 4010–4018.

Delahaye E., Welte B., Levi Y., Leblon G. and Montiel A. (2003). An ATP-based method for monitoring the microbiological drinking water quality in a distribution network. *Water Research*, **37**, 3689–3696.

Dowdell K. (2012). Trace Organic Contaminant Removal in Drinking Water Biofilters under Carbonaceous and Nitrogen-Supplemented Conditions and Evaluating Biomass with ATP and Phospholipid Methods. M.Sc. Thesis, University of Colorado at Boulder, USA.

Evans P. J., Smith J. L., LeChevallier M. W., Schneider O. D., Weinrich L. A. and Jjemba P. K. (2013a). A Monitoring and Control Toolbox for Biological Filtration. Water Research Foundation, Denver, Colorado.

Evans P. J., Smith J. L, LeChevallier M. W., Schneider O. D., Weinrich L. A. and Jjemba P. K. (2013b). Biological Filtration Monitoring and Control Toolbox: Guidance Manual. Water Research Foundation, Denver, Colorado.

Kroll J. H., Donahue N. M., Jimenez J. L., Kessler S. H., Canagaratna M. R., Wilson K. R., Altieri K. E., Mazzoleni L. R., Wozniak A. S., Bluhm H., Mysak E. R., Smith J. D., Kolb C. E. and Worsnop D. R. (2011). Carbon oxidation state as a metric for describing the chemistry of atmospheric organic aerosol. *Nature Chemistry*, **3**(2), 133–139.

Magic-Knezev A. and Van Der Kooij D. (2004). Optimisation and significance of ATP analysis for measuring active biomass in granular activated carbon filters used in water treatment. *Water Research*, **38**, 3971–3979.

Mantzavinos D., Livingston A. G., Hellenbrand R. and Metcalfe I. S. (1996). Wet air oxidation of polyethylene glycols; mechanisms, intermediates and implications for integrated chemical-biological wastewater treatment. *Chemical Engineering Science*, **51**, 4219–4235.

Marques S. M. and Esteves da Silva J. C. (2009). Firefly bioluminescence: a mechanistic approach of luciferase catalyzed reactions. *IUBMB Life*, **61**, 6–17.

Pharand L., Van Dyke M. I., Anderson W. B. and Huck P. M. (2014). Assessment of biomass in drinking water biofilters by adenosine triphosphate. *Journal of the American Water Works Association*, **106**(10), E433–E444.

Qiu J., Zhang S. and Zhao H. (2012). Nanostructured TiO_2 photocatalysts for the determination of organic pollutants. *Journal of Hazardous Materials*, **211**, 381–388.

Rieger M., Käppeli O. and Fiechter A. (1983). The role of limited respiration in the incomplete oxidation of glucose by *Saccharomyces cerevisiae. Microbiology*, **129**, 653–661.

Rittmann B. E. and Huck P. M. (1989). Biological treatment of public water supplies. *CRC Critical Reviews in Environmental Control*, **19**, 119–184.

Stoddart A. K. and Gagnon G. A. (2014). Application of photoelectrochemical chemical oxygen demand to drinking water. *Journal of the American Water Works Association*, **106**, E383–E390.

Stoddart A. K. and Gagnon G. A. (2015). Full-scale prechlorine removal: impact on filter performance and water quality. *Journal of the American Water Works Association*, **107**, E638–E647.

Stoddart A. K., Schmidt J. J. and Gagnon G. A. (2016). Biomass evolution in full-scale anthracite-sand drinking water filters following conversion to biofiltration. *Journal of the American Water Works Association*, **108**, E615–E623.

Tifft E. C. and Spiegel S. J. (1976). Use of adenosine triphosphate assay in disinfection control. *Environmental Science & Technology*, **10**, 1268–1272.

Vadasarukkai Y. S., Gagnon G. A., Campbell D. R. and Clark S. C. (2011). Assessment of hydraulic flocculation processes using CFD. *Journal of the American Water Works Association*, **103**, 66–80.

Velten S., Boller M., Köster O., Helbing J., Weilenmann H. U. and Hammes F. (2011). Development of biomass in a drinking water granular active carbon (GAC) filter. *Water Research*, **45**, 6347–6354.

Vogel F., Harf J., Hug A. and Von Rohr P. R. (2000). The mean oxidation number of carbon (MOC) – a useful concept for describing oxidation processes. *Water Research*, **34**, 2689–2702.

Zhang S., Zhao H., Jiang D. and John R. (2004). Photoelectrochemical determination of chemical oxygen demand based on an exhaustive degradation model in a thin-layer cell. *Analytica Chimica Acta*, **514**, 89–97.

Zhao H., Jiang D., Zhang S., Catterall K. and John R. (2004). Development of a direct photoelectrochemical method for determination of chemical oxygen demand. *Analytical Chemistry*, **76**, 155–160.

Section 4

Data collection and interpretation

Chapter 12

Test of sensor technologies for monitoring of microbiological drinking water quality

Ditte A. Søborg

12.1 INTRODUCTION

Monitoring microbiological water quality is necessary in order to produce drinking water that complies with the required criteria and is safe for human consumption. Microbiological water quality criteria may be exceeded as a result of various causes including growth of microorganisms in the distribution system due to e.g. temperature change or ingress of nutrients, rupture of a pipe and ingress of polluted water to the potable water, flow changes and possibly mobilization of pipe sediment, etc. (WHO, 2011). Each cause places different demands on the monitoring systems.

In Denmark, microbiological drinking water quality is traditionally measured by grab sampling followed by growth-based laboratory analyses. This monitoring strategy has the disadvantage of being time-consuming and not detecting episodic contaminations (Højris *et al.*, 2016). In the waterworks, however, physico-chemical sensors for measuring pressure, flow, temperature, dissolved oxygen, pH, conductivity and turbidity, for example, have been used by the water supply industry for decades (Corfitzen & Albrechtsen, 2011). Although these can indirectly register significant microbiological changes including sudden high-contamination events, there is a demand for sensors that can directly monitor microbiological drinking water quality.

The sensor technology field is evolving rapidly at present and several recent reviews of sensor technologies in the drinking water industry have been published (Bannaa *et al.*, 2014; Lopez-Roldan *et al.*, 2013; Storey *et al.*, 2011; Miles *et al.*,

Editors: Torben Lund Skovhus and Bo Højris
doi: 10.2166/9781780408699_0213

2011; Tatari *et al.*, 2016). Today, no microbiological sensor is available that fulfils all the waterworks' requirements of being automated, fast, sensitive, specific, cheap, producing easily interpretable results, requiring minimal maintenance, producing no false negative and only few false positive results, etc. However, several products are commercially available or are in the research, developmental or validation stage (Tatari *et al.*, 2016) which have the potential to be integrated into a functional system solution. These technologies are based on different principles such as measuring of total bacteria by optic or enzymatic methods, measuring of total cell activity by ATP, measuring of indicator bacteria by enzymatic methods, and others. Different measuring principles may be optimal for different types of contaminant or for different patterns of contamination. The optimal solution is therefore not necessarily one single sensor which can meet every requirement. Rather, the task of monitoring can be divided in an event alarm, automatic sampling and laboratory characterization. Using this monitoring strategy, a microbiological sensor can be developed with the focus of giving an event alarm while advanced automatic samplers combined with laboratory analyses can be developed with the focus of further characterization. The strategy also leaves open the possibility of using a combination of equipment types for each task, e.g. using two microbiological sensors based on different measuring principles for giving an event alarm. This is consistent with the strategy for monitoring of physical and chemical parameters which also does not rely on a single sensor being suitable for all situations.

In this study, dosing trials were carried out in laboratory scale in order to test various principles of monitoring the microbiological water quality under realistic drinking water contamination scenarios. The ability of different sensor technologies to give an event alarm were evaluated based on specificity, sensitivity, speed and user-friendliness. There will be referred to five sensor technologies in this chapter. These include an automated microbiological sensor (GRUNDFOS BACMON), an automated turbidity meter (ULTRATURB plus sc) that only indirectly measures microbiological water quality, a sensor in the developmental stage that consists of an automated high-volume sampler (HiVoSa) combined with the Colilert-18 method and two laboratory methods (ATP and Bactiquant) which both have been included in microbiological sensors. Automatic samplers were tested for the ability to collect a single water sample, a sample over time or possibly carry out a concentration of a large water sample using membrane filtration. The use of qPCR targeting the 16S rRNA gene and mtDNA was tested as a laboratory characterization method of drinking water contamination.

12.2 EXPERIMENTAL PROCEDURES

12.2.1 Experimental setup

The dosing trials were carried out in a tank containing 600 L of non-chlorinated fresh tap water, dosed with varying types of contaminant. The experimental setup is sketched in Figure 12.1. The tank was equipped with an electrical stirrer (VLA,

Peo-Tech A/S) as well as sample points, located 10 cm above the bottom of the tank, for microbiological sensors, automatic samplers and for manual water samples.

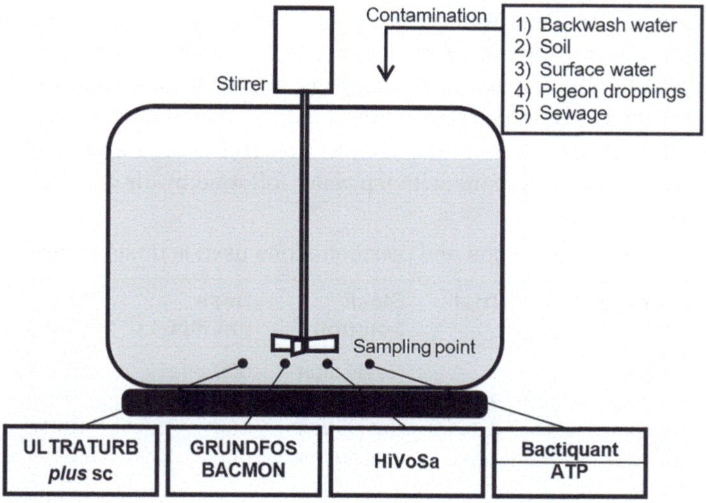

Figure 12.1 Experimental setup for dosing trials.

12.2.2 Contaminant types

Five different types of contaminant were applied during the dosing trials: (1) backwash water from Truelsbjerg waterworks, (2) humic soil from a detached garden, (3) surface water sampled from a puddle, (4) pigeon droppings from perches on a pigeon coop, and (5) raw sewage from Horsens central wastewater treatment plant. For soil and pigeon droppings, stock solutions were prepared in 1 L non-chlorinated tap water by weighing out 250 and 200 g of the contaminants, respectively and shaking vigorously to obtain the greatest degree of homogeneity. Table 12.1 provides an overview of the contaminant types and concentrations used in the dosing trials.

12.2.3 Methodology

Every dosing trial was carried out using the following methodology: The water tap was allowed to run for at least 5 minutes (which, by trial, was found to be sufficient to produce fresh water at a constant temperature), the tank was filled to 600 L, the stirrer was started and allowed to run for at least 5 minutes. Water quality of the clean water in the tank was tested to comply with Danish drinking water criteria at the tap for the following parameters: Fe, Mn, NH_4, and pH. Further, oxygen concentration and temperature were measured. The first measurements were carried out on the clean water in the tank following which the contaminant was added in the lowest concentration. Measurements were made on the contaminated water in the tank after stirring for at least 5 minutes. The whole procedure was then repeated twice; i.e. an additional quantity of the same contaminant was added, the tank was stirred and the measurements were carried out. A dosing trial, comprised of measurements on clean water and on water with three different contaminant concentrations, had a duration of approximately 5–7 hours. At the highest level of contaminant addition, the automatic samplers were activated and sampling took place over 3–15 hours, depending on the type of contamination (some types of contaminant made the filters in the samplers clog). Between each trial, the tank was cleaned by repeated rinsing with tap water followed by air drying.

Table 12.1 Contaminant types and concentrations used in dosing trials.

Contaminant Type	Trial	Stock Solution	Dose (mL/600 L)	Final Dilution/ Concentration
Backwash water (from backwashing of a waterworks pre-filter)	1	No dilution	600	10^3 times
	2	No dilution	6	10^5 times
	2	No dilution	60	10^4 times
	2	No dilution	600	10^3 times
Soil (humic soil from a detached garden)	1	250 g/L	1	0.4 mg/L
	1	250 g/L	10	4 mg/L
	1	250 g/L	100	40 mg/L
	2	250 g/L	1	0.4 mg/L
	2	250 g/L	10	4 mg/L
	2	250 g/L	100	40 mg/L
Surface water[a] (water from a puddle)	1	No dilution	1	2×10^6 times
	1	No dilution	10	2×10^5 times
	1	No dilution	100	2×10^4 times
	2	No dilution	10	2×10^5 times
	2	No dilution	100	2×10^4 times
	2	No dilution	600	10^3 times
Pigeon droppings[a] (faeces scraped from pigeon coop perches)	1	200 g/L	1	0.3 mg/L
	1	200 g/L	10	3 mg/L
	1	200 g/L	100	33 mg/L
	2	1 g/L	2	0.003 mg/L
	2	1 g/L	20	0.03 mg/L
	2	1 g/L	200	0.3 mg/L
Sewage (raw sewage from central wastewater treatment plant)	1	No dilution	0.6	10^6 times
	1	No dilution	6	10^5 times
	1	No dilution	60	10^4 times
	2	No dilution	0.6	10^6 times
	2	No dilution	6	10^5 times
	2	No dilution	60	10^4 times

[a]Note that the concentration of applied contaminant in the case of trial 1 for pigeon droppings was much higher than for trial 2, and that the concentration of applied contaminant was higher in trial 2 for surface water than for trial 1.

12.2.4 Analyses

The clean water and the contaminated water from the tank were subjected to a series of analyses.

Background measurements: Temperature, pH and dissolved oxygen were measured with the sensors SenTix® 940 and FDO® 925 and the Multi 3430 meter (WTW GmbH, Xylem Analytics). Iron, manganese and ammonium were analysed according to the manufacturer's instructions with the standard kits LCK521/LCK320, LCW532 and LCK304, respectively using the DR 3900 Spectrophotometer (Hach).

HiVoSa/Colilert-18: HiVoSa (High Volume Sampling, Amphi-Bac ApS) concentrated a 30 L water sample to 100 mL of concentrate in around 1 hour. The concentrate was withdrawn and tested manually with the Colilert-18/Quanti Tray-method (IDEXX Laboratories). The HiVoSa was manually cleaned between each trial by heat disinfection according to the manufacture's instruction. The combination of HiVoSa with Colilert-18 was under tested as a microbiological sensor solution.

GRUNDFOSs BACMON: GRUNDFOS BACMON (GRUNDFOS Holding A/S) carried out automatic measurements of bacteria and abiotic particles in the water at approximately 10 minute intervals throughout the experiment using 3D scanning by a moving digital microscope. To ensure proper mixing of contaminant and water in the tank, values determined less than 5 minutes after the introduction of a contaminant dose to the tank were excluded from the data set. The optical flow cell holding the water sample was changed between each contaminant type and performed optimally (100%) for all the measurements performed during the dosing trials (Højris, 2016).

ULTRATURB *plus* sc: The ULTRATURB *plus* sc turbidity meter (Hach) carried out automatic measurements on the water at approximately 10 minute intervals throughout the experiment. To ensure proper mixing of contaminant and water in the tank, values determined less than 5 minutes after the introduction of a contaminant dose to the tank were excluded from the data set.

ATP: Triplicate water samples were collected manually in sterile flasks and 120 mL aliquots analysed for Adenosine TriPhosphate (ATP, total active cells) within 4 hours using the test kit Quench-Gone™ Aqueous and the PhotonMaster™ (both LuminUltra Technologies Ltd.) according to the manufacture's instructions.

Bactiquant: Triplicate 250 mL water samples were collected manually in sterile flasks and analyzed for total bacteria within 4 hours using the enzymatic Bactiquant®-water method (Mycometer A/S) according to the manufacture's instructions.

Automatic water sampling: Automatic water sampling was carried out at the highest contaminant concentration using the automatic sampler from HOFOR (Døgnprøvetageren, HOFOR A/S) and the Alonda1000 (Amphi-Bac ApS). The sampling duration and the volume rate of water sampled were noted in order to calculate the filtered quantity. The automatic water sampling was supplemented with 5 L water samples that were collected manually from the tank for each contaminant concentration and vacuum filtered (Advantec MFS Inc.). The manual water samples were necessary in order to be able to characterize water of each contaminant concentration and at the same time being able to carry out a full dosing trial in a day. Nylon filters (Membrane filter Nylon 0.45 µm 47 mm white, Frisenette ApS) were used for the manual vacuum filtration as well as for vacuum filtration by the automatic samplers. DNA was extracted from the nylon filters using the RapidWater® DNA Isolation kit (Mo Bio Laboratories) according to the manufacture's instructions and stored at 4°C until qPCR analysis.

qPCR targeting the 16S rRNA gene: Total bacteria were determined by using the universal forward 5′-TCCTACGGGAGGCAGCAGT-3′ and reverse 5′-GGACTACCAGGGTATCTAATCCTGTT-3′ primers and the probe (6-FAM)-5′-CGTATTACCGCGGCTGCTGGCAC-3′-(TAMRA) for the 16S rRNA gene, as previously described (Nadkarni *et al.*, 2002). The analyses were performed with the TaqMan Universal PCR MasterMix and analysed on a Stratagene Mx3000P by Amphi-Bac ApS.

qPCR targeting mtDNA: Determination of mitochondrial DNA (mtDNA) from humans and/or mammals were analysed by using the human-forward 5′-GGCTCACTCCTTGGCGC-3′ and human-reverse 5′-CCTCGCCCGATGTG TTAGGAAG-3′ and the universal forward 5′-GCCACCGCGGTCATACGATT-3′ and universal reverse 5′-GGGTATCTAATCCCAGTTTGGGTCTTAGC-3′ primers, respectively, as previously described (Tobe & Linacre, 2008). The primer sets targeted the cytochrome b (*cytb*) and the 12S rRNA genes on the mitochondrial genome, respectively. Determination of mtDNA from pigeons (family *Columbidae*) were analysed according to a method developed by Amphi-Bac ApS (Bastholm & Starcke, 2015). All mtDNA analyses were performed with the Brilliant III Ultra-Fast SYBR® Green QPCR Master Mix and analysed on a Stratagene Mx3000P by Amphi-Bac ApS.

12.3 RESULTS AND DISCUSSION

The tested system solutions for monitoring microbiological drinking water quality consisted of combinations of five different sensor technologies, two automatic samplers and laboratory characterization. The sensor technologies covered different measuring principles: (1) total bacterial activity by ATP (LuminUltra Technologies Ltd.), (2) total bacteria based on enzymatic analysis (Bactiquant-water method, Mycometer A/S), (3) indirect measurement of microbiological water quality by turbidity (ULTRATURB *plus* sc, Hach), (4) detection of specific indicator bacteria in concentrated water samples by enzymatic analysis (HiVoSa, Amphi-Bac ApS combined with Colilert-18, IDEXX Laboratories), and (5) total bacteria and abiotic particles by optic analysis (GRUNDFOS BACMON, GRUNDFOS Holding A/S). The automatic samplers provided different types of samples: (1) 500 mL immediate water samples (Alonda1000, Amphi-Bac ApS), (2) 500 mL time-proportional water samples (Alonda1000), and (3) time-proportional membrane-filtered large (10–100 L) water samples (Alonda1000 and the automatic sampler from HOFOR). The laboratory characterization methods included qPCR targeting the 16S rRNA gene for determination of total bacterial gene copies and mitochondrial DNA (mtDNA) for detection of pigeons, humans, and mammals.

The system solutions were tested in duplicate dosing trials for which a broad spectrum of realistic contamination scenarios were selected. The experimental setup for the dosing trials is seen in Figure 12.1 and contamination types and concentrations are found in Table 12.1. The contaminations included high number of particles (backwash water and soil), few particles (surface water), high number of non-pathogenic microorganisms (backwash water), high number of heterotrophs and coliforms (soil, surface water, pigeon droppings and sewage), and high number of fecal microorganisms such as *E. coli* (pigeon droppings and sewage). The selected contaminants thus represent a very broad spectrum of types of bacteria and particles. Further, testing the response of the sensor technologies to realistic drinking water contaminants may give a better indication of their performance in a real-life situation than basing their performance on response to e.g. pure bacterial cultures.

The dosing trials comprised a series of three different contaminant dose concentrations in addition to tap-water without any dose of contaminant. Measurements on the clean tap water in the tank (with no contaminant) at the start of every dosing trial demonstrated that the water complied with the Danish quality criteria at the consumer's property for Fe, Mn, NH_4, O_2 and pH (BEK 1147, 2017), as shown in Table 12.2. The water temperature in the tank was 15.6 ± 0.6°C during the trials while the room temperature was around 20°C. The duration of the individual trials was too short for the water temperature to approach equilibrium with room temperature.

Table 12.2 Physical and chemical background measurements on the clean tap water in the tank.

Parameter	Measurement (Average)	Quality Norm (Consumer)
Fe	0.07 ± 0.02 mg/L	0.2 mg/L (max)
Mn	0.04 ± 0.04 mg/L	0.05 mg/L (max)
NH_4	0.00 ± 0.00 mg/L	0.05 mg/L (max)
O_2	9.8 ± 0.2 mg/L	> 5 mg/L (at entry to property)
pH	7.5 ± 0.1	7–8.5 (permitted range)
Water temp. (tank)	15.6 ± 0.6°C	12°C (target)

For each microbiological sensor technology, an average value and the standard deviation for measurements on clean tap water were found. The average plus two standard deviations was used as the value for which a specific sensor gave an event alarm, while the average plus one standard deviation was used to indicate warning. It is important to note that although the non-chlorinated clean tap water in the tank was expected to have a natural presence of microorganisms, no pathogens should be present. The methods based on detection of total bacteria distinguish between water samples with a normal bacterial content and water samples with abnormally high bacterial content. Thus, establishment of a baseline is required. Using this approach, the microbiological sensors targeting total bacteria therefore had a higher background level than sensors based on detection of specific microorganisms. However, this scenario reflects real-life application of the sensors for non-chlorinated drinking water treatment and distribution.

12.3.1 Specificity and sensitivity

The tested sensor technologies reacted very differently to the various contaminations. The HiVoSa combined with Colilert-18-analysis, which targeted specific indicator bacteria (coliforms and *E. coli*) rather than e.g. total bacteria, was the only method providing negative results for dosing of backwash water and positive results for all other contaminant types. As the backwash water was expected mainly to contain chemoautotrophs, i.e. beneficial bacteria from the water treatment processes that obtain energy from oxidation of inorganic compounds such as iron, manganese and ammonium, and no pathogenic microorganisms, the HiVoSa/Colilert-18 method was shown to have a good specificity with regards to monitoring the microbiological drinking water quality. Several other sensor technologies (ALARM, CALM, Coliguard, aquaBio, TECTA and Coliminder) are automated versions of the Colilert-18 analysis (Tatari *et al.*, 2016).

The advantage of providing results of presence of specific groups of bacteria, that are not normally present in drinking water, is that a higher sensitivity can be obtained since there is no background levels. This is in comparison to methods based on detection of total bacteria. The most sensitive solution was the HiVoSa/Colilert-18 method, which gave an event alarm to the lowest doses of all problematic contaminant types (soil, surface water, pigeon dropping and sewage). It was expected to observe a high sensitivity for this method, as it includes concentration of water samples. The manufacturer claims detections of 0.001 coliforms/100 mL for filtration of 100 L water (Amphi-Bac ApS, 2017), which is 1000 times lower than the Danish drinking water criterion (BEK 1147, 2017). A detection of coliforms below the drinking water criterion for coliforms was confirmed in the dosing trials with membrane filtration of 30 L water. Figure 12.2 shows the results of the HiVoSa/Colilert-18-method after dosing with surface water (trial 1, see Table 12.1). Most probable number of coliforms in the tank were calculated to 0.02, 0.11 and 3.07 coliforms/100 ml water for the three final dilutions of surface water, respectively.

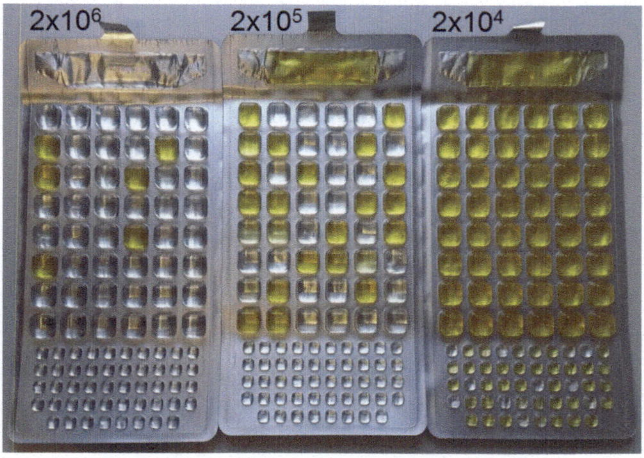

Figure 12.2 Example of the results of contaminant dosing at different concentrations using the HiVoSa/Colilert-18-method. From left to right, dosing with surface water to final dilutions of $2x10^6$, $2x10^5$, and $2x10^4$. Results were 6.3, 31.7, and 920.8 coliforms/100 mL concentrated samples (30 L water concentrated to 100 mL), corresponding to 0.02, 0.11 and 3.07 coliforms/100 ml water in the tank.

A high sensitivity was also seen for some contaminant types for the ATP method. The method gave event alarms for the lowest doses of surface water and pigeon dropping contaminations. However, the sensitivity of ATP and Bactiquant was affected by varying background levels in the experimental setup. In general, the methods based on determination of total bacteria were less sensitive. The Bactiquant method, the GRUNDFOS BACMON and the ULTATURB *plus* sc sensors only gave event alarms for the higher contaminant doses, independent of the contaminant type. Results for GRUNDFOS BACMON in the dosing with soil contamination (trial 1, see Table 12.1) can be seen in Figure 12.3.

12.3.2 Speed

When considering speed of the sensor technologies, different results were observed. In contrast to performing well on sensitivity, the HiVoSa/Colilert-18-method was not fast, having a response time of about 20 hours in the dosing trials. The methods based on determination of total bacteria or particles (i.e. Bactiquant, ATP, GRUNDFOS BACMON and ULTRATURB *plus* sc) were faster and were able to provide results within 5–30 minutes. The GRUNDFOS BACMON and the

ULTATURB *plus* sc sensors were the most rapid technologies tested. In Figure 12.3, it can be seen that a few minutes after addition of the highest concentration of soil (40 mg/L), GRUNDFOS BACMON measured bacterial counts in values giving rise to an event alarm and the high bacterial counts continued for the remainder of the dosing trial. The dosing trials therefore emphasized a general trade-off, for microbiological sensors, between specificity and speed. This observation suggested that a combination of microbiological sensors based on different measuring principles may be needed to cover different contamination scenarios. In the dosing trials, at least 2 microbiological sensor technologies were needed to detect any contamination type and concentration.

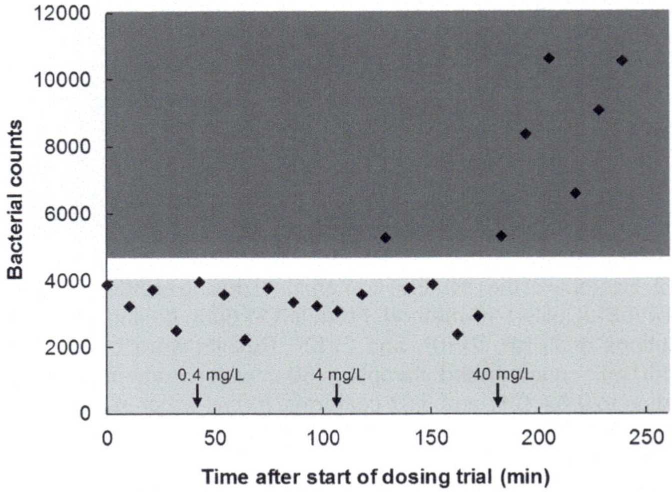

Figure 12.3 Example of the results of contaminant dosing at different concentrations using the GRUNDFOS BACMON. Dosing of soil to final dilutions of 0.4, 4 and 40 mg/L is marked with arrows. Light grey: Acceptable water quality, White: Warning, Dark grey: Event alarm.

12.3.3 User-friendliness

In addition to the ability of the microbiological sensor technologies to detect a contamination event, the user-friendliness of operating the sensors was evaluated (cleaning, maintenance, interpretation of data, etc.). For a sensor to be applicable throughout the drinking water treatment system, it should need as little maintenance as possible, should not require chemical supplies, should not produce hazardous waste (Højris *et al.*, 2016), and the data should be easily accessible and easy to interpret.

The GRUNDFOS BACMON and the ULTRATURB *plus* sc were the only fully automated sensors tested in the dosing trials. The only manual handling required for GRUNDFOS BACMON was change of the flow cell between each dosing trial. Data from this sensors was also easily accessible through different interphases (cell phone, web, SCADA, iPad). For the ULTRATURB *plus* sc, data was logged and saved on a memory card from which it was later downloaded to a PC. Both sensors can therefore be applied in the drinking water treatment system in the versions included in the study. In such a situation, in which contamination events are only rarely seen, minimal maintenance is required for the GRUNDFOS BACMON, in terms of replacing the flow cell after several months of use (Højris *et al.*, 2016). In the dosing trials, the HiVoSa functioned as an automatic concentration unit that formed part of a sensor (HiVoSa/Colilert-18). The Colilert-18 analysis was performed manually and the HiVoSa required manual cleaning between dosing trials. However, the HiVoSa is suitable for incorporation into a complete sensor as it produces a water sample, in which bacteria are concentrated, rather than a filter from membrane filtration, where the bacteria often have to be removed manually from the filter and transferred back to an aqueous phase before any further analysis can be carried out. The complete sensor, however, has the drawback that it requires addition of chemicals as all sensors based on the Colilert-18 analysis. Bactiquant and ATP analyses also required addition of chemicals. Both methods were manually performed in the dosing trials. Samples were, however, analysed regularly during the dosing trials to simulate measurements by sensors, and thereby to evaluate the applicability of the methods as microbiological sensor technologies. An automatic sensor based on the Bactiquant-water method (Bactiline, Mycometer A/S) has been tested in a previous study (Nordahn *et al.*, 2012) and a sensor based on the ATP method (EZ-ATP, AppliTek) is commercially available.

The tested sensor technologies also performed differently concerning the interpretation of data. The results from HiVoSa/Colilert-18 were easy to understand and use for evaluation of the microbiological drinking water quality as the Colilert-18 analysis is simple and part of the legislative monitoring of drinking water in Denmark (BEK 1147, 2017). In comparison, for the methods based on detection of total bacteria (GRUNDFOS BACMON, Bactiquant, and ATP), there are no limits prescribed in the regulations controlling microbiological drinking water quality, and thus the methods cannot be used for legal compliance purposes. Rather, the methods are useful in the water utilities supplemental monitoring. Due to the baseline levels of bacteria in clean drinking water that varies with time and location (i.e. in the waterworks or position in the distribution network), the interpretation of data from these microbiological sensor technologies are more complex. In the dosing trials, variations in the background levels of bacteria in the clean water due to varying total bacteria quality in the tap water at different times and other reasons also reduced the sensitivity of these methods. However, in most cases, it was unproblematic to distinguish water quality giving rise to an event alarm from the quality of the clean water in the tank. The last technology tested, the ULTRATURB *plus* sc, only indirectly measures microbiological drinking water quality and cannot be used alone to make a statement on a possible contamination scenario.

12.3.4 Automatic sampling

Even if water quality is monitored by microbiological sensors, there is still a need for regular sampling. Such samples might be used for those analyses that cannot be performed at-line, or they may be used to characterize the water that triggered an event alarm. In order to collect a sample of the water that triggered an event alarm, regardless of time of the day, automatic sampling is required. This, of course, presupposes that a microbiological sensor can communicate with a sampling unit. In the dosing trials, the two automatic samplers (Alonda1000 from Amphi-Bac ApS and the automatic sampler from HOFOR A/S) were manually activated at the highest level of contaminant addition. Using this approach, the communication between microbiological sensors and automatic samplers was not studied. Rather, the ability of the automatic samplers to collect immediate or time- or flow proportional samples was tested. Manual water sampling and filtration were used for collecting samples of water with lower doses of contaminant. An example of filters obtained after automatic and manual sampling of water from the tank with different contaminant doses can be seen in Figure 12.4.

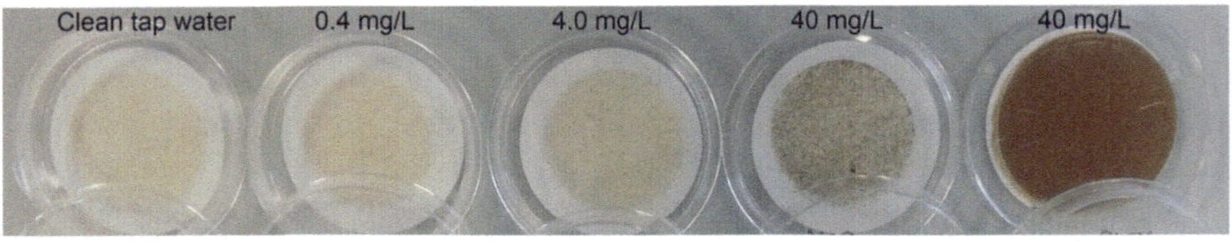

Figure 12.4 Example of filters after membrane filtration of water samples of different concentrations of contaminant dosing. From left to right, filters from manual filtration of 5 L of clean tap water and water with final dilution of soil of 0.4, 4 and 40 mg/L, respectively. Furthest to the right, filter from overnight filtration of ~ 30L water (the filter clogged after approximately 6 hours with initial flow of 85mL/min) with 40 mg/L soil using the automatic sampler from HOFOR.

It was possible to achieve concentrated membrane-filtered water samples over time by both automatic samplers. In most cased, automatic sampling were carried out with a constant flow of water overnight (up to 15 h). However, hydraulic resistance in the filters led less water through with time, and for the contaminations containing high numbers of particles (e.g. backwash water and soil), the filters clogged after approximately 3–10 hours. It was not tested if clogging of the membranes during filtration would also be a problem with lower contaminant doses, which would more likely represent the level of contamination seen in a drinking water distribution system. In addition to the membrane-filtered samples, the Alonda1000 was shown to be able to collect grab water samples and water samples over time in plastic bags. This collection of a total of four samples (grab water sample, water sample over time and two membrane-filtered samples) is advantageous, as this allows immediate analysis of e.g. a bag sample with Colilert-18, while a membrane-filtered sample can readily be stored in a freezer for later DNA analysis.

12.3.5 Characterization of contaminations

The characterization of any contamination is the final part of a system solution for monitoring microbiological contamination of drinking water. In this study, the membrane-filtered samples collected manually and by the automatic samplers were used for characterization purposes to confirm the water quality of the water in the tank. This was done to validate increasing contaminant doses as well as to determine the contamination type. At the same time, characterization by qPCR targeting mtDNA for pigeons, mammals and humans was evaluated as a source tracking method. The hypothesis behind the mitochondrial DNA method is that a contamination will contain large amounts of eukaryotic cells (e.g. intestinal cells, blood cells, plant cells etc.) along with microorganisms. Compared to detection of bacterial DNA, which does not allow source differentiation, specific mtDNA assays can be used to identify presence/absence of a given eukaryotic cell type and thereby identify the contamination type (Martellini *et al.*, 2005).

DNA was successfully extracted from all samples. The content, however, varied from 1.8–23 ng DNA/μL. As expected, large membrane-filtered water samples (up to 100 L) obtained by the automatic samplers resulted in the highest yields of DNA. The results of qPCR were analysed by plotting the C_t-values as a function of final contaminant concentration. The cycle threshold (C_t) is the number the number of cycles required for the fluorescent signal to cross the threshold, i.e. exceed background levels. In other words, the higher amounts of target (e.g. total bacteria), the lower the C_t-value. As seen in Figure 12.5, qPCR results of total bacteria indicated a relationship between C_t-value and the amount of added dose of pigeon droppings down to a final dilution of contaminant in the tank of 0.3 g/L. At this dose, the C_t-value reached the average value for water samples without contaminant dosing. For the other types of contaminant, no correlation was seen. The fact that only high doses of contaminant could be distinguished from background levels of natural presence of bacteria in drinking water is comparable to the observations obtained for the microbiological sensor technologies targeting total bacteria. Therefore, qPCR targeting total bacteria are only relevant to detect relatively high contamination events. A specific qPCR may resolve this but may require numerous assays for targeting all potential drinking water pathogens.

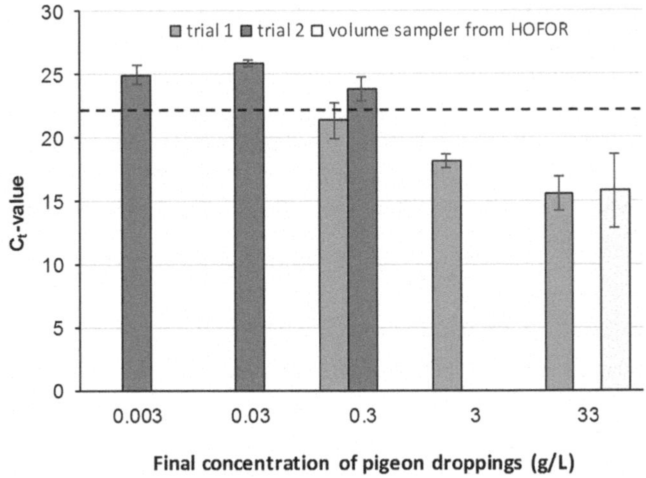

Figure 12.5 Results of qPCR targeting total bacteria (16S rRNA gene) in membrane-filtered water samples (5 L) contaminated with final dilutions of pigeon droppings of 0.3, 3, and 33 g/L (trial 1) and 0.003, 0.03 and 0.3 g/L (trial 2). Membrane-filtered water samples from the automatic sampler from HOFOR (~ 15 L, the filter clogged after approximately 3 hours with initial flow of 85 mL/min) was only included for the highest concentration of pigeon dropping. Corrected threshold cycle numbers (C_t-values) are plotted as a function of final dilution of contaminant in the tank. The dotted line shows the average C_t-value for samples without contaminant dosing.

Positive results of pigeon specific qPCR targeting mtDNA were seen for the highest dose of pigeon droppings (Figure 12.6). For the second highest concentration of pigeon droppings, the fluorescence signal only crossed the threshold value after 36–38 cycles (C_t-values = 36–38). However, for both samples, melting curve analysis showed melting points comparable to the positive control which indicates presence of pigeon mtDNA. Samples from membrane filtration of ~15 L water from the automatic sampler from HOFOR, showed C_t-values comparable to those from 5 L manual filtration. The objective of concentration of water samples is to achieve a lower detection limit. In this dosing trial, it was not possible to see an increased sensitivity by filtering higher amounts of water. The reason for this is unknown. However, parameters possibly affecting this include hydraulic resistance of the filters resulting in less water than expected being filtered with the automatic samplers or problems with transferring genetic material from the filter to a water phase prior to the qPCR. For all other contaminant types than the pigeon droppings, the pigeon specific qPCR gave negative results. Hence, the approach of targeting mtDNA from pigeons was successful as a source tracking method for high contamination events with pigeon droppings. The weak positive result for the final dilution of pigeon droppings of 3 g/L, indicated that this concentration was close to the detection limit of the method.

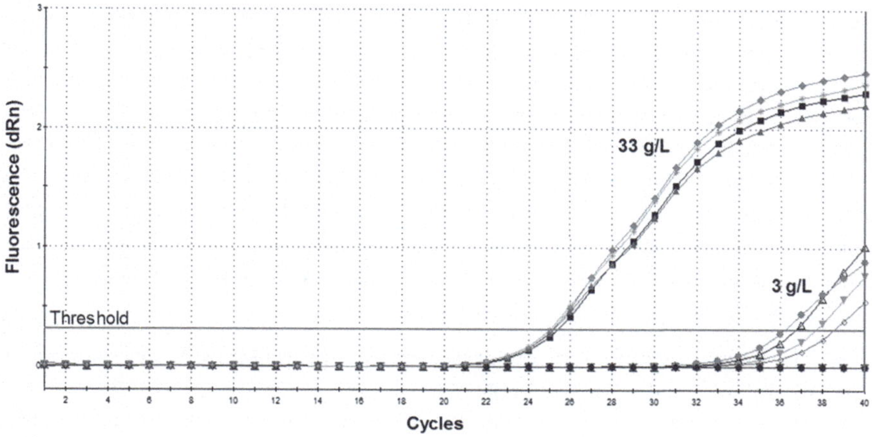

Figure 12.6 Results of pigeon specific qPCR (mtDNA from pigeons) in samples contaminated with pigeon droppings in final dilutions of 33 g/L and 3 g/L (two true replicates and two technical replicates of each concentration).

The mammal and human specific qPCR was only performed for the highest doses of contaminants. The highest dose of sewage (10^4 dilution) gave a clear positive result for the mammal test method, but a surprisingly weak result for the human test method. This was most likely due to a high content of slaughterhouse waste in the sewage used for the dosing trials which is known to be the case for Horsens central wastewater treatment plant. In general, it is expected that the composition of sewage is highly varying between wastewater treatment plants, mainly due to contents of different industrial waste in the sewage. Hence, it is important to test the mammal and human specific qPCR with sewage from a range of different wastewater treatment plants prior to using it for characterization purposes of drinking water contamination. In the dosing trials, however, all other contaminant types but sewage led to negative results in the mammal and human specific qPCR. Hence, the method was successful in the identification of the contaminant of mammal origin between the five different contaminant types included in the dosing trials.

12.4 CONCLUSIONS

On the basis of the dosing trials, it was concluded that the five different sensor technologies reacted very differently to various contaminant types. The HiVoSa/Colilert-18 method was very sensitive, but not rapid. Further, it was the only method that determined specific bacteria (the indicator bacteria coliforms and *E. coli*). The methods based on determination of total bacteria either by cell numbers (GRUNDFOS BACMON) or by ATP or enzymatic measurement (Bactiquant), were faster, but less sensitive, due to varying background levels in the experimental setup. GRUNDFOS BACMON and ULTRATURB *plus* sc were the most rapid sensors tested and were the only sensors that were fully automated in the trialled versions. The contaminant dosing trials emphasized the normal trade-off between the desire for specificity and speed of determination. These two requirements do not seem to be easy to combine in a single microbiological sensor.

The automatic samplers (Alonda1000 and the automatic sampler from HOFOR) functioned as expected, and contaminations with pigeon droppings and sewage could be identified in the filtered water samples with the mitochondrial DNA test method. The method appeared to be specific, but with a relatively high limit of detection.

Conclusions and perspectives drawn in this study:

- The study identified different strengths and weaknesses by individual microbiological sensor technologies.
- The selection of the most suitable sensor therefore depends on the specific purpose of monitoring, including legislative and supplemental monitoring.
- For giving an event alarm, the waterworks will, at the moment, need to rely on several types of microbiological sensor technologies to cover the broad range of different potential types of contaminants and patterns of contaminations that the industry faces. This is consistent with the strategy for monitoring of physical and chemical parameters which also does not rely on a single sensor being suitable for all situations.
- In the dosing trials, at least 2 microbiological sensor technologies were needed to detect any contamination type and concentration.
- In the future, a microbiological sensor that is fast, sensitive, automatic and specific may be developed and applied in the water industry, e.g. based on PCR and other molecular methods such as hybridization.

This study showed that with today's technologies, it is possible to divide the task of microbiological monitoring in three: event alarm by microbiological sensors, automated sampling and laboratory characterization. With this combined approach, any contamination event can be detected, sudden or slow, and handled accordingly to secure the water quality for the consumer.

12.5 REFERENCES

Amphi-Bac ApS 2017. http://www.amphi-bac.dk/da/produkter-og-services/drikkevand/hivosa (accessed 7 May 2018).

Banna M. H., Imran S., Francisque A., Najjaran H., Sadiq R., Rodriguez M. and Hoorfar M. (2014). Online drinking water quality monitoring: review on available and emerging technologies. *Critical Reviews in Environmental Science and Technology*, **44**, 1370–1421.

Bastholm S. and Starcke S. (2015). DNA kit til drikkevand. Report 7238.2011, Vandsektorens Teknologiudviklingsfond, Denmark.

BEK 1147 (2017). Statute on water quality and supervision of water supply systems. Danish Ministry of the Environment. 24 October 2017.

Corfitzen C. B. and Albrechtsen H. (2011). On-line kontinuert måling af drikkevandskvalitet. Naturstyrelsen.

Højris B., Christensen S. C. B., Albrechtsen H. J., Smith C. and Dahlqvist M. (2016). A novel, optical, on-line bacteria sensor for monitoring drinking water quality. *Scientific Reports*, **6**, 1–10.

Lopez-Roldan R., Tusell P., Cortina J. L., Courtois S. and Cortina J. L. (2013). On-line bacteriological detection in water. *TrAC - Trends in Analytical Chemistry*, **44**, 46–57.

Martellini A., Payment P. and Villemur R. (2005). Use of eukaryotic mitochondrial DNA to differentiate human, bovine, porcine and ovine sources in fecally contaminated surface water. *Water Research*, **39**, 541–548.

Miles S. L., Sinclair R. G., Riley M. R. and Pepper I. L. (2011). Evaluation of select sensors for real-time monitoring of *Escherichia coli* in water distribution systems. *Applied and Environmental Microbiology*, **77**, 2813–2816.

Nadkarni M., Martin F. E., Jacques N. A. and Hunter N. (2002). Determination of bacterial load by real-time PCR using a broad range (universal) probe and primer set. *Microbiology*, **148**, 257–266.

Nordahn P., Mogensen S., Miller M. and Iversen S. (2012). Kobling af at-line målinger med realtidsmodel på ledningsnet. Naturstyrelsen.

Storey M. V., van der Gaag B. and Burns B. P. (2011). Advances in on-line drinking water quality monitoring and early warning systems. *Water Research*, **45**(2), 741–747.

Tatari K., Corfitzen C. B., Albrechtsen H. J. and Christensen S. C. B. (2016). Sensors for microbial drinking water quality. Technical University of Denmark, DTU Environment, Kgs. Lyngby.

Tobe S. S. and Linacre A. M. T. (2008). A technique for the quantification of human and non-human mammalian mitochondrial DNA copy number in forensic and other mixtures. *Forensic Science International: Genetics*, **2**(4), 249–256.

WHO (2011). Guidelines for Drinking-Water Quality, 4th edn. World Health Organization, Geneva, Switzerland.

Chapter 13

From sensor to decision – augmented and automated decision-making based on real-time data

Mathis Dahlqvist

13.1 INTRODUCTION

Drinking water plays a central role in human life, and for decades, thousands of water professionals have worked hard to keep it safe and drinkable, without the general public taking notice. In an attempt to always stay one step ahead and reduce risk of major incidents occurring, data logging, systems monitoring, and remote control have for a long time been standard. Thus, combining sensors, computers, and networks for systems monitoring is not a new idea.

However, in order to keep utilities secure, the majority of systems are based on private self-contained networks with only limited access to sensors, data, and information. But as the world becomes more connected, networks turn to IP-based operation, sensors and computing power become cheaper, electronics are miniaturized, data analytics become more advanced, and cloud computing is being adopted, our lives become more digitalized with new expectations and new opportunities arising together with a new generation of water professionals.

Increasingly we see consumer products, cars and trucks, as well as industrial components like sensors, actuators, and entire systems, being internet connected, and combined with computing capabilities that allow them to include powerful data analytics; which transforms these connected objects into a new breed of smart objects capable of generating, exchanging, and consuming data with only minimal human interactions – this emerging trend of hyper-connectivity combined with computing capability is also popularly known as The Internet of Things (IoT).

Editors: Torben Lund Skovhus and Bo Højris
doi: 10.2166/9781780408699_0231

Rose *et al.* (2015) anticipates that by 2025 as many as 100 billion connected IoT devices will be in operation, which without question will alter the way water utility systems are monitored. In essence, we need to prepare ourselves for a connected world that is shaped by indirect engagement with connected devices instead of active interactions with electronic content. But before this transformation can take place and the potential benefits can be reaped a set of serious challenges needs to be addressed. Some of the key challenges lie within security, privacy, and standardization, as well as regulatory and financial matters.

To date, most water utility networks have relied upon private self-contained networks, so IoT brings a new set of security challenges that needs to be addressed to ensure that the future internet connected system monitoring solutions will be as secure as the private networks are.

Poorly secured sensors and devices serve as potential entry points for cyber-attacks and leave the entire network vulnerable for exploits and data thefts. The interconnected nature of IoT devices amplifies this challenge which means that IoT suppliers needs to be extra aware of the security responsibility they have, in particular when IoT devices are combined with traditional systems, which are often based on security paradigms that more or less solely relies on a private network strategy. In these networks, there is an increased risk of operating dated, un-patched software that is extremely vulnerable in case the privacy of the network is short-circuited by IoT devices that sends data on both the secure private network, and via the internet.

Hence IoT developers, system integrators, and users have a shared obligation to ensure the integrity of the network when installing connected devices, in order to ensure that they do not expose anyone to potential harm. It is a fundamental priority that water utilities and their customers can trust that IoT devices and the data they produce are secure, in particular when they become deeply integrated in the utility systems.

IoT is redefining relations between suppliers, operators, and customers, and unlocks incredible and unique potential by allowing data to be shared between different stakeholders, which completely changes the ways data can be collected, analysed, enriched, and used, hereby enabling a much more personal and detailed experience, but at the same time give rise to concerns about privacy considerations, which might hinder full adoption of this technology.

Implementing extensive systems monitoring solutions will also raise many regulatory and legal questions. The rapid evolution of IoT technology will often render current rules and legislation outdated. Since data sharing between multiple stakeholders is common in IoT, these solutions are often challenged around legislation covering cross-border data flows, where different data protection laws could be in place. Other things like balance between law enforcement surveillance and civil rights, data retention and destruction also needs to be considered.

Extensive data collection and analysis might even intensify the feeling of being under surveillance and concerns about not being able to "opt-out" of certain monitoring schemes, hence it is key to ensure trust and confidence in these solutions by respecting privacy rights and by being transparent and honest about data usage, to have fully aligned privacy expectations.

Even though IoT based system monitoring solutions are not dramatically different to the older private network based solutions, IoT based solutions can lead to significant advantages due to the data sharing element, which redefines the relationships between objects and people, but a series of elements needs to be in place to harvest the full potential. In the end, solutions that harvests the full potential of IoT will be built through open and honest dialogue by collaborating with a whole range of stakeholders that all hold a "win-win" perspective.

13.2 CONNECTING SENSORS – LESSONS FROM INTERNET OF THINGS

Since data forms the basis of any calculation, analysis or reasoning, it is one of the key ingredients in forming new valuable relations; hence one of the first challenges of IoT is to get data, i.e. get devices connected. Since one of the key benefits from IoT comes from data sharing, proprietary and fragmented implementations of IoT and system monitoring will diminish the value created. Thus appropriate standards and architectural reference models are needed to guide the implementation of IoT. In 2015 Tschofenig *et al.* (2015) released a report that outlines four different IoT communication models used by IoT and smart objects, each with different advantages. The four models illustrate the flexibility of ways that IoT can be configured, and include: Device-to-Device, Device-to-Cloud, Device-to-Gateway, and Back-End Data-Sharing.

13.2.1 Device-to-Device connection

The direct Device-to-Device communication model is illustrated in Figure 13.1 and shows two or more devices communicating with each other using the same communication protocol via either cabled or wireless network. This model is well known in the water industry and is often seen with actuators taking direct cabled input from sensors such as pressure, temperature, or pH typical using analogue or traditional fieldbus communication.

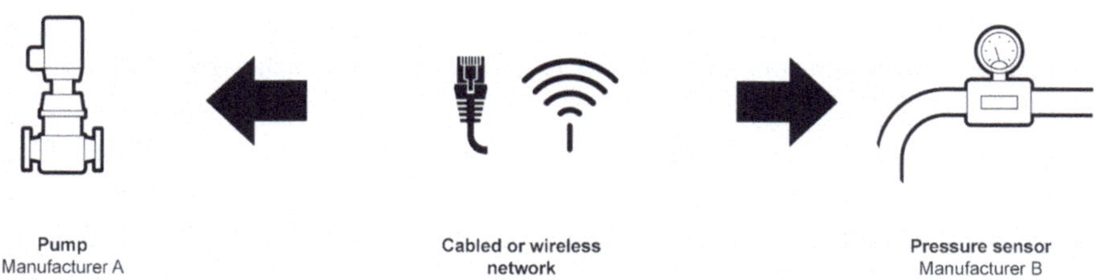

Figure 13.1 Example of Device-to-Device communication. Pump communicating with a pressure sensor via either cabled or wireless network. Adapted from Tschofenig et al. (2015).

Device-to-Device communication can happen over many different networks, but when data rates are low and data packets are small like in home automation and residential IoT it is more and more seen that wireless protocols such as Bluetooth, Z-wave, ZigBee, and Thread as the anticipated new protocol, are used for direct device-to-device communication to avoid direct cabling. Since devices in the direct Device-to-Device communication model are communicating directly with each other they are highly dependent on interoperability, i.e. their ability to "speak the same language". In a completely interoperable setup any IoT device would be able to communicate with any other, in practice it is however more complex and not always feasible, and several incompatible protocols exist, meaning either users are essentially limited to a specific family of devices that use the same protocol once a protocol has been selected, or alternatively device manufacturers need to account for multiple protocols by including redundant capabilities in the devices, adding both cost and complexity.

Some companies see a competitive advantage in building proprietary systems and protocols, but from an overall perspective this may lead to a constrained solution that lowers the value of the IoT implementation. So, it is important to choose protocols and ecosystems carefully, since it might make it impracticable to select the devices with the most suitable features in the future.

13.2.2 Device-to-Cloud connection

The Device-to-Cloud communication model is illustrated in Figure 13.2, and is used when IoT devices connect directly to an internet connected server like a cloud-based application service provider.

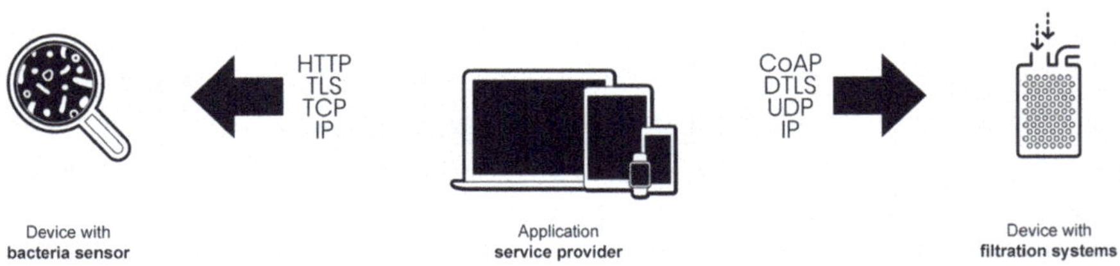

Figure 13.2 Example of Device-to-Cloud communication. Bacteria sensor and filtration system with built-in cloud-connection. Adapted from Tschofenig et al. (2015).

This is often seen in smart homes with more advanced devices like Smart TVs, fridges, and some thermostats; and in industry with advanced water quality sensors, and OEM-built sub-systems like filtration skids and pressure boosting systems. This type of communication often relies on Ethernet, Wi-Fi, or cellular based technologies.

Cloud connected devices enable users and applications to remotely access data, and often supports pushing software updates from the cloud to the device. Data can often be consumed from a multitude of interfaces like smartphones, tablets and desktop browsers, making usage of cloud connected devices extremely flexible.

It is seen more and more that cellular capabilities are combined with positional tracking either through GPS or cell-tower triangulation which can be extremely convenient for tracking assets or devices that are moving, like trucks or portable equipment used for temporary installations.

Performance of cloud connected devices are often enhanced by the remote computing capabilities, i.e. microphones can use powerful cloud computation capabilities for voice recognition used to turn them into hands-free virtual assistants, cameras can recognise items, and performance sensors like accelerometers can recognise and predict faults like a failing bearing on rotating equipment by analysing patterns in the vibration signals.

As with the Device-to-Device communication interoperability can be a challenge with Device-to-Cloud connected devices if devices from multiple manufacturers are integrated in the same solution. It is often seen that device and cloud service are from the same service provider which moves the integration interface from the device to the cloud. However, since Device-to-Cloud communication often relies on at least two parties a service provider and a network provider e.g. a cellular company, it can be a little more complex to ensure stability, and end-to-end security must be in place to ensure full data integrity.

In general, Device-to-Cloud connectivity can add value and capabilities that would be otherwise impractical or expensive to obtain, and which far surpasses the native capabilities of the un-connected device.

13.2.3 Device-to-Gateway connection

Device-to-Gateway communication models is essentially an extension to the Device-to-Cloud model. In the Device-to-Gateway model, as shown in Figure 13.3, the device connects to an intermediary device that supplies the device with access to the cloud service.

This model often relies on local execution of application software, and could provide functionalities like data transcoding, device to device, or device to cloud communication handling, as well as, data aggregation, preliminary data-analysis ("Edge Computing") and data security.

A gateway is often able to bridge some of the interoperability challenges that exist in the Device-to-Device model, since gateways are often able to communicate on different standards and fieldbuses, making them very attractive when retrofitting already installed equipment with remote management capabilities.

This model is applied a lot in smart-homes e.g. Philips HUE that uses a gateway to connect smart bulbs to the internet, as well as with fitness trackers that uses a smartphone to establish connection between the device and the internet. In the water industry, this model is also slowly emerging with vendors supplying gateways to connect already installed equipment to cloud-based remote management software.

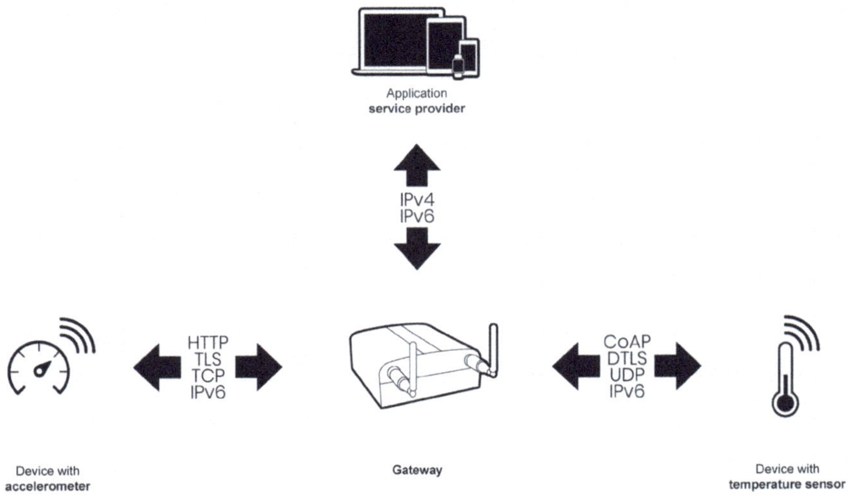

Figure 13.3 Example of Device-to-Gateway communication. A gateway provides cloud connectivity to a temperature sensor and an accelerometer. Adapted from Tschofenig et al. (2015).

13.2.4 Cloud-to-Cloud connection

Since practical application of both Device-to-Cloud and Gateway-to-Cloud communication models often relies on devices or gateways being from the same service provider as the cloud service to ensure interoperability, the integration-point moves from the device side to the cloud side, as illustrated in Figure 13.4.

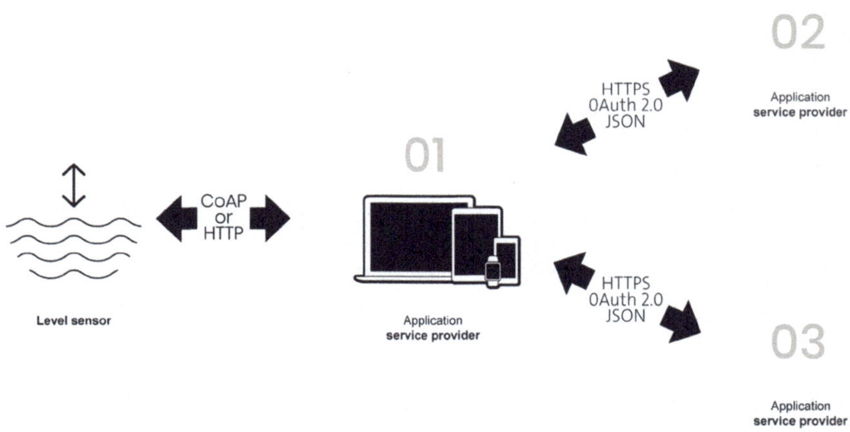

Figure 13.4 Example of Cloud-to-Cloud communication. An application with a cloud connected level sensor is enhanced by integrating with other cloud applications. Adapted from Tschofenig et al. (2015).

Cloud-to-Cloud connection enables sharing of data from IoT devices at the "back-end" of cloud connected systems and is used to tie multiple authorised cloud-based services together to enhance each other, and the user experience.

This could e.g. be sensor manufacturers, actuator manufacturers, and weather forecast providers that share data to create a solution that distributes water in an intelligent way, according to current or expected weather.

Typically cloud solutions provide well defined interfaces for exchange of data, this could be through REST APIs which enables easy and secure exchange of data between services. The advantage of Cloud-to-Cloud communication is that often a lot of devices can be integrated in reasonably straight forward ways, with a single well defined and standardized interface; this means that in an IoT inspired future Cloud-to-Cloud communication will help break down data silos and improve interoperability, since all providers have a chance to enrich data from the devices they deliver.

13.3 STORAGE STRATEGIES

Recent research conducted by Plummer *et al.* (2017) predicts that 95% of all new products created in 2020 will contain IoT functionalities. So data and internet connected devices can also be expected to enter the Water Utility industry in massive scale.

In 2013 SINTEF (2013) estimated that 90% of the world's data was created within the previous two years; further many predict exponential data growth moving forward, with Ffoulkes (2017) observing that there is agreement among researchers that at least a doubling of data quantities every two years will take place. He further predicts that machine generated data will out-grow human generated business data by a growth-rate factor of 50, as illustrated in Figure 13.5. Data acquisition has clearly entered the scene and will only increase going forward.

Figure 13.5 Exponential growth of data towards 2020, adapted from Ffoulkes (2017).

With that in mind it is important to have a solid storage strategy. Historically speaking data has typically been stored in structured or relational databases based on rows and columns, essentially like giant spread sheets. The advantage of relational databases is that on simple datasets it is fast to find specific values since everything is structured, and stored according to a pre-determined schema. The disadvantage is however that in order to establish this schema all columns must be defined before any data is entered into the database, and all rows must contain data for all columns. This means that if the schema needs to be extended with new measurement, e.g. a new type of sensor that has been procured, then the entire database needs to be altered to accommodate the new data type, which can be time-consuming and expensive.

The dot-com era in the late 90s and early 00s gave rise to alternative storage technologies: the unstructured or non-relational databases which describes a different breed of databases with the main ones being document, graph, key-value, and columnar. In general, unstructured data can be much more than numbers and simple text; it can also be complex data like images, video, e-mails, social media data, documents, etc., so it does not necessarily conform neatly into a spread sheet or database. In order to cope with this complexity, unstructured databases are based on dynamic schemas where new information-types can be added on the fly, and in contrary to the structured databases, rows do not need to contain data in all columns.

The unstructured databases have a further advantage when considering the rapidly growing data quantities, because they scale horizontally, meaning when you run out of resources you can start up a second, third, etc., server to help cope with the load. Structured databases mainly scale vertically meaning the server that is hosting the database needs to be bigger and bigger as data grows, which can quickly get very expensive when it is no longer built from standard components.

A fundamental challenge of unstructured data is however that they are not intuitive for nontechnical users to consume. Apart from the obvious lack of structure which can be time- and energy-consuming to establish, the sheer quantity that can be gathered of this type of data can be a challenge. Hence it can be beneficial to define a process where the cost of analysing unstructured data can be reduced. Because of this the two database types are often used together, unstructured databases for collection and storage of bulk data and structured databases for preparation and presentation of specific data.

When it comes to data storage, choices have consequences, and there is no "one size fits all" solution so it is important to make conscious decisions when selecting data storage technologies. If you are expecting high quantities, rapid growth, or changing data types in the future, then unstructured databases should be considered, on the other hand if nothing is changing and the business is predictable then structured database could yield the best performance.

13.4 DATA ANALYTICS

The devices are connected and sending data, you have stored the information, but what does it tell you? Court (2015) reports that retailers who leverage the full power of big data could increase their operating margins by as much as 60%, but it is not an easy task.

Throughout the last ten years data analytics have been through four major data analytics approaches. Starting from a period where descriptive analytics was the main approach, meaning creating visualizations and reports that describe and explain what had happened in the past. It is of course important to be able to understand performance and the underlying reasons, but a complete retrospective view does not help you move forward and make good decisions.

Silicon Valley encountered the limitations of descriptive analytics in the early 00s when big software companies like Google, eBay, and LinkedIn tried to analyse all the clickstream data they had in order to create better products and experiences. This data was very different, high volume, and high speed, so new technologies were needed in order to extract meaning from them. So in 2006 Hadoop, an open-source solution for storing large quantities of data, was created by Doug Cutting and Mike Cafarella based on ideas published in a research paper by Google. Along with Hadoop came a whole range of new open-source tools like Spark, Pig and MapReduce for data processing and analysis which sparked a new approach centred on experimentation with analysing big data-sets. The analyses created were typically not that sophisticated but it helped develop a way of dealing with big data.

The third approach emerged when descriptive analytics and big data analytics was combined. Typically, bigger companies have a need to both understand descriptive data-sets in combination with larger and more complex datasets like behavioural data from social media or vast machine data from internet of things enabled sensors. An example could be that it would be beneficial for a water utility to combine contextual understanding of what people in the area are saying and doing, and how the network is running, with the number of complaints received; this could e.g. help pin-point quality problems before they escalate. This combination of descriptive and predictive analytics is also known as '*operational analytics*' since it becomes an integrated part of the production and operational processes. Companies that excel at operational analytics are often highly dependent on making great operational decisions, but they also realise that these analyses can help them make better products and services in the future.

Recently a new approach in data analysis has surfaced – the three first approaches are heavily dependent on humans – data scientists – to formulate hypotheses, models, and conduct investigations in order to unveil the knowledge hidden in the data. But the most recent analytics trend is to remove or at least limit the efforts needed by human data scientists. Artificial Intelligence and cognitive technologies have undergone massive investments from companies like IBM, Google, Microsoft, and Amazon, and have the potential to truly change the way the world works. One of the core elements in the change is the use of machine learning, which in its essence is statistical analysis, but it is the machine that creates the models and tests them not the data scientist. This allows for many more models, and more complex models to be created and tested in the same time as a human could create one, and since the machine is not limited by biased human thinking machine learning can often uncover trends that the human data scientist could not find. This does however also create a challenge, since machines are able to create complex models that are not necessarily possible for human data scientists to interpret and understand, machine learning can deliver results that are coupled with uncertainty or mistrust, because the algorithm that has produced the results is essentially a black box that cannot be understood. Hence data scientists need to keep evaluating how good the machine learning models are at predicting the outcomes that are being tracked, and in the case of discrepancy it might be time to correct them.

13.5 DECISION MAKING – AUGMENTED OR AUTOMATED

Regalado (2013) estimates that only 0.5% of all data is ever analysed and used, just imagine the potential for improving your business if all the knowledge and insights the collected data contain could be utilized.

According to Herbert Simon, former Nobel Prize Winner and Professor at Carnegie Mellon University, decision-making roughly involves three stages: the first stage is about prospecting and understanding the situation, the problem is identified, desirable outcomes are formulated and information is gathered and analysed. The second stage is the conception where the solution is created by weighing options. The third stage is the actual decision where one solution is chosen and implemented by taking actions accordingly. What is left is to experience the consequences of the decision, which is essentially the product or outcome of the decision process. Outcomes are then often used for evaluating the effectiveness of the decision, which could feed back to framing a new decision process. In real life decision making does however not follow a standardized process, and inconsistencies often increase with increasing hierarchical level of the decision maker. Decisions are often made by taking two aspects into consideration: a rational and analytical aspect, and an emotional and personal aspect. The higher a leader is placed in an organisation the higher probability that self-confidence and experience is biasing the decisions towards the emotional and personal aspect, in particular if the decision has to be taken under time pressure.

Data-driven decision making has been evolving in parallel with the evolution of data analytics. In the early days of analytics data-driven decisions often required a data analyst to spend weeks on gathering data analysing, and creating visualizations and reports often just to realize that all the hard work was ignored when decisions were to be made, because the decision maker "knew better" either because of experience or because of lack of trust in data. As the data grow bigger and the analyses more complex, it becomes harder and harder to ignore the analyses, but even today many decision-makers and operators feel more comfortable with making their own decisions because they are the ones that have to live with them. When wanting to move decision-making more towards the rational and analytical side, trust becomes important.

The decision system needs to display a degree of transparency and predictability for the decision maker to learn to accept the advice and behaviours the analytical setup is providing. One way to build-up trust between decision maker and analytical service is to build augmented decision processes, where the analytical system delivers transparent actionable advice that the decision maker then can implement. By tracking the decision makers tendency to follow the communicated advice the process can slowly be automated, by letting the system take over decisions where the decision maker has a high tendency to agree.

Decisions can typically be classified into three types. Scheduled decisions that are repetitive and have a high degree of routine, these are typically the first ones that gets fully automated. Unscheduled decisions often appear under time pressure hence they are often made using experience, intuition, and creativity, and hence harder to automate because the process is dominated by the emotional and personal aspect meaning higher discrepancy between the decision process and the analytically suggested solution. The third type is a combination of the two, where automated responses exists side-by-side with decision makers judgement.

So in order to harvest the value of all the unused data it is important to build a flexible analysis and information system that can provide agile, fast, and reliable advice and actions.

13.6 REFERENCES

Court D. (2015). Getting big impact from big data, McKinsey Quarterly, January 2015.

Ffoulkes P. (2017). The Intelligent Use of Big Data on an Industrial Scale. Whitepaper. InsideBIGDATA. https://insidebigdata.com/white-paper/guide-big-data-industrial-scale/ (accessed April 4, 2018).

Plummer D. C., Andrews W., Marquis H., McGuire M., Chesini F., Leow A., Reynolds M., Lovelock, J.-D., Hill J. B., Furlonger D., Smith D. M., Golvin C. S., Velosa A., McCoy D. W., Kutnick D., Sicular S., Perkins E. and Revang M. (2017). Top Strategic Predicitions for 2018 and Beyond: Pace Yourself, for Sanity's Sake. Report ID: G00340316, Gartner.

Regalado A. (2013). The Data Made Me Do It, *MIT Technology Review*, *Business Report*, May 3, 2013. https://www.technologyreview.com/s/514346/the-data-made-me-do-it/ (accessed April 4, 2018).

Rose K., Eldridge S. and Chapin L. (2015). The internet of things: An overview, Understanding the issues and challenges of a more connected world. Report, Internet Society (ISOC).

SINTEF (2013). Big Data, for better or worse: 90% of world's data generated over last two years. *ScienceDaily*, 22 May 2013. www.sciencedaily.com/releases/2013/05/130522085217.htm (accessed April 4, 2018).

Tschofenig H., ARM Ltd., Arkko J., Thaler D. and McPherson D. (2015). Architectural considerations in smart object networking. Report #7452, Internet Architecture Board (IAB).

Section 5

Water safety, education and legislation

Chapter 14

HACCP in drinking water systems

Finn Mollerup, Carsten Vejergang, Bo Vægter and Anne Esbjørn

14.1 INTRODUCTION

The application of the principles behind the HACCP (Hazard Analysis and Critical Control Points) system has improved drinking water safety in many Danish water utilities. The present chapter introduces HACCP principles in drinking water and their applicability in relation to the use of on-line sensors and microbiological sensors in the production and distribution of drinking water.

In Denmark, the supply of drinking water is based on simple water treatment, which means that drinking water is not subject to secondary treatment, e.g. chlorination or UV treatment. Drinking water utilities are vulnerable to extraneous microbiological contamination. This means that water utilities are dependent upon their production and distribution systems being completely closed – and this requires monitoring and management of both control systems and the water quality. The HACCP principles constitute the template for designing the necessary monitoring and control systems in Denmark.

When looking at water supply in general, monitoring of the quality of drinking water is conducted through random test sampling. This alone is not adequate in relation to the HACCP standards adopted by the utilities. Water utilities have issued requests for the development of sensors that can be applied in relation to direct monitoring of the quality of drinking water, and they have petitioned for

Editors: Torben Lund Skovhus and Bo Højris
doi: 10.2166/9781780408699_0245

expansion of the usage of sensors monitoring control measures, so that these can be implemented as management tools in production.

14.2 WATER SUPPLY IN DENMARK

The geology of Denmark provides us with large groundwater sources that are suitable for drinking water. In general, the quality of groundwater in Denmark is very high, and can be used as source water for drinking water, requiring only simple water treatment.

Groundwater in Denmark is, however, vulnerable to pollution, and in order to maintain the minimum treatment policy, action must be taken to protect the future water supply. At national levels, large investments have been made in the groundwater mapping of areas of particular interest as regards drinking water. The location of the groundwater source, its volume and degree of vulnerability have been mapped, and initiatives have been taken to protect the groundwater, and to ensure that future generations also have access to groundwater sources that are clean to be used as sources for drinking water. Figure 14.1 shows a map of Denmark, indicating the areas of interest in relation to drinking water.

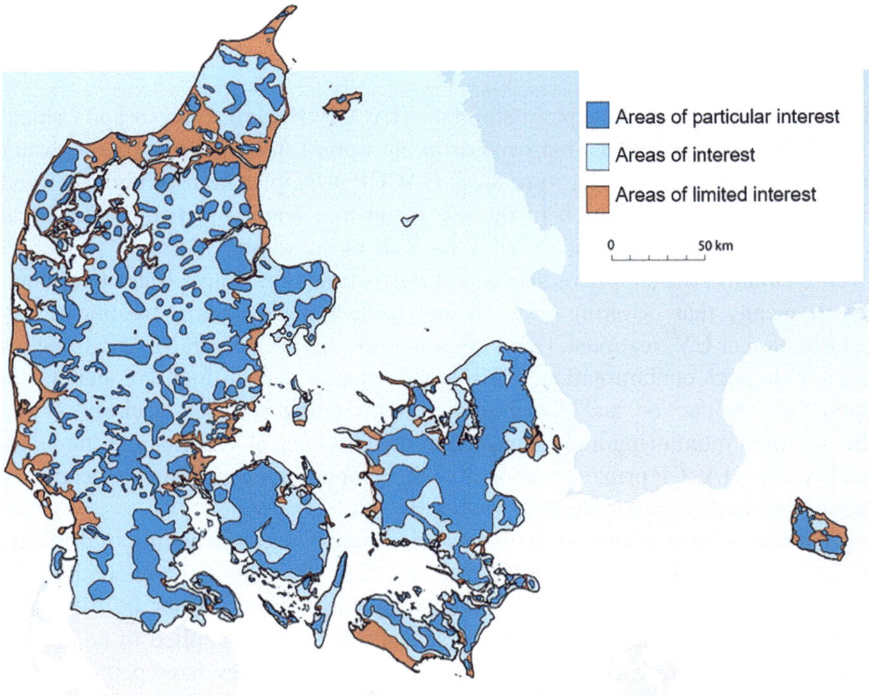

Figure 14.1 Areas in Denmark of particular interest in relation to groundwater resources suitable for drinking water (geus.dk, undated).

The tradition for simple treatment of drinking water goes back to the middle of the 18th century. The first waterworks in operation (1853) was located in Odense (Berthelsen, 2003). Since the very beginning, there has been a political consensus to restrict treatment of drinking water to simple/minimum treatment, where this is in any way possible. Historically, and at present, this is possible in most geographical areas of Denmark.

In Denmark treatment of water for human consumption is, as mentioned above, usually a simple process: the raw water is aerated and filtered through rapid down-flow sand filters. The water is generally not treated with UV-filters or chlorine, or any other barriers that change the microbiology of the water quality of the water. Since there is no secondary treatment for eliminating undesired bacteria, the minimum-impact approach requires that hygiene and process control are maintained at the highest possible levels throughout the water's journey from the source to the consumer: i.e. from abstraction, through treatment and to distribution to the consumer.

The simple aeration process serves two purposes: to remove undesired gasses (i.e. methane and hydrogen sulphide) and to increase the oxygen content in the water. Filtering through a rapid sand filter removes particles from the water (i.e. ferrous material and manganese that precipitate during the aeration process), and ammonia is oxidised into nitrate (nitrification). Precipitation occurs both via chemical and biological processes (Arvin, 2002).

Drinking water is distributed in a closed, pressurised network of pipelines. Booster stations and elevated water tanks ensure that water is distributed to the consumer at the desired pressure. This serves to provide consumers with a steady water supply, and to prevent pollutants from percolating or seeping into the drinking water, through leaks in the pipe system. Figure 14.2 shows a flow chart of the water treatment process, from groundwater to consumer.

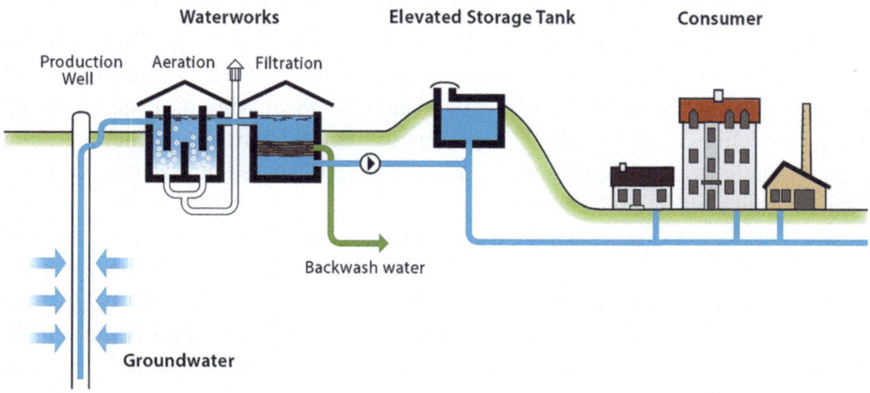

Figure 14.2 Flow chart shows treatment and transport of drinking water from groundwater to consumer (VCS Denmark, undated).

Production facilities and the distribution system for drinking water differ from other food manufacturing facilities as they are not cleaned and disinfected. Such treatment would disrupt the ecological balance (benign, indigenous water bacteria) that ensures the safety and high quality of the drinking water supplied to the consumers.

The issuing of statutory and regulatory instruments and guidelines is carried out at the national level. The municipality constitutes the supervisory water authority, and this means that the local municipality carries out the supervisory functions relating to the water utility.

14.3 HISTORY AND INTERNATIONAL STANDARDS

Food safety, as a concept, originated in the US, where it was initially linked to space sciences. The HACCP concept was developed as the result of a collaboration between NASA and a food manufacturing company (NASA, 1991). In 1959, the company was commissioned to manufacture nutritional pollutant-free products, with no direct or indirect – microbiological, chemical or physical pathogenic properties. The goal was to develop foods/nutrition that would not make the astronauts sick, while travelling in space. At that time, quality control focussed on end-products. For this type of quality control to be 100% reliable, testing would need to be carried out on 100% of the end-products – which is impractical, as the testing itself would consume the entire production, and no products would find their way to the astronauts. It was clearly necessary that focus for quality control be shifted towards the *system* that produces the food or nutrients, in order to obtain the highest attainable level of food safety. This was the process that led to the creating of HACCP.

In 1994, the use of HACCP was tested for applicability in drinking water safety (Havelaar, 1994). Over subsequent years, the concept developed and was finally realised in the Water Safety Plans. In 2004, IWA (International Water Association) and WHO (World Health Organisation) issued the Bonn charter (IWA/WHO, 2004), which established a high-level framework for drinking water risk management. In addition, WHO published the third edition of the Drinking Water Guidelines (WHO, 2004). Together, the Bonn Charter and the Drinking Water Guidelines promoted the use of Water Safety Plans as integrated procedures for managing drinking water safety.

Since the beginning, HACCP has been further developed and fine-tuned and in 1997 FAO (Food and Agriculture Organisation) and WHO jointly adopted the Codex Alimentarius Commission's collection of standards, codes of practice etc. – and it is under these auspices that HACCP has become the internationally recognised operating method that helps companies in the food and beverage industries identify food safety risks, prevent food safety hazards and address statutory compliance.

DS/EN ISO 22000:2005 is an international standard that specifies the requirements for food safety management systems, that focusses on the ability of the given enterprise to conform and maintain the required food safety regulations. The management system provides a logical method for analysing the food related processes in the enterprise, and it identifies possible risks and critical control parameters in the production processes that must be observed, in order to prevent distribution of food that is harmful to the consumers.

The new international standard ISO 22000 was introduced in 2005. This standard was a Danish initiative directed at improving food safety in food manufacturing industries across the globe. ISO 22000 was the first universal standard dedicated to food manufacturing industries. The objective was to replace the multitude of national standards and regulated production methods defined by national boards, industrial boards and consumer retail chains in the food manufacturing industries with an internationally recognised and globally uniform standard.

ISO 22000 is a management system that is designed to promote continual documentation, risk management and quality control. The standard is supervised by the International Organisation for Quality Control (ISO), an organisation consisting of 153 national standard bodies.

14.4 HACCP – INTRODUCED TO DANISH UTILITIES

In August 2002, an *E. coli* contamination was detected in the drinking water supply, affecting 100,000 households in Denmark. The contamination lasted approx. 3 weeks, and during this period, all of the affected households had to boil any and all tap water to be used for human consumption.

Historically, there have been many major or minor incidents of contamination or pollution of drinking water, that have required the instigation of emergency measures, and in connection with which the authorities have recommended that consumers boil their tap water, before they use it for human consumption (Miljøstyrelsen, 2009).

The above occurrences generated a change in the way that the water utilities view their product. The industry began researching how other providers and manufacturers of products for human consumption were handling food safety. The new approach was underscored by a re-categorisation of water: Water was now subject to the same standards as other food products.

The food manufacturing industries have been subject to statutory regulations for many years. Compliance with the HACCP principles in the processing of food products served as assurance of the quality of the manufactured products, and as documented proof of the food manufacturer's focus on food safety.

In 2004, the Danish EPA (Environmental Protection Agency) conducted a pilot project focussing on safe drinking water in the Danish water utilities

(Miljøstyrelsen, 2005). The project was based on WHO's "Water Safety Plans" (WHO, 2005), and the principles of HACCP – which by now had become a well known management system in food manufacturing industries (FAO/WHO, 1996; EN ISO 22000, 2005). The objective of the project was to determine whether the HACCP principles were applicable in the Danish drinking water sector.

The outcome of the pilot project was that the systematic approach to uncovering risks, and the establishing of management tools provided in the HACCP system were, indeed, highly applicable in the water utilities. The project also showed that a lot of work would be required for HACCP to be fully implemented in the water sector.

The pilot project prompted a number of Danish major water utilities to initiate the process of implementing HACCP. More or less concurrently, the Danish EPA issued guidelines for the implementation of HACCP in the Danish water sector. As a consequence, many of the Danish water utilities have now adopted the HACCP approach. The guidelines issued by the EPA were publicised under the title: Guidelines for drinking water quality (Miljøstyrelsen, 2005).

The concept of "Documented Drinking water Safety" (DDS) was incorporated in Danish legislation in 2013 (BEK nr. 132, 2013; Naturstyrelsen, 2014), at which time a new statutory order was issued requiring quality control of public waterworks. With the new order, it was incumbent on water utilities supplying in excess of 750,000 m^3 of drinking water per year to implement quality control systems, either by adopting the ISO 22000 system, or by introducing other systems based on the HACCP principles, e.g. DDS or other comparable systems. Water utilities supplying between 17,000 and 750,000 m^3 of drinking water were required to implement a simpler system that also included a number of HACCP principles (BEK No. 132, 2013). Several water utilities have decided to obtain formal certification in food safety by implementing the ISO 22000 standard (EN ISO 22000, 2005).

14.5 HACCP – THE DANISH APPROACH

In the Danish water utilities, the HACCP efforts focus on minimising risks in the core production processes (i.e. abstraction, treatment and distribution of water), in order to ensure that drinking water supplied to consumers is of a high quality. This focus is process based rather than end-product based. Quality control focusing on water coming from consumer taps poses an inherent risk in itself: by the time that problems are detected in tap samples, the problem may potentially already have spread to the entire system. In other words, tap sample testing is a poor management tool, as it is generally too late for remedial action by the time a problem is detected. The HACCP principles are applicable in all types of drinking water utilities, but the specifics of the HACCP approach depend on the type, size and organisation of the given water utility.

Microbiological barriers such as UV treatment, ozonisation, microfiltration and chlorination are used as part of the water treatment process in many areas of the world. In Denmark, however, these types of barriers are not permitted, and this means that the practical implementation of HACCP in Denmark is significantly different to the practical implementation of HACCP in countries using e.g. microbiological barriers. Implementation of HACCP requires that close attention be paid to the given local conditions, e.g. type of source water, type of water treatment and methods of distribution.

The implementation of HACCP serves as consumer assurance and is instrumental in ensuring drinking water of the best quality. It is essential that utilities gain and maintain the consumers' trust. As an end-user, one has very little possibility to control the water quality oneself. One must therefore be able to trust the utility with one's life.

14.5.1 Risk assessment

In order to identify possible risks in relation to a given water utility, the first step is to gain a clear overview of the key parameters in the water supply process (i.e. detailed insight into the processes of abstraction, treatment and distribution), and to collect detailed knowledge about the utility's consumers e.g. pressurised systems in industry, waste water treatment plants and rain water collecting systems. A flow chart is developed for each key parameter, detailing each process stage. The flow chart in Figure 14.2 is an example, of such a general overview of the water processes – the required level of detail for each process stage in HACCP flow charts is, however, much higher.

The flow charts form the basis for the risk assessments. All risks are included – also those for which mitigating measures have already been taken. Risk assessment takes a point of departure in a hypothetical scenario, in which no preventive / mitigating measures whatsoever, have been taken to protect the quality of drinking water. Risks can be divided into 3 general categories: biological, chemical and physical. Examples are listed below:

Biological: animals/insects, pathogenic bacteria, fungi, viruses and other xenobiotics

Chemical: natural undesired substances, pesticides and other xenobiotic substances

Physical: metal and plastic fragments / particles, grains of sand etc.

As a minimum, the HACCP risk assessments are correlated with water quality requirements, i.e. the WHO requirements for drinking water and national legislation. Other requirements could include internal policies regarding water quality as adopted by water utilities. Figure 14.3 shows the nature of requirements related to drinking water quality.

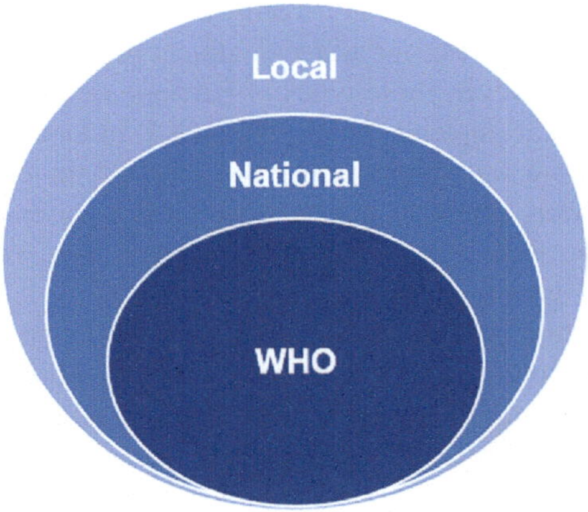

Figure 14.3 Quality goals for drinking water – general overview.

Risk factors are assessed based on their degree of potential severity (consequence) and the estimated probability of occurring. This ensures relevant prioritisation and clarification as regards to the specific risks that must be controlled. A risk assessment tool is used to categorise risks. An example can be found in Figure 14.4.

Risk assessment		Probability			
		Very low	Low	Moderate	High
Consequences	High	PRP	OPRP	OPRP/ CCP	OPRP/ CCP
	Moderate	PRP	PRP	OPRP	OPRP
	Low	PRP	PRP	PRP	PRP

Figure 14.4 Example of risk assessment tool. Green: very low to low risk; Yellow: moderate risk; Red: moderate to high risk. PRP: Pre-Requisite Procedures; OPRP: Operational Pre-Requisite Procedures; CCP: Critical Control Points.

It is important that clear definitions are established to distinguish between low, moderate and high consequences – and this applies similarly to distinguishing between very low, low, moderate and high probability. The assessment of risks varies from utility to utility, due to varying local conditions for water production and other local factors. The distribution of Probability and Consequences may also vary, and fewer or more assessment categories may be used. Based on the chosen criteria, risks are divided into categories, e.g.:

Very low risk: No control measures are required.

Low risk / PRP: Risks in this category are controlled via supporting programmes (e.g.: education, establishing of work procedures, maintenance plans, guidelines, demands in relation to materials, distribution of information material, etc.). Some processes are subject to control – and others cannot be controlled – and this means that all staff must have the necessary competences to be able to minimise risks through their actions and work routines. This aspect has been described in more detail in section 14.5.3.

Moderate to high risk/OPRP and CCP: Provisions must be made to actively address these risks – and efforts must be documented. Such measures include activities that eliminate, prevent or reduce risks to a tolerable level in relation to the established targets for water quality. It must be ascertained that the implemented control measures do, in fact, generate the desired results. Standard water sampling (random test sampling) does not suffice, as results are not immediately available. Standard sampling results are only available after 1–4 days (due to the incorporation time, transport and preparation of sampling) after the tested water has been used by consumers.

It might be preferable to use on-line monitoring in relation to the specific parameters posing a critical factor in a given situation. Direct monitoring of critical parameters may not be tenable financially or technically, but in some cases indirect monitoring of the given parameter may be feasible.

Risk assessment is conducted to identify risks that require the implementation of control measures. These measures are implemented in a plan for safe drinking water, which includes:

- *Process stages*: Identify the water supply process in which the critical risk belongs
- *Adverse events/injury*: What is the probability of occurrence?
- *Risk factors*: Under which harmful/dangerous parameter (biological, chemical or physical) does the risk belong?
- *Control measures*: Which measures have been initiated to minimise the given risk?
- *Critical limit and point of required action*:
  - At which values are indications of potential risks detectable/observable?
  - Set up a clear threshold for when values become critical.
  - Identify the values at which preventive action be initiated to prevent values from becoming critical?
  - Identify the demarcation line between values for which preventive actions can suffice and the value at which immediate corrective action is required.
- *Monitoring*:
  - Set up a monitoring plan (i.e. *what* is monitored, and *when* is monitoring carried out?)
  - Make sure it is clear who is responsible for monitoring.

- *Corrective actions*: Clarify the number and nature of measures for dealing with values that reach the critical limit (pre-emptive) and the measures for handling values that reach the critical point (immediate action).
- *Documentation/demonstration*: Set up follow-up procedures for preventive and critical actions to determine whether they generate the desired effect.

Corrective actions may not always suffice – other control measures may be required to ensure drinking water safety. HACCP includes provisions for this, as it is an iterative process with continued assessment of the applicability and efficacy of initiated measures (Miljøministeriet, 2006).

14.5.2 HACCP as a management tool

HACCP is a management tool that contributes to handling risks related to quality, and that serves to support management's prioritisation of activities and managing of company finances.

When HACCP is to be introduced in a utility, it is essential that management provide its full support and commitment to both the principles behind it and to the full implications of the implementation process that follows. A formally appointed HACCP team leader is the driving force and authority in charge of the establishment, implementation and maintaining of the system – and the team leader constitutes the communication link to management.

It is essential that a strong cross-organisational and cross-disciplinary HACCP team be appointed to carry out the implementation process, together with the HACCP team leader. The team must cover all areas of expertise and all parts of the organisation, including operations – and the appointed team members must also possess the required personal skills and competences for the job.

Management makes provisions to ensure that the team is provided with the necessary competences and instruction (whether this be formal education or participation in supportive preparatory courses). In order for the team to be able to successfully advocate for the changes relating to the new system, all employees must understand the relevance and importance of their own activities in relation to the new system.

The practical execution of decisions is allocated to the staff. Consequently, many Danish enterprises have adopted the principles of value-based management. Value-based management provides staff with the authority to make work-related decisions, without having to seek prior approval from their immediate superior. This means that the staff's degree of comprehension of HACCP, and dedication towards upholding its principles, are pivotal to drinking water safety. Staff must experience ownership of the HACCP principles – and for their implementation in daily operations. This makes it of paramount importance that staff are included in the follow-up procedures, documenting or demonstrating the effects of the measures taken to minimise risks.

The implementation of HACCP in water utilities has already yielded a number of significant improvements. Among the benefits are a general broadening of the staff's insight into the functions of the water utility – at all levels – and the natural incorporation of "hygienic design" in all projects, including restoration projects and construction. The many benefits achieved have greatly improved the identification of potential risks to the quality of drinking water.

The vastly improved quality control in the water sector also serves to uphold the long-term general trust that consumers have in their water utility. Despite the fact that HACCP, admittedly, requires a lot of allocated resources, it is clearly a highly efficient management tool that improves quality control in daily operations. The HACCP system makes it easier to document the efficiency of the water utilities, both to society as a whole, and to the relevant authorities. A good example of the benefits of improved documentation is found in the improved collaboration between utility and authority, where HACCP is instrumental in ensuring transparency and overview.

Local authorities support the HACCP standard, and a tangible example of the positive effect of this is the changed inspection frequency of waterworks in a number of municipalities. Inspections are no longer the authority's only assurance of insight into the activities of water utilities; information and documentation are readily available, and to a much larger extent than before, and in much more detail than was possible earlier. Inspections of each individual plant used to be carried out once a year, each requiring the presence of representatives from the municipality and a senior staff member. The new procedure only requires such on-site inspections to be carried out every second year.

14.5.3 An efficient tool: Education and HACCP training programmes

To a large extent, Danish water utilities base the maintaining of drinking water safety on staff knowledge and competences – both regarding the staff's specific professional areas of expertise and their knowledge about HACCP. It is an essential element in the Danish approach that utilities prepare and support staff with tools and knowhow so that they are able to make competent risk assessments on location, and are able to handle unexpected events in close collaboration with their colleagues. This approach is preferred as compared to simply issuing work instructions that staff must follow. Basing operations purely on work instructions is inadequate as a method for dealing with all the contingencies that staff experience on a daily basis. Instructions only cover the handling of a limited number of pre-defined situations, and they are not exhaustive of the full range of potential contingencies in the water utility. Water utilities ensure that staff members maintain focus on the HACCP principles by requiring staff participation in courses, meetings, audits (both internal and external) where compliance with the HACCP system is reviewed. The requirement that staff continue developing their knowledge and know-how concerning HACCP, is extended to contractors who are also required to comply with the HACCP principles in their work for water utilities.

In 2008, the Danish Water and Waste Water Association (DANVA) took the initiative to develop a course on drinking water safety based on the HACCP principles. The course was aimed at staff in water utilities. The first course was held in 2009, with course participants from nine different water utilities. The course included teaching of various topics including: basic knowledge about micro-organisms and bacteria in drinking water systems; risks (from the drilling of boreholes to the end-user's tapped water) relating to bacteriological, chemical and physical influences and the maintaining of required hygiene in daily operations.

The new statutory order regarding quality control in public water utilities (BEK nr. 132, 2013) issued in 2013, resulted in an upgrade of water safety in the water sector – specifically, in the private small to medium-sized waterworks with an output of less than 300,000 m^3 per year. The new regulations require that the person in charge of operations in a public water plant must complete a course on general operations relating to the supplying of drinking water, and a course on basic drinking water safety requirements in water plants. Regionally, major water utilities (in collaboration with DANVA) offer half-day courses on drinking water safety. The target groups include those who collaborate with water utilities (e.g. contractors and craftsmen) but also local attendants of waterworks and board members in small and medium-sized waterworks, who are often volunteers and often unpaid.

Several of the major water utilities currently require, as a minimum, that all associated craftsmen and contractors participate in a half-day hygiene course, before they undertake assignments for the utility. It is a requirement that these courses be followed up every three years. This is enforced by making HACCP training a requirement, e.g. in connection with the bidding for tenders.

14.5.4 Procedures: Drinking Water Safety brochure

Almost all major water utilities have issued a Drinking Water Safety brochure to support the HACCP efforts. The brochure is intended as information for staff, external craftsmen, contractors and visitors, so that everyone is provided with basic knowledge about how to conduct themselves on site.

The brochure contributes to minimising the risks of polluting drinking water with biological, chemical or physical contaminants. The brochure provides examples of the risks that may occur in connection with direct contact with the drinking water. It explains the conduct that must be observed in the undertaking of daily assignments for a water utility – in particular, when such assignments entail direct contact with drinking water. In general, the brochure addresses issues such as:

- personal hygiene, dangers of infection and work wear
- handling, storing and transporting of materials, chemicals, tools and equipment
- procedures for repairs and commissioning of new plants or pipes
- general requirements as regards designing projects for the water plant
- conduct and passing through a water plant, when working on site

Concurrently with the developing of the Drinking Water Safety brochure, a number of water utilities have also divided their waterworks into water safety zones. This is done to ensure that all parties on site are kept aware that they are visiting or working in a food manufacturing plant. The closer visitors or workers get to the drinking water stream, the more precautions and attention must be paid to personal conduct and the handling of materials and equipment. Figure 14.5 shows an example.

Figure 14.5 An example of water safety zones.

14.6 MONITORING OF THE QUALITY OF DRINKING WATER

Monitoring of drinking water quality is often carried out through random test sampling. Testing of water samples involve processes that are very time-consuming, and it means that results are not available until 1–4 days after the tested water has reached the consumers. This poses a risk that can be alleviated by implementing microbiological sensors for detection of pathogens in the water supply, as this generates on-line data that provides basis for significantly reducing response time.

In Denmark, test sampling of drinking water is subject to statutory regulation. However, regulations only require that very few test samples are taken. For a water utility supplying 1 million m^3 per year (the equivalent of the water supply for 8,500 standard households) only 10 water samples per year are required from consumers' installations and no further samples are required from the distribution system or waterworks for microbiological testing (BEK no. 1147, 2017). From a drinking water safety perspective, this is not adequate to ensure the quality of drinking water. Consequently, many of the major water utilities in Denmark obtain far more samples than are strictly required. Water utilities such as VCS Denmark and Aarhus Vand typically carry out weekly test sampling from several parts of the distribution network – and they take test samples once or twice every month from the outflows from waterworks and elevation tanks. The extra test sampling is introduced as a direct consequence of the implementation of HACCP. Furthermore, other additional test samples are also collected in connection with operations and maintenance of plants and distribution networks. The results of the tests are included in the assessment of whether measures to ensure the quality of drinking water and minimising risks are sufficient or require further action. The sampling procedures contribute to the documentation and verification of the results obtained through HACCP management.

In cases with elevated risks, and in which drinking water quality could be threatened, the frequency of test sampling is increased – and microbiological flow sampling may also be conducted (i.e. continual collecting of samples from a side stream over a given period, e.g. 24 hours). Flow sampling may also detect contamination values that fluctuate over time. Fluctuating concentrations of contaminants may occur in a number of cases, e.g. due to tank leakage or if the concentration of contaminants is very low, and only detectable when samples are concentrated.

On-line solutions are used to monitor daily operations as a supplement to direct monitoring of the water quality via random test sampling. The figure 14.6 below shows an example.

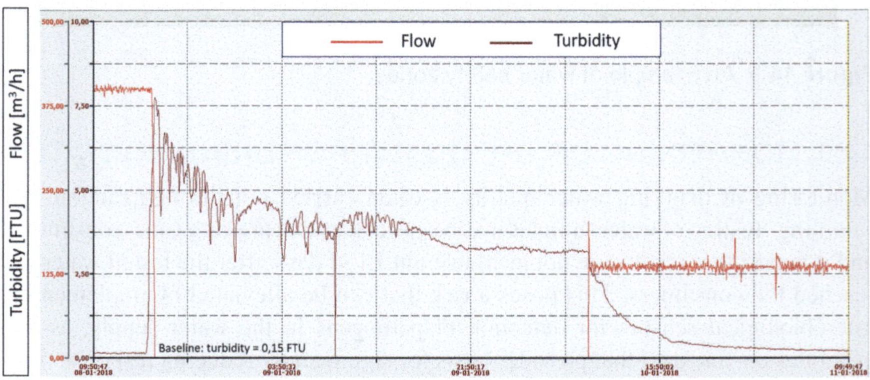

Figure 14.6 On-line monitoring at the waterworks Holmehaveværket, located in Odense.

Any deviation from the baseline indicates that something unusual has happened. The normal baseline for turbidity is 0.15 FTU (threshold = 1 FTU). In the figures the turbidity suddenly increases rapidly over a few minutes. This indicates an abnormal situation and triggers an alarm. In this particular case, the pumps were immediately stopped and the service team was dispatched to the waterworks to handle the situation. After 48 hours normal operations were recommenced. Figure 14.6 is a good example of how we are able to ensure drinking water of good quality for our consumers.

14.7 ON-LINE MICROBIOLOGICAL SENSORS

Water authorities and utilities alike, increasingly inquire into having active on-line measurements/readings approved as a management tool. However, many of the available sensors only indicate threats to consumer safety indirectly. In particular, there is a strong call for solutions that make detection of indicator microorganisms – preferably, pathogenic bacteria – instantaneous, precise and specifically directed at discerning between various types of microorganisms. Such smart microbiological sensors would be a valuable tool in relation to HACCP. Faster detection of undesired bacteria would make it possible to initiate the necessary adjustments in production to avoid the spreading of contaminated drinking water at an earlier stage than what is possible today, and, furthermore, it would make it possible to initiate prompt measures to detect the cause of a contamination.

Danish water utilities have conducted on-line monitoring of the bacterial count in the drinking water system. As regards the on-line systems, a commonality is that the measuring methods significantly deviate from the methods applied to conduct the statutory analyses. Another common finding is that the on-line systems do not include provisions for distinguishing between harmless bacteria and pathogenic bacteria. The tested on-line results do not lend themselves to direct application in relation to the statutory, microbiological control measures (Tatari *et al.*, 2016).

Microbiological sensors that are not as precise in (nor specifically dedicated to) the detection of indicator bacteria or pathogenic bacteria may, however, still be applicable in water utilities. Such microbiological sensors could be applied to the monitoring of the bacterial count in general – but they would not be adequate as replacements for traditional random test sampling. It would seem that microbiological sensors as management tools for preventing pathogenic bacteria from reaching consumers are not currently in the cards. It would seem, however, that on-line systems are applicable as optimisation tools in operations – making them indirectly applicable to obtaining improved HACCP management.

14.8 HANDS-ON EXPERIENCE FROM TWO WATER UTILITIES

In 2006 and 2007, respectively, VCS Denmark (9 million m^3 distributed drinking water per year) and Aarhus Vand (15 million m^3 distributed drinking water per year) were among the first in Denmark to obtain the ISO 22000 certification for food safety management systems. The certification entailed an increased focus on HACCP in both utilities.

The obtaining of the ISO 22000 certification has prompted several water utilities to adopt integrated management systems, where certification to obtain other ISO standards (e.g. quality, work environment, energy and CSR) is an integral part. These management systems have generally generated added value in the water sector, and, in particular, these systems have strengthened Danish HACCP thinking and practices significantly.

The figure 14.7 below shows microbiological test results from Aarhus Vand, for the last decade. The graph clearly shows that targeted efforts invested in drinking water safety generate positive results in the quality of drinking water.

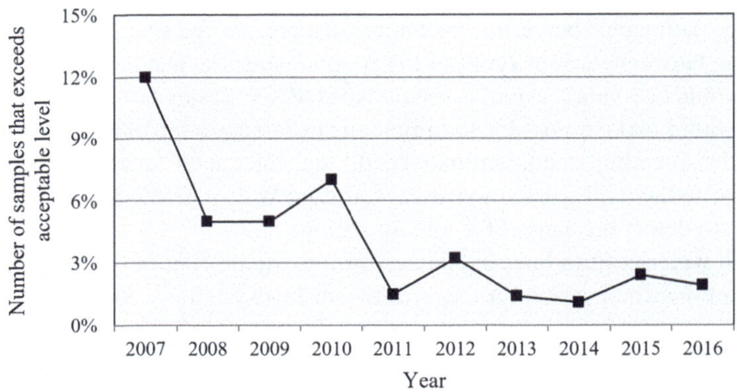

Figure 14.7 Microbiological test results from Aarhus Vand shows the percentage of drinking water samples that exceeds acceptable level for bacterial counts at 22 and 37°C, Coliforms or *E. coli.*

It is possible to improve the water quality by using the principals from HACCP, but there is a wide range of initiatives that need to be implemented. Several examples can be found in the following chapters 14.8.1, 14.8.2 and 14.8.3.

14.8.1 HACCP and maintenance of pipe systems – an example

Repairs in connection with pipe bursts and pipe-related projects constitute the greatest risk factors for water utilities. The risks are especially severe in cases of pipe bursts, as soil or surface water may enter the pipe system, if procedures for repairs are not followed carefully. The conditions of the burst itself, or the surroundings, may vary significantly. This means that, in addition to the general policies relating to pipe bursts, staff must be highly competent as regards making on-site risk assessments, and they must have the required skills and competences to initiate the necessary remedial actions. General policies from the utility Aarhus Vand might include:

- Maintaining pressure in the pipe system while the excavation to reach the burst is being carried out
- Ensuring that excavation is defined as complete removal of soil around the burst pipe
- Making certain that pumps are used to drain the excavation and keep it dry before repairs are carried out

It is important that situations in which the burst gets out of hand (e.g. if water from the excavation is about to enter the pipe system) are reported as HACCP non-conformances. The incident will then be included as documented experience that can be useful in other similar situations in the future. Reported incidents can also be studied at a later stage, and contribute to the development of better control systems for future occurrences of the same nature.

Specific procedures have been established for commissioning new pipes – or recommissioning pipes that have been subject to bursts, or that have been cut off temporarily to accommodate other projects. Such procedures include flushing and random test sampling for laboratory testing before the pipes are re-opened for operation. These procedures serve to maintain drinking water safety. Experiments and collecting of experiences have made it possible to establish procedures that ensure a high degree of drinking water safety in connection with the commissioning or re-commissioning of pipes for operation. Other benefits include ensuring that commissioning runs smoothly and that the resources required are reduced to a minimum.

14.8.2 Case study: Clean water tanks

When inspecting and cleaning a drinking water tank, HACCP procedures ensure that focus is afforded to all aspects of the process (i.e. through risk assessment). Drinking water tanks are cleaned and inspected regularly to remove sediments. and to ensure that the structural integrity of the tanks is intact.

The HACCP approach at VCS Denmark entails that it is possible to distribute drinking water to consumers, as soon as the cleaning of the tanks is completed, without the need for disposing of large volumes of drinking water, and without the use of disinfectants. Best practice procedures are based on HACCP, and include procedures for inspection, cleaning and start-ups.

Elevated bacterial counts are often registered after the re-commissioning of drinking water tanks that have been subject to inspection or cleaning. A study at the waterworks Hovedværket located in Odense has examined whether elevated bacterial counts are indicative of extraneous bacteria from the surrounding environment, or indicative of naturally occurring, indigenous, drinking water bacteria, that do not pose a health risk. The results showed that bacterial counts did not exceed guideline values at any point during the investigations. No pathogens or indicator bacteria were detected in the water, sediment or biofilm (Christensen, 2013).

This was not the only good news for the management team and the operators; they could also clearly see that by following the procedures carefully when entering the drinking water tank, the staff involved posed no risk to the drinking water system. This has been a greatly encouraging finding and a major motivating source for their continued HACCP efforts.

A very important aspect of implementing HACCP in the Danish drinking water utilities is that management entrusts staff with a high degree of autonomy and responsibility. It is important that follow-up sampling is perceived as a control measure to ensure that the HACCP system is providing the desired results, and not as a means to control the performance of the staff.

14.8.3 Case study: Truelsbjerg waterworks

When water utilities build new waterworks, the collected HACCP experiences from the entire water sector are included in the design considerations. Truelsbjerg Waterworks that was commissioned for use in 2015 by Aarhus Vand illustrates this. The combination of 10 years of HACCP experience and the establishing of a new waterworks proved an excellent basis for the designing of a plant that took drinking water safety to the cutting edge.

General considerations regarding the developing of a design supporting the strong focus on the quality of drinking water included the demand for full accessibility to the processing plant – and that it be fully accessible for on-site inspection. The design was required to support early and easy detection of leaks.

Another design requirement was that water and air must be pressurised throughout the production processes. This eliminates the majority of risks, as leaks in the production plant would result in water or air leaking from the system and into the surroundings, instead of having potentially contaminated water or air leaking from the surroundings and into the production plant.

The design requirements were instrumental in the development of the following technical solutions:

- piping on the well field is always pressurised
- the well field is divided into two production lines
- aeration of raw water using pure oxygen
- air used in production (e.g. back-flushing of filters, pressurising of the water tanks) is purified using an active carbon filter
- pressurised water treatment and pressurised water tanks
- water tanks in stainless steel, and with full accessibility for visual inspection
- after back-flushing of filters, the produced drinking water is recirculated through the filter until the turbidity is at an acceptable baseline level, only then is the water released to production
- service of sensors can be carried out while the plant remains pressurised

Figure 14.8 below show the two steal water tanks at Truelsbjerg Waterworks.

Figure 14.8 Water tanks in stainless steel that can be visually inspected.

Truelsbjerg Waterworks produces the cleanest drinking water of the eight waterworks operated by Aarhus Vand, and it is tangible proof that HACCP thinking actually works.

14.9 HACCP – AN ONGOING PROCESS

Since the implementation of HACCP, the water sector has taken a lot of initiatives that have improved the water safety level in the water utilities significantly. There is absolutely no doubt that HACCP is here to stay.

HACCP is fully integrated in the managerial functions. To keep HACCP procedures relevant and up-to-date, they are reviewed at least once a year. Reviews must be carried out for each process stage in each individual area and compared to the general development in the utility. Reviews include: new projects or changes in working procedures, introduction of new products or materials in the plant, new contractors and craftsmen etc. that are assessed for the undertaking of assignments for the utility etc. Frequent audit inspections are carried out (similar to safety inspections) with the objective of determining whether procedures match reality, and whether we actually "practice what we preach".

Reality is that members of staff come and go, and it is therefore important that the HACCP system is continually and fully updated – and that development of the system is considered an ongoing project. New insights and ideas for the maintaining and improvement of HACCP are fixed items on the agenda, which is an efficient and easy way of ensuring that the system is always updated.

HACCP is an on-going development process – it must never be used as a pretext for inaction.

14.10 CONCLUDING REMARKS

From a food safety perspective, drinking water is a product that differs from other products for human consumption. Water utilities cannot recall contaminated water, or, in the event of poor water quality, prevent sub-standard drinking water from reaching consumers, as quality control is conducted as random test sampling. Test results are only available after the tested volumes of water have been supplied to the consumers. The implementation of HACCP in Danish water utilities constitutes a paradigm shift. The primary focus has shifted from security of supply towards the quality of drinking water, and the water sector has also adopted a much more systematic approach to product control.

It is necessary that both production and distribution systems be subject to continuous control measures. Management systems based on HACCP principles (e.g. ISO 22000) are well-suited as means for ensuring the establishing and maintaining of the desired levels of quality control. HACCP includes a number of key elements that must be prepared and developed before the control systems can be implemented. Such key elements include a full description of the production facilities, a risk assessment of all production processes and a plan for eliminating or minimising risks. HACCP provides an excellent framework for the documentation of control measures.

Random test sampling in water utilities is primarily used as a tool for documenting controlled production. The intervals between testing can be quite long, and this means that the only measure for ensuring the quality of the drinking water during such intervals is an efficient and fully functional HACCP system.

The application of bio-sensors could prove to be the next developmental step in water utilities. Bio-sensors are going to be able to provide continuously updated data that is much more detailed than the data generated from random test sampling. In other words, bio-sensors can contribute to the documentation needed to meet the requirements of HACCP control systems. This means that bio-sensors could prove conducive to ensuring continual improvements in the production and distribution of drinking water – which is a prerequisite for securing and maintaining safe drinking water and continued production of high-quality drinking water.

14.11 REFERENCES

Arvin E. (2002). Filtration of drinking water; Environment and Resources. Technical University of Denmark.

BEK nr 132 af 08/02/2013 (2013). Bekendtgørelse om kvalitetssikring på almene vandforsyningsanlæg (Executive order on quality assurance at public water supply facilities). Miljø- og Fødevareministeriet.

BEK nr 1147 af 24/10/2017 (2017). Bekendtgørelse om vandkvalitet og tilsyn med vandforsyningsanlæg (Executive order on water quality and oversight at water supply facilities). Miljø- og Fødevareministeriet.

Berthelsen A. W. (2003). Vand og vækst – om vandforsyningen i Odense gennem 150 år (Water and growth - about water supply in Odense through 150 years). Odense Vandselskab, Odense, Denmark. ISBN 87-98771-42-6.

Christensen S. C. B., Esborn A. and Mollerup F. (2013). Mikrobiel vandkvalitet i rentvandsbeholdere efter inspektion og rensning (Microbial water quality in clean water tanks following inspection and cleaning). DTU Miljø, Kgs. Lyngby.

EN ISO 22000 (2005). Food safety management systems – Requirements for any organization in the food chain.

FAO/WHO (1996). Codex Alimentarius Commission. Report of the Twenty-Ninth Session of the Codex Committee on Food Hygiene. Washington D.C.

geus.dk (undated) Areas in Denmark of particular interest in relation to drinking water. http://www.geus.dk/media/15359/gv03_osd_large.jpg (accessed 8 May 2018).

Havelaar A. H. (1994). Application of HACCP to drinking water supply. *Food Control* **5**, 145–152.

IWA/WHO (2004). Bonn Charter For Safe Drinking Water. IWA, London, UK.

Miljøministeriet – by og Landskabsstyrelsen (2006). DANVA Vejledning nr 72. Vejledning i sikring af drikkevands-kvalitet (Dokumenteret Drikkevands-Sikkerhed - DDS) (Guidance on quality assurance of drinking water).

Miljøstyrelsen (2005). HACCP – et værktøj til risikostyring i vandforsyningen, Miljøprojekt nr. 989 (HACCP - A tool for risk management in the water supply sector, project nr. 989).

Miljøstyrelsen (2009). Mikrobiologiske drikkevandsforureninger 2000–2002 omfang, årsager, aktion og sygdom (Microbial drinking water contamination 2000–2002 extent, causes, action & disease).

NASA (1991). A Dividend in Food Safety. https://spinoff.nasa.gov/spinoff/database/spinoffDetail.php?this = /spinoff//jsc/JSC-SO-165.

Naturstyrelsen (2014). Kvalitetssikring på almene vandforsyninger. Naturstyrelsens vejledning, July 2014.

Tatari K., Corfitzen C. B., Albrechtsen, H.-J., & Christensen S. C. B. (2016). Sensors for microbial drinking water quality. Technical University of Denmark, DTU Environment, Kgs. Lyngby.

VCS Denmark. (undated) Flow chart: http://www.vandcenter.dk/-/media/vandcenter/illustrationer-og-figurer/vandets-vej_boring-hane.jpg?la=da&hash=FC579C08FD76A0A95BC29747AABE7884019597E6 (accessed 8 May 2018).

WHO (2004). Guidelines for Drinking-water Quality, 3rd edn, Vol. **1**. WHO Geneva, Switzerland.

WHO (2005). Water Safety Plans. Managing Drinking-Water Quality from Catchment to Consumer. WHO Geneva, Switzerland. WHO/SDE/WSH/05.06.

Chapter 15

Enhancing knowledge of microbiological sensors through education

Torben Lund Skovhus

15.1 BETTER EDUCATION OF WORK FORCE IN THE WATER SECTOR

In most European countries, there is a general lack of well-educated and qualified people with technical and scientific skills, such as engineers, scientists and technicians; the demand is simply too high. This shortage is mainly due to the limited interest in science and technology (STEM) in the school system as compared to the demands of society and industry, combined with the low birth rate of children in many European countries. In Denmark, this has been a general trend for the past few decades, and has led to several intensive campaigns to get young people interested in science and technology through the educational system; from primary up to high school level (Engineer the Future, 2018).

In Denmark, the future is looking more optimistic as the uptake numbers have increased by 60% (for 3.5 years BSc level) and 73% (for 5 years Masters level) from 2010 to 2017 (Ministry of Education). In the same period, the general uptake at the higher education level in Denmark was averaged 25% for all academic subjects.

The water sector in Denmark consists of highly specialised and qualified professionals trained to make critical decisions during the daily operation of the water utility system. As in any other industry in Europe, there is a high demand of newly educated young engineers in the water sector.

 Microbiological Sensors for the Drinking Water Industry
Editors: Torben Lund Skovhus and Bo Højris
doi: 10.2166/9781780408699_0267

When clean and safe drinking water is delivered at the tap and the sewer system functions well, we do not hear much about the water sector and the excellent work of water professionals in the daily news. On the other hand, when the streets are flooded due to cloudburst events and wastewater enters houses resulting in a lack of clean drinking water, water companies and water professionals are suddenly the focus of the media and public attention and qualified answers are required.

An example of such negative awareness was the water contamination case from Køge south of Copenhagen, Denmark in January 2007. Over one weekend, approximately 27 m^3 of sewer water contaminated the drinking water system in Køge through backflow from the wastewater treatment plant (WWTP) into the drinking water system (Vogt-Nielsen *et al.*, 2007). Two valves were malfunctioning and technical staff at the WWTP had illegally coupled the pressurised internal water system at the WWTP containing sewer water with the drinking water system of the city. Several hundred people became ill from drinking the contaminated water in the days following the contamination. The source of the contamination was traced back to the WWTP using a newly developed DNA-based fingerprinting technique applied in samples collected from the drinking water distribution network (Saunders *et al.*, 2007). Technical failures of components and incompetent operational management of the WWTP, combined with lack of microbiological sensors in the system, were the main causes of this tragic disaster.

Another example of the ways in which water handling can easily go wrong is during the commissioning of new sections of drinking water pipes. Before two sections of new and old polyethylene (PE) pipe can be connected, the new pipe section must be cleaned by running a sponge-pig through the pipeline to remove plastic fragments from the cutting and welding process, and flushed with clean water for several days before commissioning. The flushing is done to provide biologically stable water with a mature biofilm on the inner surface of the pipe (Skovhus *et al.*, 2018). During commissioning of a new section of pipe in Aarhus, Denmark in September 2015, the installation process failed due to staff oversight (Figure 15.1). At the time the new pipe section was scheduled for pigging and flushing, the contracting staff forgot to install a pump in the ditch where the coupling was to take place. Hereby dirty water from the flushing started to rise and eventually entered the newly cleaned pipe after the sponge-pig had arrived. The flushing period of the newly installed pipe section had to be extended because the dirty water in the drinking water system resulted in several cases of increased levels of Heterotrophic Plate Count (HPC) bacteria at 22°C above the allowed limit of 200 cfu/ml (BEK Nr 1147, 2017). As some coliforms were also introduced into the system with the dirty water, the citizens of the area had to boil the water over a long period before it could be used for cooking and food preparation.

Several things went wrong during the planning of the pipe commissioning; e.g., the contracting staff were not aware of the consequences the contamination would have on public health; they were not well prepared on the procedure they were instructed to follow, and they had a strict time schedule so they rushed to finish the work. After this episode, the procedure was strengthened and further optimized by the water company to meet a higher risk management standard.

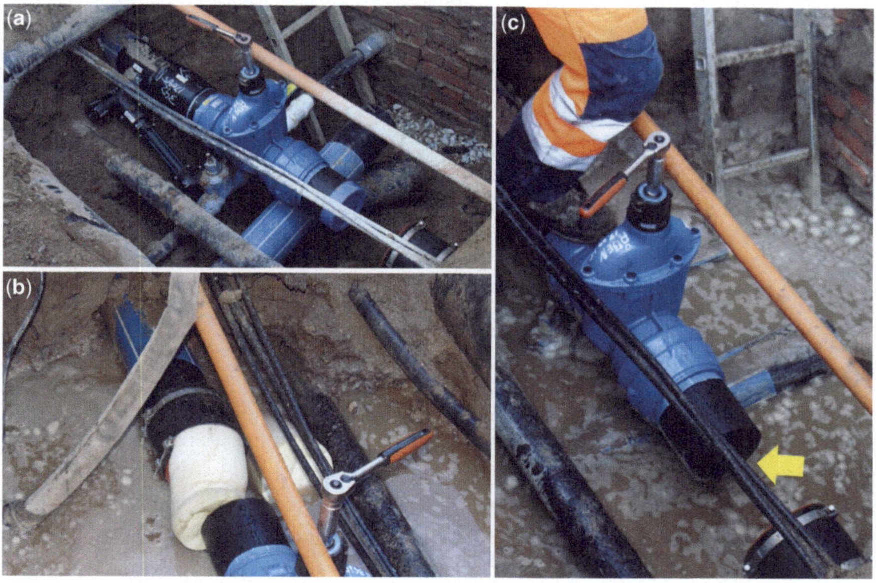

Figure 15.1 Contamination of new clean drinking water pipe section. The new PE pipe section (with the valve) is ready to be flushed with a sponge-pig (a). The two sponge-pigs comes out of the clean pipe section along with several 100 litres of water (b). The contracting staff forgot to install a pump and therefore dirty water entered the clean pipe section as indicated by the yellow arrow (c).

From these two examples, it is obvious that better understanding and education of technical staff in the consequences of microbiological contamination would have been a great help to prevent these failures. Furthermore, it is postulated that a better education level of the staff on water safety, combined with online monitoring systems based on advanced sensor technology, would advance our understanding of contamination events so that they can be prevented or avoided in the future.

Below are listed some of the benefits of introducing (microbiological) sensor technology in addition to having formally educated staff to interpret the sensor data:

- Being able to follow the (microbiological) processes in real time for optimization purposes
- Being able to react on events/peaks if needed in real time
- A need for fewer grab samples to be analysed with limited value (with low statistical relevance)
- Fluctuations can be followed and analysed over long time intervals
- Manpower is mainly needed during installation and maintenance

- Manpower can be focused on data interpretation and optimization instead of sampling
- Combining both abiotic and biotic sensors will increase the awareness of water safety

With the many benefits and possibilities of introducing sensor technology in the water sector, one might ask how they are being integrated and applied in the educational landscape today? This is discussed further in the next section that deals with the introduction of microbiological sensors at different levels of the educational system.

15.2 INTRODUCTION OF MICROBIOLOGICAL SENSORS IN THE EDUCATIONAL SYSTEM

The following provides an overview of how microbiological sensors are introduced in the educational landscape in Denmark: for students at the BSc engineering education level, industry professionals at industry courses, and during industry specialist workshops. Based on a survey among international water organizations and utilities, some interesting insights on how microbiological sensors are applied internationally in the water sector are provided. The short answer is that sensors are not well utilized and there is significant room for improvement among lawmakers, educators and technologists to work together to improve water quality standards and strengthen risk mitigation programs.

15.2.1 Supply engineering education

Bachelor of Supply Engineering is focused on drinking water, wastewater and district heating systems and is the only qualification of its kind (VIA University College, 2017). The Supply Engineering course is a 3.5 year degree provided by VIA University College in Horsens, Denmark. During the first four semesters, the students learn about geology, groundwater, construction management, climate change, hydraulics, process technology, materials, corrosion and the distribution network. Besides the engineering courses, the students have several relevant science subjects such as mathematics, thermodynamics, physics, chemistry, statistics and microbiology. Each of the first four semesters also contains a practical project that relates to the theme of the semester. During the fifth semester, the students work as interns at a water utility, district heating plant or in a private company, with a relevant topic from their specialization. During the sixth and seventh semesters, the students specialise in one of the main directions of the educational program: drinking water, wastewater or district heating. The sixth semester is concluded with a practical project and the last semester is concluded with the BSc project. During the degree at least three specific courses contain microbiology and advanced sensor technology as outlined in Table 15.1.

Table 15.1 Courses in the Supply Engineering degree containing microbiology and sensor technology.

Course (Semester)	Technical Course Content
Chemistry and Microbiology (3rd semester)	Scientific background, general microbiology principles and brief introduction to online microbiological sensor technology. Practical laboratory work. Methods: HPC, ATP, Microscopy, Colilert-18, etc.
Process Technology of the Water Sector (3rd semester)	Instrumentation and unit operations. Methods to measure pressure, flow, turbidity, oxygen, corrosion, nutrients, particles, gases, vibrations, noise, etc.
Operation and Process Management of the Drinking Water System (7th semester)	Sensor technology and application in the supply industry. Online and at-line sensors. Physical, chemical and biological sensors; data management; risk assessment; optimization and asset management theory. Practical laboratory and field work.

Besides these three specific courses, the students are encouraged to apply state-of-the-art sensor technology in their semester projects during their education and in the final BSc project.

15.2.2 A representative student BSc project on Ice Pigging of drinking water pipes

A recently concluded BSc project investigated the cleaning of drinking water pipes with slush ice using a technique called Ice Pigging (Mikalainis, 2017). The project was a collaboration between Centre of Applied Research and Development in Building, Energy & Environment, VIA University College and TREFOR, Aarhus Water, Envidan and SUEZ. The Ice Pigging method had never been applied in Denmark and the aim of the project was to demonstrate and document potential benefits that could be obtained from the cleaning method in a non-chlorinated drinking water distribution network. The method is straight-forward and consists of crushed ice (a salt brine) that is pumped into the drinking water pipe section that needs cleaning (Ice Pigging, 2018). The ice is discharged from the distribution system through an inserted fitting or a fire hydrant where samples also can be obtained (Figure 15.2).

The project involved a group of volunteer students from the Supply Engineering specialization taking samples during the project period that lasted for one week in April 2017. The students applied several types of sensors during the project; these included turbidity sensor for measuring particles, ATP (adenosine triphosphate) for measuring bacterial activity, and Coulter counter for measuring particle size. Further, several additional parameters were evaluated by the students, such as organic matter, nutrients, HPC, *E. coli*, coliforms and microbiological abundance and diversity in the samples using Next Generation Sequencing (NGS).

Figure 15.2 The Ice Pigging project with involvement of students and industry professionals. Measuring skid downstream the cleaned pipe section with connections and instrumentation of sensors (a). Sampling of "slush ice" coming out after cleaning the pipe (b). Student and teacher working with samples for further analysis in the laboratory (c). Time series of discharge samples representing the removed pigging debris and biofilm from the inner surface of the pipe wall (d).

Overall the project concluded that Ice Pigging was a fast and efficient cleaning method for drinking water pipes in a non-chlorinated distribution system. However, additional time was needed for practical implementation of the method in a system that was not prepared for the pigging procedure beforehand (Ramsay *et al.*, 2017).

15.2.3 Industry courses and workshops

Besides the traditional engineering education levels (BSc, MSc and PhD) there are also a number of venues for industry professionals to be educated and updated on state-of-the-art microbiological sensors in the water sector. It is here we differentiate between industry courses and workshops; the first being regular courses that are provided annually with a curriculum and a test. Workshops are defined as meetings that are more ad hoc with a specialized topic aiming to bridge the gap between the latest academic knowledge and industry practice. Below, three courses are described that are provided by the Danish water organization (DANVA, 2018).

15.2.3.1 Water quality

DANVA, the Danish water and wastewater association, has for several years offered a two-day course in water quality (11). The target groups for the course are the operating staff in waterworks as well as operating managers. The aim of the course is to give the participants knowledge about the physical, chemical and microbiological water quality so the participants can prioritize, as well as solve, practical water quality problems that may occur at the waterworks. The course gives a deeper understanding of water quality, in order to understand the protection of the aquifer, how to decide the best place to establish a borehole, adapting the pumping strategy, evaluating the risk of deterioration of the raw water quality over time and improving the health and safety aspect of the drinking water. The water quality course takes the participants through:

- Status of the Danish drinking water quality
- Legal Framework (water quality criteria, authority's rights and how to respond to quality problems)
- Analysis parameters (physical, chemical and microbiological)
- Measurements (in-line, at-line, field and laboratory)
- Introduction to geology (types of aquifer and different environments)
- Introduction to water chemistry (acid–base conditions, redox and water types)
- Introduction to microbiology (beneficial, pathogenic, and indicator bacteria, microbiological sampling, microbiological analysis methods (culture, enzymatic and DNA))

15.2.3.2 Practical hygiene for people working with drinking water

DANVA also offers a two-day course in practical hygiene for people working with drinking water (DANVA). This course is a legal requirement for everyone responsible for operations at waterworks. Therefore, DANVA holds this course several times every year and potential participants include more or less everyone employed at waterworks.

The aim of the course is to give the participants a good understanding of microbiology in the drinking water system, as well as good waterworks hygiene. The course gives the participants a mind-set that leads to good working habits and the skills to see when behaviour can be critical for the water quality (hygiene). The course gives the participants a basic understanding of microbiology, which makes it logical to follow the rules of hygiene in the daily work and easy to react if good hygiene practice departs from the normal. The course takes the participants through:

- Basic knowledge of microorganisms
- Bacteria in the drinking water system
- From the well to the water meter – when/where is the risks of contamination.
- What can be done to prevent contamination?
- HACCP (Hazard Analysis and Critical Control Points)

- Good hygiene rules in daily work with drinking water
- How to discover contamination

15.2.3.3 Practical hygiene for external contractors working for the waterworks

Most waterworks in Denmark require that external contractors working for the waterworks pass a 4 hour course in practical hygiene every 3 years. The course gives the participants an understanding of the risk factors in the drinking water system and an insight in good working habits to assure good hygiene practice. The course takes the participants through:

- Overview of the water – from the well to the consumers – and where you can find sources of contamination
- Physical, chemical and microbiological risk factors
- Hygiene at the waterworks, the distribution system and hygiene zones
- Preparedness and HACCP

15.2.3.4 Ad hoc industry workshops

Besides the regularly held industry courses by DANVA, there are also a number of workshops held by private companies, academic institutions and technical societies. On the topic of microbiological sensors, there have been three such workshops for 2015–2016 in Denmark as outlined in Table 15.2.

All three events were well attended (around 50 participants per event) and some stimulating discussions followed the presentations. The workshop by DANVA in 2016 was combined with a technical exhibition of instruments and sensors that gave a good practical perspective to the case studies.

In a survey, more than 20 international were asked water organizations and utilities in Europe and North America if they were hosting any courses or workshops on the topic of *microbiological sensors*. The answer was rather devastating, as none of the organizations facilitate such initiatives in their current portfolio of courses and workshops. Only the Finnish Water Utilities Association (FIWA) answered that they had a course similar to that mentioned above from DANVA called "Obtain the license to work with drinking water". Here, hygiene and traditional microbiological parameters were included, but not sensors.

So the general conclusion of the survey was that as long as the legislation does not accept data from microbiological sensors in the monitoring strategy for water safety purposes, the water companies and their organizations do not forge ahead and educate their staff about available sensor technologies. This seems like a classical "Catch-22" situation where everybody is waiting for the other to make the first move (Heller, 1961). As mentioned earlier in this section, there is significant room for improvement in this field worldwide for lawmakers, educators and technologists to work together on common ground for the sake of higher water quality standards and risk mitigation.

Table 15.2 Recent industry workshops about microbiological sensors in the Danish water sector

Workshop Title	Date and Location	Workshop Focus	Audience	Host Organization
Drinking water sensors 2016 – expo and knowledge sharing	27 January 2016 Skanderborg, Denmark	Three presentations on industry needs, market survey and technical overview. Exhibition with 10 different companies showing and demonstrating their sensor products.	Water industry professionals	DANVA danva.dk
Biological sensors for the water industry – what is possible, what is needed, and how do we apply it?	4 September 2016 Copenhagen, Denmark	Five speakers discussed the following: Current technologies and needs of the drinking water industry. Development phases of a specific microbiological sensor. Case stories on how Big Data from microbiological sensors can be managed for improved decision-making.	Water industry professionals and academics working in the field	MEWE mewe2016.org
Sensor technologies in the water sector	7 September 2017 Vejle, Denmark	Broader view on sensors including sensors for wastewater treatment plants and storm water. A majority of presentations was about online monitoring of microorganisms in drinking water.	Industry professionals from the water sector and consultants	Danish Water Forum and IDA Environment www. danishwaterforum.dk

15.3 GAP ANALYSIS AND WAY FORWARD

Question: *What moves our civilization forward*? (Hint: answer at the end of the chapter).

Interviewing two university professors in Denmark about the integration of microbiological sensors in the educational system gave the same impression as the international survey previously highlighted. Their viewpoint was that it is interesting to work with sensors from the educational and technical perspective, but as long as it is not part of the legislative framework, it is a "nice to have" rather than a "need to have" tool in the toolbox.

One way to move forward would be to get the right parties (lawmakers, educators and technologists) to discuss the benefits and limitations of applying (microbiological) sensor technology and "big data" management in the water sector through the traditional educational system, training courses and industry workshops (Figure 15.3).

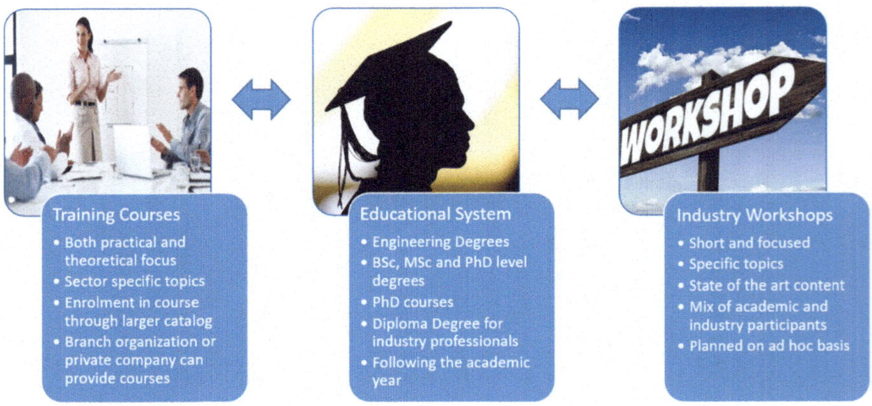

Figure 15.3 Collaboration between the traditional educational system and other operators in the educational market will improve the knowledge level of the current and next generation of water professionals worldwide related to applying sensor technology and big data management.

The traditional educational system will need to work closely together with other stakeholders to improve the knowledge level of the current and next generation of water professionals worldwide. By increasing the general awareness in the population, politicians and lawmakers may lead to a new or improved legislative framework based on the best available practice.

So, returning to the question of what moves our civilization forward, the answer is this: creativity, innovation and knowledge sharing. These start with improving education at all levels.

15.4 ACKNOWLEDGEMENTS

Thanks to colleagues and collaborators in the water sector for giving technical input to this chapter and for giving valuable feedback during the review process. A special thanks to Britt Cramer Dalén from DANVA for helping with the industry survey and providing up to date details on the DANVA courses.

15.5 REFERENCES

BEK nr 1147 af 24/10/2017 (2017). Bekendtgørelse om vandkvalitet og tilsyn med vandforsyningsanlæg. (Executive order on water quality and oversight at water supply facilities); Miljø- og Fødevareministeriet.

DANVA (2018). https://www.danva.dk/arrangementer/liste-over-alle-danva-arrangementer/ (accessed 7 May 2018).

Engineer the future (2018). http://engineerthefuture.dk (accessed 7 May 2018).

Heller J. (1961). Catch-22. Simon & Schuster, New York, USA.

Ice Pigging (2018). https://www.ice-pigging.com/en (accessed 7 May 2018).

Mikalainis M. (2017). Ice Pigging and Beyond. BSc project at VIA University College, Horsens, Denmark.

Ministry of Education (2017). *https://ufm.dk/uddannelse/statistik-og-analyser/sogning-og-optag-pa-videregaende-uddannelser/2017/notat-5-ingenior.pdf* (accessed 7 May 2018).

Ramsay L., Andresson G., Juhl A. B. and Bærentzen R. (2017). Ice Pigging: Ny metode til rensning af drikkevandsledninger. (Ice Pigging: A novel method for cleaning drinking water pipelines). *Vand & Jord*, **24**, årgang nr. 4, 160–163.

Saunders A. M, Wejse P. and Skovhus T. L. (2007). Kilden til forureningen af drikkevandet i Køge blev opsporet med hjælp fra ny kvantitativ DNA-metode (The drinking water contamination source in Køge was detected using a new quantitative DNA method). *Spildevandsteknisk Tidsskrift* nr. 1.

Skovhus T. L., Søborg D. A., Ramsay L., Andreasen J. O., Kristensen K. B., Hansen K. L., Jensen K. V., Jepsen T. L., Hansen J. B. and Højris B. (2018). Characterization of biofilms that develop during the installation phase of new drinking water pipes in Aarhus, Denmark. *Poster Presented at the Nordic Drinking Water Conference NORDIWA 2018*, Oslo, Norway.

VIA University College (2017). http://en.via.dk/programmes/technology-and-construction/supply-engineering-bachelor (accessed 7 May 2018).

Vogt-Nielsen K., Hagedorn-Rasmussen P. and Larsen, I. (2007). *Evaluering af håndteringen af forureningen af drikkevandet fra Vandværket Lyngen i Køge Kommune* (*Evaluation of the handling of the contamination of drinking water from Lyngen waterworks in Køge municipality*). CASA. ISBN 978-87-91795-57-2.

Chapter 16

Challenges of regulatory compliance using sensor technology for drinking water distribution systems

Syed A. Imran

16.1 INTRODUCTION

Drinking water quality regulations define in quantitative terms water that is 'safe' for human consumption by imposing acceptable treatment standards and contaminants levels. Currently, the United States Environmental Protection Agency (USEPA) National Drinking Water Regulations, the World Health Organization's (WHO) Guidelines for Drinking Water Quality, the European Union (EU) Drinking Water Directive, and the Guidelines for Canadian Drinking Water Quality are the most comprehensive and transparent sources of the history, science, and human-health significance of regulated contaminants. Similar drinking water quality guidelines are implemented for other countries based largely on scientific principles and guidelines recommended by WHO.

In most countries, the acceptance of any monitoring technology by practitioners is dependent on whether the regulatory agencies accept the generated data for demonstrating compliance to water quality standards and guidelines. Monitoring of water quality is essential in the protection of the health and safety of its users. However, because of the complexity of the distribution system, traditional compliance strategies are inadequate in minimizing the risks for the consumer. Currently drinking water regulations have stringent requirements for sampling, transport, storage, analytical and reporting methods. Though the regulatory agencies allow for alternate methods, these methods are mostly validated through comparison with historical reference methods. Therefore, most current water

Editors: Torben Lund Skovhus and Bo Højris
doi: 10.2166/9781780408699_0279

quality sensors try to mimic laboratory techniques and provide an indication of levels/ranges of the parameters of regulatory concern.

Many recent advances in sensor hardware, data mining and computational techniques promise increased economy of costs and operations in water quality monitoring and process control. Future sensors will use a variety of input to 'sense' the health and suitability of water quality for its intended use – without particularly trying to quantify contaminant concentrations. However, these methods have not gained the expected uptake mainly due to lack of acceptance from the regulatory viewpoint. Practical implementation of sensor technology would require not only the development of robust sensors for detecting changes in water quality, but also the use of the generated data for demonstrating regulatory compliance.

In this chapter, we will discuss the role of water quality regulations and guidelines and future directions in regulations that could impact the development and use of sensor technologies for regulatory compliance. We will also discuss some current trends that are likely to have a positive impact on the adoption of sensor technology in future.

16.1.1 USEPA drinking water standards

Generally, the term Acts, in the US context, refers to laws enacted and codified in the United States Code. Regulations or Rules refer to specific interpretations of the Acts that are recorded in the Code of Federal Regulations and subsequently implemented and enforced. The first regulation related to safe drinking water in the US was the abolition of *common cups* on carriers in interstate commerce, promulgated under the 1893 Interstate Quarantine Act in the year 1912. The 1914 Treasury Standards were established, to regulate drinking water quality, by the United States Public Health Service (USPHS), and later revised in 1925, 1942, 1946 and 1962. Through the 1974 Safe Drinking Water Act and its subsequent amendments in 1986 and 1996, the US Congress authorized the USEPA to regulate contaminants that present health risks and which are known to occur in some public drinking water supplies.

The USEPA sets legal limits for the regulated contaminants in drinking water through promulgating MCLs (Maximum Contaminant Levels) and by suggesting treatment technologies (techniques). The MCLs are intended to be as protective of human health as is "feasible". Feasibility is dependent on the contaminant concentrations, removal efficiencies using best available treatment technologies (economically achievable) and the accuracy of analytical measurements. In cases where available analytical methods or normal monitoring schemes are inadequate, the USEPA may impose treatment techniques to be followed under specified conditions. The Safe Drinking Water Act gives individual states the authority to set and enforce their own drinking water standards if these are at least as stringent as the USEPA's national standards. The 1986 Amendments to the Safe Drinking Water Act required the USEPA to set MCL Goals (MCLGs) and National Primary Drinking Water Regulations for 83 specific contaminants. The Amendments included 15 provisions for periodic review of the data and studies affecting the basis of the MCLGs and MCLs. The 1996 Amendments of the Safe Drinking Water Act resulted in several provisions related to the Small Systems Compliance Technology Lists.

The MCLs are not the only regulatory constraints on a public water service. Many other checks and balances exist in the form of source-water consumptive use permits, protection of wetlands and drainage areas, process waste discharge permits, use (restriction of use) of construction material and treatment chemicals etc. For this project, only the regulatory drivers for maintaining safe drinking water are considered (Imran *et al.*, 2009).

16.1.2 Guidelines for canadian drinking water quality

The Guidelines for Canadian Drinking Water Quality are established by the Federal-Provincial-Territorial Committee on Drinking Water and published by Health Canada (2017). These guidelines are based on current, published scientific research related to health effects, aesthetic effects, and operational considerations. Health-based guidelines are established based on comprehensive review of the known health effects associated with each contaminant, on exposure levels and on the availability of treatment and analytical technologies. Aesthetic effects (e.g., taste, odour) are considered when these play a role in determining whether consumers will consider the water drinkable. Operational considerations are factored in when the presence of a substance may interfere with or impair a treatment process or technology (e.g., turbidity interfering with chlorination or ultraviolet disinfection) or adversely affect drinking water infrastructure (e.g., corrosion of pipes).

The Federal-Provincial-Territorial Committee on Drinking Water establishes the Guidelines for Canadian Drinking Water Quality specifically for contaminants that meet all the following criteria:

1. Exposure to the contaminant could lead to adverse health effects in humans;
2. The contaminant is frequently detected or could be expected to be found in many drinking water supplies throughout Canada; and
3. The contaminant is detected, or could be expected to be detected, in drinking water at a level that is of possible human health significance.

Older guidelines are systematically reviewed to assess the need to update them. Science-based documents published as part of the Guidelines for Canadian Drinking Water Quality (i.e., Guideline Technical Documents, Guidance Documents) are developed through a documented process which includes a literature review, internal and external peer-reviews, public consultations and Federal-Provincial-Territorial approval processes (Health Canada 2017).

16.1.3 EU drinking water directive

The first wave of EU water legislation took place in the latter part of the 1970s and culminated in the Drinking Water Directive (CD 80/778/EEC). The second wave of

EU legislation took place in the early 1990s. Eventually in 1998, in order to upgrade CD 80/778/EEC to scientific and technological progress, the EU Drinking Water Directive – CD 98/83/EC on the quality of water intended for human consumption was adopted (Bernasconi *et al.*, 2003). A total of 48 microbiological, chemical and indicator parameters were established as standards that were monitored and tested regularly.

In general, WHO's guidelines for drinking water and the opinion of the Commission's Scientific Advisory Committee are used as the scientific basis for the quality standards in the drinking water. When translating the Drinking Water Directive into their own national legislation, Member States of the EU can include additional requirements, e.g. regulate additional substances that are relevant within their territory or set higher standards. Member States are not allowed, nevertheless, to set lower standards as the level of protection of human health should be the same within the whole EU. The Directive also requires providing regular information to consumers. In addition, drinking water quality has to be reported to the European Commission every three years. The scope of reporting is set out in the Directive. The Commission assesses the results of water quality monitoring against the standards in the Drinking Water Directive and after each reporting cycle produces a synthesis report, which summarizes the quality of drinking water and its improvement at a European level (Council of EU 1998).

16.1.4 World Health Organization (WHO) guidelines

The Guidelines for drinking-water quality provide the recommendations of the WHO for managing the risk from hazards that may compromise the safety of drinking-water. The Guidelines are intended to support the development and implementation of risk management strategies that will ensure the safety of drinking-water supplies through the control of hazardous constituents of water. These strategies may include national or regional standards developed from the scientific basis provided in the Guidelines.

The Guidelines describe reasonable minimum requirements of safe practice to protect the health of consumers and derive numerical "guideline values" for constituents of water or indicators of water quality. When defining mandatory limits, it is preferable to consider the Guidelines in the context of local or national environmental, social, economic and cultural conditions. The Guidelines should also be part of an overall health protection strategy that includes sanitation and other strategies, such as managing food contamination. This strategy would also normally be incorporated into a legislative and regulatory framework that adapts the Guidelines to address local requirements and circumstances.

The main reason for not promoting the adoption of international standards for drinking-water quality is the advantage provided using a risk–benefit approach (qualitative or quantitative) in the establishment of national standards and regulations. Further, the Guidelines are best used to promote an integrated preventive management framework for safety applied from catchment to consumer. The Guidelines provide a scientific point of departure for national authorities to develop drinking water regulations and standards appropriate for the national situation (WHO 2011). Despite its reluctance to provide hard standards for various drinking water contaminants, the WHO Guidelines form an integral part of water quality standards for many countries around the world.

16.2 STRATEGIES FOR REGULATORY COMPLIANCE

The task of practically implementing the drinking water standards often falls on the Drinking Water Systems (DWSs). DWSs are typically drinking water utilities run by public or private agencies that are tasked with processing the source water to applicable standards. Depending on their mandate, the DWSs can also be responsible for the distribution and monitoring of the treated drinking water. An exact definition of a DWS is out of scope of this chapter as each jurisdiction classifies DWSs differently based on size, population served etc. Depending on their size and water source, DWS may not have to comply with all the drinking water regulations. Most regulations are limited to DWSs of particular size and source. Therefore, where an effort has been made to include all the relevant drinking water regulations, DWSs should determine the regulations applicable to them based on their size and water source.

Different countries also have different levels of comfort with treatment technologies for drinking water treatment. Many factors affect the appropriateness of a given technology for treatment, distribution and monitoring. These factors include; seasonality and reliability of source water; economic considerations of centralized DWSs; energy requirements; and public expectations. In the interest of conciseness, the drinking water regulations are discussed in context of the USEPA regulations throughout the rest of this chapter.

16.2.1 Microbiological contaminants

The acute and chronic risk to human health from water borne pathogens has been the primary impetus towards the development of drinking water regulations around the world. Water-borne pathogens are different from other chemical contaminants in that they can cause acute as well as chronic health effects. Pathogens are often aggregated or adherent to suspended solids in water, and the pathogen concentrations vary in time, so the likelihood of acquiring an infective dose cannot be predicted from their concentration in water. Pathogens can rapidly multiply within the infected host and some pathogens can grow and multiply even outside their hosts (WHO, 2011). Due to this uncertainty, it is necessary to employ a holistic view of compliance (rather than just at point of treatment). Regulatory agencies therefore favour a multi-barrier approach where the risks of microbiological intrusion, proliferation and exposure are minimized from the source water all the way to the consumer's tap.

Control measures for microbiological contaminants differ from other organic and inorganic contaminants in that their levels, occurrence and sources cannot be precisely predicted in time for proactive mitigation measures. Under the Surface Water Treatment Rules and the Ground Water Rule, utilities are expected to estimate the baseline level of pathogens in their source water and provide removal measures based on the source water levels. Filtration is the treatment of choice for the removal of microorganisms from source waters. However, filtration is often practiced in conjunction with other unit processes such as coagulation, lime softening or disinfection. Filtration is almost always followed by disinfection due to the increased potency of disinfectants in filtered waters.

Researchers in the US have made a compelling argument for maintaining disinfectant residuals for North American water supplies (Haas, 1999; LeChevallier, 1999). Disinfection, especially chlorination, has been the treatment of choice in many water systems to provide primary and secondary disinfection. Primary disinfection is provided at the treatment plant to mitigate pathogens occurring in source waters. Before introduction into the distribution system, a small residual dose of disinfectant is added to limit microbiological growth in the pipes as well as to mitigate intrusion of pathogens at breaks. However, concern over chlorinated disinfectant by-products has prompted the need for alternate disinfectants. Chloramination, ozonation, chlorine dioxide, ultraviolet irradiation and advanced ozone/peroxide processes have been gaining ground as the alternate treatments. However, some (chloramination) lack potency, some (ozonation) do not provide residual protection and most are complex in application compared to chlorination.

In addition to the primary role of disinfection at the treatment stage, the secondary role as a residual is important to avoid regrowth or intrusion problems. Additional protection after treatment is provided, by requiring covered reservoirs, minimum disinfectant residuals and monitoring for pathogens in the distribution system (under the Total Coliform Rule). The choice of the disinfectant is also determined by the source water quality and the potential for forming by-products that might impact consumer health (USEPA 2001).

16.2.2 Contaminants specific to water sources

Though all the USEPA drinking water regulations attempt to mitigate the health risk at the point of use, from a DWS's perspective some contaminants are easier to control. These contaminants are easily removed by conventional technology in place or by slight modification of existing processes (conventional coagulation vs. enhanced coagulation) and do not re-occur after removal. These contaminants are best monitored at the point of entry into the distribution systems. Most organic and inorganic contaminants can be classified under this category. Included in this category are contaminants listed in the Phase I, II and V Rule, the Arsenic Rule and the Revised Radionuclide Rule. A significant change made by the USEPA is that some contaminants (Radionuclide Rule contaminants) need to be in compliance at all points of entry into the distribution system, where previously only a representative point in the distribution system was monitored for compliance.

Some contaminants can also be classified based on their ease of removal by certain treatment processes. For instance, granular activated carbon is effective in removing more than 80% of all regulated organic contaminants from drinking water sources and can be termed as an organic contaminant-specific technology. Similarly, coagulation and filtration can be termed as effective treatment technologies for most heavy metals.

Microorganisms in source waters can present another challenge for compliance due to uncertainty of occurrence and complexity of sampling and quantification. The USEPA attempts to minimize this risk by providing minimum level of treatment for certain classes of microorganisms categorized with respect to its prevalence in certain source waters (groundwater, surface water or groundwater under the direct influence of surface water). Treatment techniques, complemented by rigorous monitoring, can reduce the endemic and acute health risks associated with pathogenic microorganisms.

The various Surface Water Treatment Rules, the Total Coliform Rule and the Ground Water Rule attempt at mitigating the most endemic and difficult to treat pathogenic microorganisms present in many source waters by establishing minimum treatment criteria instead of having treatment goals. This is due to the large variability in the level of microorganisms in the source water due to seasonal variations or during epidemics (Ashbolt *et al.*, 2005). The proposed Long Term 2 Enhanced Surface Water Treatment Rule is the most comprehensive microbiological control regulation for surface water and groundwater under the direct influence of surface water sources where the required level of treatment is linked to the prevalence of pathogenic microorganisms in the source-waters. Reliance on a single strategy of treatment (either filtration or disinfection) is discouraged to provide a more comprehensive safety barrier in case of occasional treatment deficiencies.

When criteria of compliance for different contaminants are in conflict, the USEPA promulgates the conflicting regulations in pairs. For instance, the various Surface Water Treatment Rules require different levels of treatment for different pathogens and one way of complying would be to disinfect with high disinfectant concentration or provide long contact times. However, this may lead to problems of either higher exposure levels to disinfectants or to higher disinfectant by-products due to long contact times. This conflicting objective is balanced by providing the counter-regulation to the Surface Water Treatment Rules in the form of the Disinfectant/Disinfectant By-product Rule where the disinfectant and disinfectant by-products are regulated (Imran *et al.*, 2009).

16.2.3 Contaminants generated within the water distribution infrastructure

Certain contaminants are generated in the distribution system due to the interaction of bulk water constituents or the interaction of bulk water with pipe surfaces. Trihalomethanes formation is an instance where bulk water constituents (organic matter and disinfectant) interact to form unacceptable by-products. The disinfectant by-products of chlorine were the first to be regulated. As utilities began using alternate disinfectants; other potentially harmful disinfectant by-products associated with ozone and chlorine dioxide came to be recognized and regulated under the Stage1 Disinfectant/Disinfectant By-product Rule and the proposed Stage 2 Disinfectant/Disinfectant By-product Rule.

Residual disinfectant in the distribution system presents a unique challenge for both regulatory agencies as well as the public water systems. Though residual disinfectant may be required for control of microorganisms in the distribution system, a high concentration may render the water unacceptable for human consumption. The health problems associated with high residual disinfectant are mitigated under the Disinfectant/Disinfectant By-product Rules by regulating the maximum concentration of chlorine, chloramines, ozone and chlorine dioxide in the distributed water. In contrast, the regulation dealing with microbiological control in the distribution system requires a minimum residual disinfectant concentration at all points within the distribution systems.

The interaction of the bulk water with the pipe surfaces may lead to corrosion and leaching of pipe material into the transported water. Lead and copper corrosion, biofilm sloughing, and red water problems are instances where the bulk water interacts with the pipe surfaces. The Lead and Copper Rule is unique among USEPA rules in that it regulates the contaminants that are derived from the corrosion of the distribution system. Therefore, monitoring is done at the consumers tap and not the treatment plant. If 10% of the samples exceed the action levels for lead (15 μg/L) and copper (1.3 mg/L) the utility has to take documented action to mitigate the lead and copper levels. The Lead and Copper Rule monitoring requirements presents a challenge to utilities in determining the location and frequency of sampling as well as collaborating with the consumers to get the samples at their taps. However, the Lead and Copper Rule is very difficult to enforce due to diverse age, lengths, residence times, locations, state of corrosion and materials of household plumbing. Some of these variations are mitigated by following a standard sampling protocol, while others are inevitable due to the complex nature of the distribution network.

Contaminants generated in the distribution system are due to the interaction among bulk water constituents (residual chlorine and organic matter react to form trihalomethanes and haloacetic acids, interaction with pipe surfaces (corrosion products, biofilm slough-off) and hydraulic effects.

The internal surfaces of distribution pipes can act as a sink and/or source of chemical and biological contaminants (Schock *et al.*, 2008; Lytle *et al.*, 2004; Valentine & Stearns, 1994). Changes in water quality can trigger the release of these contaminants to an extent where the water quality becomes unacceptable for use even though the water quality at the treatment plant was acceptable (Imran *et al.*, 2005).

Table 16.1 Drinking Water Regulatory Standards for Microbiological Parameters.

Parameter	Canadian Guideline (1)	USEPA Standard (2)	WHO Guideline (3)	EU Directive (3)
Heterotrophic plate counts	–	n/a Treatment: <500 colonies/mL	–	(colony count @ 22°C) No abnormal change.
Cryptosporidium	Treatment: min 3 log removal	MCLG (0 mg/L) Treatment: watershed control provisions	–	–
Giardia	Same as Cryptosporidium	MCLG (0 mg/L) Treatment: >99.9% removal/inactivation	–	–
***Clostridium perfringens* (including spores)**	–	–	–	0/100 mL
Total coliforms	0/100 mL	MCLG (0 mg/L) MCL: a mix of monitoring and treatment requirements under the Revised Total Coliform Rule	0/100 mL	0/100 mL (same as *E. coli*)
E. coli	0/100 mL	Same as total coliforms	0/100 mL	0/100 mL
Enteric viruses	Treatment goal: min 4 log reduction	MCLG (0 mg/L) Treatment: >99.99% removal/inactivation	–	–
Enterococci	–	–	–	0/100 mL
Legionella	–	MCLG (0 mg/L) Treatment: same as Cryptosporidium	–	–
Turbidity	Treatment limits for individual filters	Treatment limits for individual filters	–	–

(1) The methods and monitoring guidelines for these contaminants are collected and published as technical monographs by Health Canada (Health Canada 2017).
(2) The methods and monitoring requirements for these contaminants are specified in 40 CFR 141.402(c)(2). Additional methods are listed in Appendix A to Subpart C of Part 141. (USEPA 2017).
(3) The methods and monitoring guidelines for these contaminants is listed in ISO 9308-1, 9308-2 and 9308-3 (WHO 2011).

All physical, chemical and microbiological processes that take place in the distribution system are related in some way to the quality of the water being distributed. Over a period of historical usage any source-water will establish an "apparent" equilibrium of physico-chemical and biological reactions in the distribution system. This apparent equilibrium makes it possible to predict and maintain a specific quality for the water being distributed. Recent trends in source-water diversification, regulatory restrictions and demand dynamics may result in distribution systems receiving variable water quality. These water quality changes may be either gradual or sudden depending upon the operational conditions. Any change in the finished water quality will disrupt the existing equilibrium resulting in a transitional state until a new equilibrium is achieved. This new equilibrium may either improve or deteriorate the water quality at the consumers tap (Taylor *et al.*, 2005). Distribution systems that have maintained an historic single source for drinking water supplies can have problems maintaining the distributed water quality when alternate sources are used (Imran *et al.*, 2006).

16.3 REGULATORY SHORTFALLS

Out of the 14,028 reported violations for health-based standards in 2002, the Total Coliform Rule violations (69%) far exceeded the chemical contaminants based violations (10%), Lead and Copper Rule violations (5%) or the Surface Water Treatment Rule violations (15%) (USEPA 2004). The violations must also be viewed in terms of applicability of regulations. The Total Coliform Rule applies to all water systems, while the Lead and Copper Rule and Surface Water Treatment Rule are limited by the size of the system classification or type of source water. The higher incidence of Total Coliform Rule violations underlines the importance of maintaining distribution infrastructure in a good condition (Roberson, 2003).

In Canada, the worst modern water-borne outbreaks occurred in Walkerton in 2000 and in North Battleford in 2001. These outbreaks resulted in seven deaths and over 7000 illnesses (Bereskie *et al.*, 2017). While there were no major waterborne disease outbreaks since 2001, there were roughly 1766 drinking water advisories in communities across Canada in 2008. By 2015, although the locations of the water-quality advisories had shifted, the number of boil water advisories was 1838. Though such advisories are expected to be transitory, among First Nations communities these advisories have been in place for up to 5 to 15 years (Eggertson, 2015).

Distribution system deficiencies in community water systems caused 31% of the reported waterborne outbreaks during 1971–2000. During the same period, distribution system contamination was responsible for 8% of the outbreaks in non-community water systems. The remaining outbreaks were attributed to treatment breakthrough and disinfection deficiencies (Craun *et al.*, 2003). Contaminants entered the distribution system through cross-connections, back-siphonage, corrosion and leaching of metals, broken or leaking mains, storage deficiencies and construction and repair of mains. Craun *et al.* (2003) reported that the most important chemical contaminant causing distribution system outbreaks was copper. Acute outbreaks associated with copper plumbing were primarily due to corrosion and plumbing deficiencies.

Iron in drinking water networks arises from two sources; breakthrough from the treatment plants and corrosion and subsequent release of corrosion by-products from iron pipes. The aesthetics of the water deteriorate at a much lower iron level than its health-related maximum level. Therefore, iron levels are managed under the National Secondary Drinking Water Regulations. Some states and utilities have adopted these secondary regulations as mandatory water quality goals. However, the usual point of monitoring for iron is at the point of entry to the distribution system and does not consider the iron contributed by the distribution system itself. Iron contributed from iron pipes is a major source of turbidity and coloured water within the distribution system and has been linked to increased disinfectant demand, microbiological-shielding from disinfectant action and in microbiological proliferation.

With one notable exception of the Lead and Copper Rule, the USEPA does not require the water quality to be protective of the distribution system. Other regulations like the Total Coliform Rule might require integrity of distribution systems for compliance (intrusion through compromised distribution infrastructure may cause microbiological failures). This compliance is ultimately achieved through water quality criterion (residual chlorine) and by infrastructure rehabilitation.

Utilities traditionally have relied on single source-waters and therefore had to deal with minimum variations in water quality and in the type or concentration of contaminants. This practice afforded the utilities to establish a single compliance strategy for the specific regulated contaminants in their source water. However, over-reliance on single sources is being abandoned due to adverse environmental impacts associated with unsustainable withdrawals. Today, utilities (especially in communities with high growth) may have multiple source-waters and hence complex compliance strategies. For instance, a utility having separate treatment of groundwater and surface water (and complies with regulations individually), does not have a guidance on how mixing of the waters within the distribution system and its related problems will affect its compliance with regulations. Such mixing may have a complex impact on the pH, disinfectant by-product precursors and formation potentials, and corrosion related problems in distribution systems.

Since the primary objective of the drinking water regulations is the protection of human health, distribution system deficiencies should be taken seriously or even regulated. As stated earlier, the first drinking water regulation in the US was the abolition of *common cups* on interstate carriers. The spirit of this regulation that "*the quality of drinking water is as good as the container in which it is served*" should be followed in the case of distribution systems as well (Imran *et al.*, 2009).

16.3.1 Rise of class action lawsuits

Increasingly communities that have been affected by drinking water crisis are taking to the courts to have their grievances redressed. In the US, various groups of plaintiffs have filed class-action lawsuits alleging due process violations and torts as well as Safe Drinking Water Act claims. In January 2016, officials and the City of Flint were sued for various Safe Drinking Water Act violations. The 2017 settlement from the Flint, Michigan, Safe Drinking Water lawsuit requires the City of Flint and the State of Michigan to replace Flint's lead and galvanized steel service lines within three years, while, in the meantime, providing water filter support and education as well as extensive tap water testing. The agreement requires the state to provide $97 million to fund the replacement of the lead and galvanized steel water pipes (Cagle & Broadaway, 2017).

Chicago residents filed a class action lawsuit accusing the city of Chicago of knowingly contaminating the city's water supply when replacing lead water pipes in the city. The elevated, unsafe levels of lead in the drinking water have persisted since 2008 when the city began construction on the old lead pipes, delivering water to 80 percent of Chicago (Tuttle, 2016).

In the US, Legionella-contamination settlements and jury awards over the past decade range from $307,000 to $5.2 million, and mostly targeted owners of buildings, cruise ships, hot tubs, hospitals and cooling towers. There are no statutes or laws requiring testing of building water systems for Legionella, but a 2015 Legionella outbreak in the South Bronx resulted in the City of New York adopting regulations requiring routine testing of cooling tower water for Legionella.

In Flint, in addition to lead-contamination lawsuits, residents have filed Legionella-contamination suits against the State of Michigan, which was managing the city's water system at the time of the outbreak. The suits claim failure to add corrosion-control chemicals allowed Legionella bacteria to breed in the city's pipes. At least 91 people reportedly contracted Legionnaires' disease in Flint, and 12 died from it (Espinola, 2017).

In Canada, four Alberta First Nations are suing the federal government over its alleged inaction to address the deplorable condition of drinking water on nearly three-quarters of aboriginal reserves in Canada. Many aboriginal communities experience water shortages, cross-contamination and boil-water advisories. The group also claims lack of clean drinking water violates constitutional rights to life, liberty and security of aboriginal people and that legislation passed last year to address the issue does nothing but protect the government from liability. The allegations have not been tested in court (Barrett, 2014).

In 2016, the Government of Saskatchewan and the City of North Battleford set compensation for young people who became sick in the city's 2001 tainted water scandal. Earlier, a class-action lawsuit worth $3.3 million was approved by the courts. Anyone under the age of 18 who got sick between March 20 and April 25, 2001 was offered financial compensation, as long as they hadn't already been paid (Cowan, 2016).

16.4 ALTERNATE REGULATORY METHODS

In most countries, drinking water regulations are viewed as legally binding obligations of the DWSs. Consequently, activities that demonstrate compliance are strictly defined in unambiguous legalese. The timing, frequency of collection, transportation, storage, preparation, and analytical methods for monitoring are well-defined and publicly available. This strict definition of compliance actions results in a uniform application of the laws across different jurisdictions.

DWSs are typically amenable and do in fact use various sensors for measuring physical and chemical parameters in their day to day business (for instance, flow, pressure, temperature, chemical dosing control via SCADA systems). However, the acceptance of microbiological sensors has been painstakingly slow. The regulatory requirement for comparison with standard methods raises the testing and validation bar for the adoption of innovative sensor technologies for compliance monitoring.

When regulatory agencies set a monitoring requirement for a contaminant, they also establish in the regulations at least one reference analytical method for analysis of the contaminant. For instance, under USEPA's Drinking Water Alternate Test Procedure Program, organizations can submit applications for the evaluation of modified versions of existing test methods, and new test methods as equivalent alternatives to test methods included in the drinking water regulations. Equivalent alternative methods can be demonstrated by either a side-by-side comparison test with a reference method or through a QC-acceptance based criteria similar to the reference method.

Side-by-side comparison of methods uses the following parameters (USEPA 2010):

1. Recovery. Does the new method have similar, better or worse recoveries of the target organism as the reference method?
2. Precision. Are the recoveries by the new method significantly less or more variable than the reference method?
3. False positive rate/specificity. Is the new method significantly more likely or less likely to detect non-target organisms or other sample constituents that would be reported as the target organism by the analyst when compared to the reference method?
4. False negative rate/sensitivity. Is the new method significantly more likely or less likely to exhibit non-detects for samples with the target organism or to exhibit results that are biased low when compared to the reference method?

To generate these parameters, samples are analysed by a single laboratory. During this evaluation, true positives are those positive results from the alternate methods that are confirmed by the reference method. False positives are those not confirmed as detected by the reference method. Conversely, true negatives are those negative results confirmed as negative and false negatives are negative results that were shown as positive on the reference method. The number of samples (and matrix types) used in the study are determined using a historical EPA standard (USEPA 2010).

QC acceptance criteria-based comparison studies are generally conducted to demonstrate that the alternate test or new method can meet the QC acceptance criteria of the EPA-approved reference method. A minimum of three laboratories should be used for a QC acceptance criteria-based comparison study. All three laboratories should meet all QC acceptance criteria in the EPA-approved reference method. For all QC acceptance criteria-based comparison studies, there should be one matrix per laboratory. For ambient water studies, the turbidity of at least one matrix should be greater than 10 NTU. The number of replicates per matrix and laboratory should be specific to the QC criteria to which the results will be compared. Generally, the number of reagent results should be four and the number of source water results per matrix should be two (USEPA 2010).

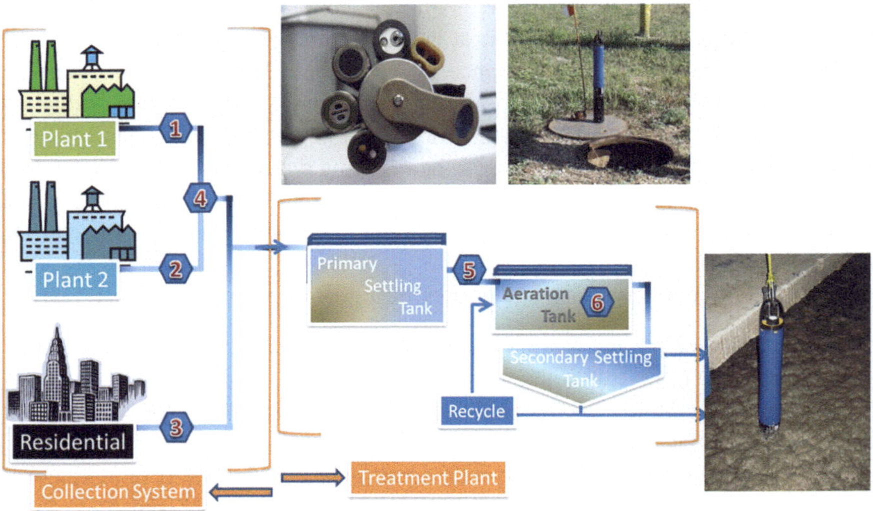

Figure 16.1 Sensors that are robust, self-calibrating, self-cleaning, with on-board data storage and retrieval systems are now available off the shelf. Many of these common water quality sensors have a small form factor and watertight for deployment in pipes and clearwells. (Adapted from Imran *et al.*, 2014)

For microbiological sensors, the requirement of specificity and sensitivity comparable to the reference methods is often an insurmountable barrier. Although sensors are increasingly appearing on the market, effective implementation in water utilities has not been realized for several reasons. Sensors for monitoring of microbiological drinking water quality have seen great development in the recent years, but still the 'ideal sensor' as defined by the utilities (with total coliforms and *E. coli* in focus) is not yet available (Tatari *et al.*, 2016).

16.5 ADVANTAGES OF SENSOR TECHNOLOGY FOR REGULATORY COMPLIANCE

The slow adoption of water quality sensors is due to the perception that sensors do not meet practical utility needs. The ubiquity of different sensors with different claims requires independent tests and verification before acceptance by utilities. Also, the sensors require changes in DWS practices that are outside the comfort zone of most field practitioners. Additionally, there is the burden of managing large data quantities and translating them into meaningful information for operational processes. Field practitioners are typically used to standardized methods/guidance for data analysis and interpretation (Raich, 2013).

However, the current generation of water quality sensors is far different and versatile than those that were commercially available even a few years ago. Commercial sensors that are robust, self-calibrating, self-cleaning, with on-board data storage and retrieval systems are now available off the shelf. In contrast, current monitoring methods are costly and time-intensive, and limitations in sampling and analytical techniques exist there as well. Ho *et al.* (2005) reported that Looney and Falta (2000) report that the Department of Energy Savannah River Site requires manual collection of nearly 40,000 groundwater samples per year, which can cost between $100 to $1,000 per sample for off-site analysis. Wilson *et al.* (1995) report that as much as 80% of the costs associated with site characterization and clean-up of a Superfund site can be attributed to laboratory analyses (Ho *et al.*, 2005).

Sensor Monitoring Systems include the sensors, data retrieval and storage, data analysis algorithms, and decision support systems. This is a significant cost to the drinking water provider. While typically the costs of monitoring through an on-site or off-site certified laboratory are viewed as ongoing operating costs, installation of the Sensor Monitoring System is initially a significant capital investment. A business case for sensors, in the absence of a mandatory regulatory requirement, is extremely difficult to make in an age of tightening public funding.

Figure 16.2 shows the costs of implementing a simple pH sensor system into a process stream compared to the cost of analysis at an accredited laboratory. The costs of sensor implementation are initially quite high compared to laboratory costs. However, due to the low costs of operation of a pH sensor (once installed) results in a break-even with laboratory costs after just 30 days. Where sensor technology shines is in the high-frequency analysis (300 samples/day), where the break-even with laboratory costs occur in just under 3 days. Another important aspect of sensor systems is that the incremental costs due to high frequency analysis are nearly similar to that for a low-frequency analysis.

According to den Berg *et al.* (2015), current regulatory sampling programs with around 200 samples a year for an area the size of Leeuwarden only have a 5–10% detection likelihood for a short-lived contamination (2 hours) with pure sewage. Taking more samples means higher detection likelihoods. However, as more and more samples are taken, the added value of each sample decreases. Even for sampling programs with a 10 times increased sampling frequency compared to a typical regulatory program the detection likelihood remains relatively low at 35%.

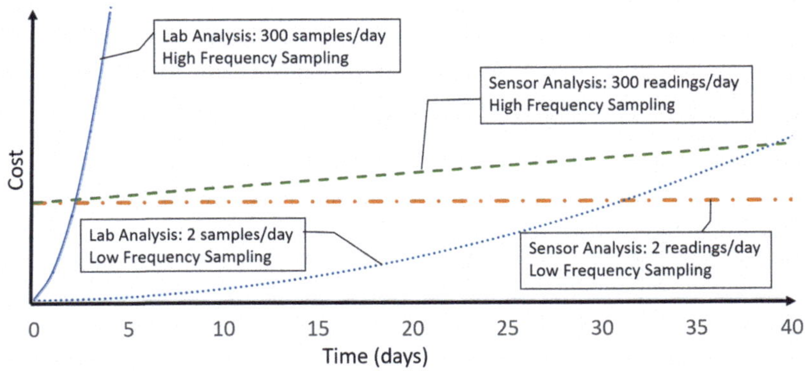

Figure 16.2 Comparison of laboratory analytical costs to a simple pH sensor system implementation. The costs of sensor implementation are initially quite high compared to laboratory costs.

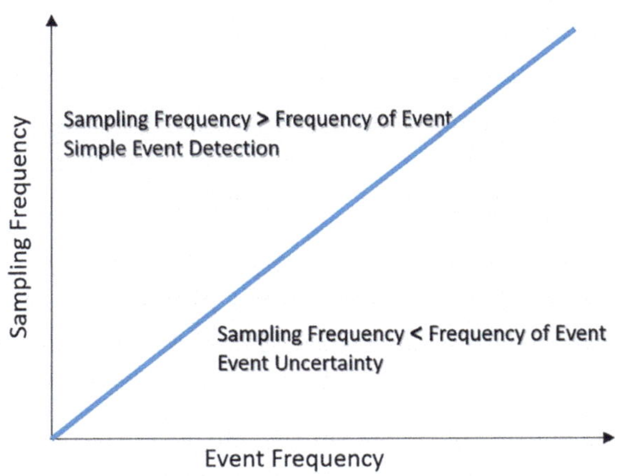

Figure 16.3 Increasing sampling frequency helps in detecting anomalies in water quality with greater confidence. Online monitoring is especially useful in the detection of unusual, short-lived contamination events.

Detection likelihoods for typical regulatory sampling programs are predominantly low because of the importance of time: short-lived contaminations can travel a long distance through the drinking water network, but are only detectable at any one location for a limited amount of time. As samples are taken at very specific times this reduces the detection likelihood. Sensors measure (semi-) continuously, eliminating the effect of time from the detection likelihood. As a result, a network of only 10 continuously measuring sensors performs far better than a sampling program of 2500 locations per year. Sensor networks showed detection likelihoods of 50–65% depending on the optimization goals instead of 35% of the theoretical high frequency sampling program of 2500 locations, let alone the 5–10% detection likelihood of a typical regulatory sampling program (den Berg *et al.*, 2015).

This indicates that while current regulatory sampling programs themselves are unable to reliably detect short-term contamination events, regulatory requirements still place an inordinate burden of validation on the adoption of Sensor Monitoring Systems. In addition, the integrity of the off-site laboratory analyses can be compromised during sample collection, transport, storage, and analysis, which can span several days or more. Clearly, a need exists for accurate, inexpensive, long-term monitoring of environmental contaminants using sensors that can be operated on site or in situ. A primary consideration that still remains to be addressed is the performance of these sensors in each of the field applications. Features such as sensitivity, stability, selectivity, speed, size, and cost need to be tested and evaluated under actual operating conditions. Harsh and fluctuating environmental conditions can degrade the performance of many of these sensors. However, the ability to deploy and use emerging sensors for these applications is uncertain due to both cultural and technological barriers.

16.6 THE DATA ANALYTICAL APPROACH FOR MICROBIOLOGICAL SENSORS

The regulatory challenge to the use of sensor data for demonstrating compliance with microbiological standards is significant and will not be easy to overcome. Most faecal indicators used in the regulatory standards have been developed, used and validated over a period of several decades. The concept behind the regulation of faecal indicators is that though not pathogenic as such, they could provide an indication of whether the drinking water is contaminated with sewage. Many sensors are also being developed on similar concept – that it might not be necessary to detect specific microorganisms as long as a valid indicator of faecal contamination is available.

The main complaint against microbiological regulations is that any compliance issue can only be detected once the drinking water has already been consumed by the consumer. In response to this, regulatory agencies and public water systems are investing in advance warning systems using commonly available non-microbiological sensors. These sensors can correlate physical and chemical indicators or co-contaminant with possible microbiological breakthrough and provide real-time warning of contamination.

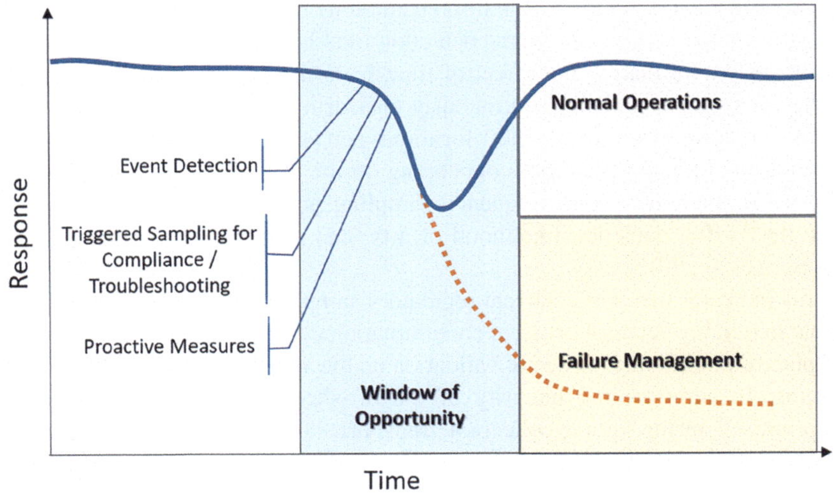

Figure 16.4 Implementation of sensors in contaminant warning systems can alert operators to potential contamination. Proactive measures taken within the small time window of opportunity will decrease the likelihood of having to manage failure and communicate to consumers.

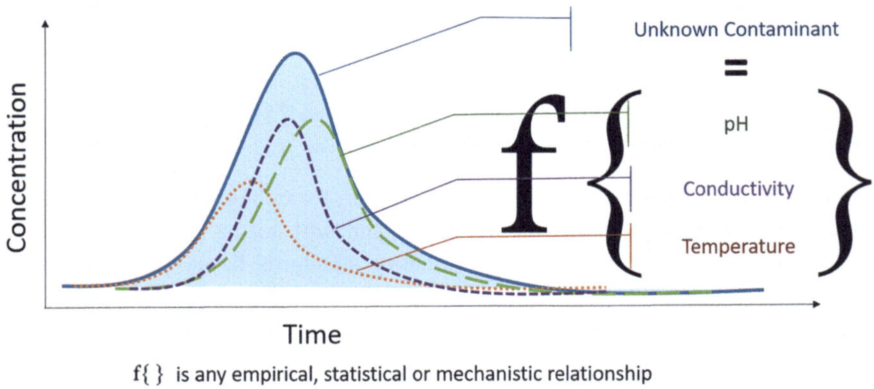

Figure 16.5 Real-time data from sensors can correlate physical and chemical indicators or co-contaminant with possible microbiological breakthrough and provide real-time warning of contamination.

According to Hall *et al.* (2009), EPA's concept for contamination warning systems is designed to extend this monitoring approach to multiple locations within a water distribution system. Therefore, in addition to evaluating online water quality monitoring and sensor technologies, EPA has collaborated with various research entities to develop two key software tools that provide these functionalities, described below. Threat Ensemble Vulnerability Assessment – Sensor Placement Optimization Tool (TEVA-SPOT). TEVA-SPOT can be used to determine the optimum number and locations for monitoring stations within a water distribution system. The CANARY algorithm evaluates water quality sensor responses and identifies changes in water quality that could indicate a contamination event. The name CANARY is not an acronym, but suggests a parallel with the historic "canary in the coal mine" event detection approach in which the coal miners used canaries to detect poison gas events. Similarly, the CANARY software evaluates real-time water quality data obtained from various instruments and uses mathematical and statistical techniques to identify the onset of anomalous water quality events (Hall *et al.*, 2009).

16.7 CONCLUSIONS

Regulatory compliance requires that newly developed sensor technology be comparable to laboratory methods either in terms of their results or in terms of the quality control measures used to develop the reference methods. In such a case, the only successful sensors would be those that either miniaturize, speed-up or automate existing laboratory techniques. These sensors have the same inherent drawbacks as the laboratory methods. Tatari *et al.* (2016) suggest that a contributing factor can be that the principle of the concept is not suited for application in drinking water. Drinking water is characterized by a large number of bacteria with high diversity living in an oligotrophic environment, thus being constantly starved with a low energy turnover. It is therefore not always possible to transfer a technology developed for other fields e.g. food industry or medical diagnostics, as these environments are often characterized by high nutrients levels and growth of single or few bacterial strains.

Hall *et al.* (2009) suggest that baseline water quality conditions can be monitored continuously in real-time such that a sudden change in water quality parameter(s) can trigger a contamination warning. Monitoring baseline water quality parameters within the distribution system will also provide multiple benefits of improved water quality closer to the point-of-use and additional security for detecting intentional or unintentional contamination events within the system. The capital, operational, and maintenance costs for contamination warning system will be difficult to sustain unless multiple benefits are identified. For water utilities, it is important to first maximize the security benefits by strategically placing the selected online monitors in the network and utilizing suitable techniques to evaluate the online sensor responses.

DWSs are mostly responsible for maintaining the safety and aesthetic quality of drinking water from source intake through treatment, storage and distribution. In some cases, the DWSs' obligations extend right into the consumers premises and internal plumbing systems (as in the case of Lead and Copper Rule and *Legionella*). With the increase in the number of class-action lawsuits against erring DWSs and water providers, it is increasingly urgent for them to implement some form of contamination warning system. The implementation of contamination warning system will help in proactive management of any system anomalies and also might help in demonstrating due diligence in courts of law.

The future of microbiological sensors therefore does not lie in their development as a regulatory compliance tool. Sensor developers seeking approval as an alternate method for regulatory compliance monitoring may face decades of delay, potential disclosure of proprietary information and costs to demonstrate comparability. On the other hand, the developments in implementation of sensors in contamination warning system will accelerate the adoption of this technology.

16.8 REFERENCES

Ashbolt N., Davies C., Kaucner C., Altavilla N., Ferguson C., Krogh M., Hijnen W., Medema G. and Deere D. (2005). *Fate and Transport of Surface Water Pathogens in Watersheds*. AWWA Research Foundation, Denver, CO, USA.

Barrett J. (2014). 'National disgrace': Alberta First Nations sue federal government over drinking water. *Calgary Herald*. http://www.calgaryherald.com/news/alberta/alberta/9944060/story.html (accessed 19 Jan 2018).

Bereskie T., Rodriguez M. J. and Sadiq R. (2017). Drinking water management and governance in Canada: an innovative Plan-Do-Check-Act (PDCA) framework for a safe drinking water supply. *Environmental Management*, **60**, 243–262.

Bernasconi C., Daverio E. and Ghiani M. (2003). Microbiology dimension in EU Water Directives. European Commission, JRC, IES, Inland and Marine Waters Unit. EUR 20787 EN. TP 300, 21020 Ispra (VA), Italy.

Cagle M. and Broadaway S. R. (2017). Toxic taps: Flint litigation and drinking water infrastructure. *Trends*, **48**(2), https://www.americanbar.org/publications/trends/2016–2017/november-december-2016/toxic_taps.html (accessed 19 Jan 2018).

Council of the European Union (1998). Council Directive 98/83/EC of 3 November 1998 on the quality of water intended for human consumption. *Official Journal of the European Communications*, **L330**, 32.

Cowan M. (2016). $3.3M settlement for Saskatchewan cryptosporidium outbreak gets preliminary approval. http://www.cbc.ca/news/canada/saskatchewan/cryptosporidium-settlement-preliminary-approval-infants-1.3812395 (accessed 19 Jan 2018).

Craun G. F., Calderon R. L. and Craun M. F. (2003). Waterborne outbreaks in the United States 1971–2000. In: Drinking Water Regulation and Health, F. W. Pontius (ed.), Wiley-Interscience, New York, USA.

den Berg J. V., Blokker E. J. M. and de Graaf B. (2015). Evaluation of the ability of sensor networks to detect *E. coli* events compared to sampling programmes. *Procedia Engineering*, **119**, 337–346.

Eggertson L. (2015). Canada has 1838 drinking-water advisories. *Canadian Medical Association Journal*, **187**(7), 488.

Espinola A. (2017). Litigation trend could snag utilities. https://www.awwa.org/legislation-regulation/leg-reg-news-details/articleid/4474/litigation-trend-could-snag-utilities.aspx (accessed 19 Jan 2018).

Haas C. N. (1999). Disinfection. In: Water Quality and Treatment: A Handbook of Community Water Supplies, R. D. Letterman (ed.), McGraw Hill, New York, USA.

Hall J. S., Szabo J. G., Panguluri S. and Meiners G. (2009). Distribution System Water - Quality Monitoring Sensor Technology Evaluation Methodology and Results. A Guide for Sensor Manufacturers and Water Utilities EPA 600/R-09/076 https://www.epa.gov/sites/production/files/2015-06/documents/distribution_system_water_quality_monitoring_sensor_technology_evaluation_methodology_results.pdf (Accessed 19 Jan 2018)

Health Canada (2017). Guidelines for Canadian Drinking Water Quality—Summary Table. Water and Air Quality Bureau, Healthy Environments and Consumer Safety Branch, Health Canada, Ottawa, Ontario.

Ho C. K., Robinson A., Miller D. R. and Davis M. J. (2005). Overview of Sensors and Needs for Environmental Monitoring. *Sensors*, **5**, 4–37.

Imran S., van der Gaag B., Neubert R., Peesker R., and Buchholzer M. (2014). Smart wastewater monitoring and diagnostics. Western Canada Water Conference and Exhibition, September 23–26, 2014. Regina, Saskatchewan.

Imran S. A., Dietz J. D., Mutoti G., Taylor J. S., Randall A. A. and Cooper C. D. (2005). Red water release in drinking water distribution systems. *Journal – American Water Works Association*, **97**(9), 93–100.

Imran S. A., Dietz J. D., Mutoti G., Xiao W., Taylor J. S. and Desai V. (2006). Optimizing source water blends for corrosion and residual control in distribution systems. *Journal – American Water Works Association*, **98**(5), 107–115.

Imran S. A., Sadiq R. and Kleiner Y. (2009). Effect of regulations and treatment technologies on water distribution infrastructure. *Journal – American Water Works Association*, **101**(3), 82–95.

LeChevallier M. W. (1999). The case for maintaining a disinfectant residual. *Journal – American Water Works Association*, **91**(1), 86–94.

Looney B. B. and Falta R. W. (eds.) (2000). Vadose Zone: Science and Technology Solutions. Battelle Press, Columbus, OH, USA.

Lytle D. A., Sorg T. J. and Frietch C. (2004). Accumulation of Arsenic in Drinking Water Distribution Systems. *Environmental Science & Technology*, **38**(20), 5365–5373.

Raich J. (2013). Review of sensors to monitor water quality. Work Programme Task No 1. Thematic group 'Chemical and biological risks to the water sector'. EUR 26325 EN.

Roberson J. A. (2003). Complexities of the new drinking water regulations – everything you wanted to know but were afraid to ask. *Journal – American Water Works Association*, **95**(3), 48–9.

Schets F. M., Nobel P. J., Strating S., Mooijman K. A., Engels G. B. and Brouwer A. (2002). EU Drinking Water Directive reference methods for enumeration of total coliforms and *Escherichia coli* compared with alternative methods. *Letters in Applied Microbiology*, **34**(3), 227–231.

Schock M. R., Hyland R. N. and Welch M. M. (2008). Occurrence of contaminant accumulation in lead pipe scales from domestic drinking-water distribution systems. *Environmental Science & Technology*, **42**(12), 4285–4291.

Tatari K., Corfitzen C. B., Albrechtsen, H.-J. and Christensen S. C. B. (2016). Sensors for MMicrobial Drinking Water Quality. Technical University of Denmark, DTU Environment, Kgs. Lyngby.

Taylor J. S., Dietz J. D., Randall A. A., Hong S. K., Norris C. D., Mulford A., Arevalo J. M., Imran S. A., LePuil M., Mutoti I., Tang J., Xiao W., Cullen C., Heaviside R., Mehta A., Patel M., Vasquez F. and Webb D. (2005). Effects of Blending on Distribution System Water Quality. AWWA Research Foundation and Tampa Bay Water, Denver, CO, USA.

Tuttle K. (2016). Class action lawsuit filed over lead in Chicago drinking water. *Jurist.* Friday 19 February 2016. http://www.jurist.org/paperchase/2016/02/class-action-filed-over-lead-in-chicago-drinking-water.php (accessed 19 Jan 2018)

USEPA (2017). Analytical Methods Approved for Compliance Monitoring under the Ground Water Rule. EPA 821-F-17–003.

USEPA (2001). Controlling disinfection byproducts and microbial contaminants in drinking water. EPA 600/R-01/110.

USEPA (2004). Providing safe drinking water in America – 2002 national public water systems compliance report. EPA 305-R-04–001.

USEPA (2010). EPA Microbiological Alternate Test Procedure (ATP) Protocol for Drinking Water, Ambient Water, Wastewater Monitoring Methods. EPA-821-B-10–001.

Valentine R. L. and Stearns S. W. (1994). Radon release from water distribution system deposits. *Environmental Science & Technology*, **28**(3), 534–4.

WHO (2011). Guidelines for Drinking-Water Quality, 4th ed. http://www.who.int/entity/water_sanitation_health/publications/dwq-guidelines-4/en/index.html (accessed 19 January 2018).

Wilson L. G., Everett L. G. and Cullen S. J. (eds.) (1995). Handbook of Vadose Zone Characterization & Monitoring. CRC Press, Boca Raton, FL, USA.

Index